DESIGNING

TRANSISTOR I.F. AMPLIFIERS

By the same author:

TRANSISTOR BANDPASS AMPLIFIERS

The theory of analysis and design of selective amplifiers as used in the I.F. parts of radio, television and radar receivers, is here dealt with, especially in relation to the application of transistors.

Single-stage amplifiers as well as multi-stage amplifiers, with arbitrary types of interstage or terminating networks are treated in detail as regards stability, power gain, amplitude response curve and envelope delay curve; also neutralization of the transistor internal feedback and problems associated with spreads in transistor parameters.

DESIGNING
TRANSISTOR I.F. AMPLIFIERS

W. Th. H. HETTERSCHEID

1966

SPRINGER-VERLAG BERLIN HEIDELBERG GMBH

This book contains xiv + 330 pages and 138 illustrations and 96 pages two-coloured design charts, 3 folding-out tables.

U.D.C. nr. 621.375.121 : 621.382.3

Library of Congress Catalog Card Number: 66-22377

Additional material to this book can be downloaded from http://extras.springer.com

ISBN 978-3-662-38672-9   ISBN 978-3-662-39544-8 (eBook)
DOI 10.1007/978-3-662-39544-8

© Springer-Verlag Berlin Heidelberg 1966
Originally published by N.V. Philips' Gloeilampenfabrieken, Eindhoven,
The Netherlands, in 1966
Softcover reprint of the hardcover 1st edition 1966

Trade marks of N.V. Philips' Gloeilampenfabrieken

PREFACE

In most cases the design of I.F. amplifiers for radio, television and radar receivers involves a large amount of computational effort. This is due to the numerous requirements imposed on such an amplifier which, in most cases, are contradictory. As a consequence, a compromise must usually be found which will result in an optimum performance of the amplifier with regard to the most important specification points.

The above comments are applicable to all typts of I.F. amplifier, irrespective of whether they are equipped with electron tubes or transistors. The design of transistor I.F. amplifiers, however, imposes even more problems than that of the tube counterpart. This is due to the internal feedback of the transistors, which is not negligibly small.

A transistor I.F. amplifier can, therefore, not be treated on a stage-by-stage basis as in the case of amplifiers equipped with modern penthode tubes, operating at a not too high frequency. The stage-by-stage design of a transistor I.F. amplifier is not possible because the internal feedback of the transistor in each stage affects the performance of all the other stages.

This implies that an I.F. amplifier equipped with transistors must be treated as a whole. A new parameter is therefore added to the variety of parameters which already exists with tube equipped amplifiers: this new parameter being the amount of feedback present in each stage. In this book, a method is presented which facilitates that the design of transistor I.F. amplifiers can be carried out with great ease and accuracy. The key of this design method is formed by a large number of computed "design charts" which are used in connection with a step-by-step design procedure.

These design charts contain amplitude response, envelope delay and gain characteristics, taking into account a certain set of parameters. One of these parameters is related to the amount of feedback of the transistors in the various amplifier stages. The step-by-step method of design leads to the choice of a particular set of design charts. Consultations of this set enables an accurate determination of the performance of the amplifier to be made. The design charts present moreover the information necessary for constructing the bandpass filters used as interstage networks.

The subject-matter contained in this book is based on work carried out in the Philips Semiconductor Application Laboratory at Nijmegen, The Netherlands.

The computed design charts are based on the analytical approach to amplifier design presented in the author's book "Transistor Bandpass Amplifiers".

The author wishes to express his gratitude towards his colleagues for the many discussions and the helpful suggestions. In this respect he especially wishes to mention Mr. E. J. Hoefgeest and Mr. A. H. J. Nieveen van Dijkum. For correcting the manuscript the author is indebted to Mr. W. H. Cazaly, Ilford, England.

March, 1966   The Author

CONTENTS

CHAPTER 1

INTRODUCTION

An intermediate frequency amplifier as used in radio, television and radar receivers provides the greater part of the high-frequency gain of the whole system and almost entirely determines the amplitude and phase response characteristics. The I.F. amplifier therefore largely governs the total performance. For these reasons a great deal of attention is usually given to the design of such an amplifier.

Often the design of an I.F. amplifier is laborious because of the large number of requirements and conditions that must be taken into account. This applies especially to amplifiers in which transistors are used as amplifying elements, when problems arise associated with the internal feedback of the transistors. (In amplifiers equipped with modern electron tubes, interelectrode feedback offers few difficulties provided that no extra feedback is introduced by the arrangement of components and wiring.)

In this book a systematic method is developed which facilitates the design of transistor I.F. amplifiers by making use of a large number of computed amplifier design charts.

The heart of the design method lies in the determination of proper values of the *regeneration coefficients* of the transistors in the various stages of the amplifier. The "regeneration coefficient" is a quantity dependent on the admittance parameters of a transistor and on the admittances of tuned circuits at its input and output terminals. The relative values of the regeneration coefficients of the transistors in the various stages largely affect the performance of the amplifier.

The various phases of the design procedure are schematically illustrated in the block diagram in Fig. 1.1. From the specification of performance requirements of the amplifier and data concerning the transistors to be employed, a choice is made of the amplifier configuration (arrangement of transistors and single-tuned and/or double-tuned bandpass filters) and the number of stages needed.

With the amplifier configuration and the number of stages known, use can be made of a series of *design charts*. From these design charts amplitude response and envelope delay characteristics can be derived. Taking into account the relevant items of the design specification a certain group of curves

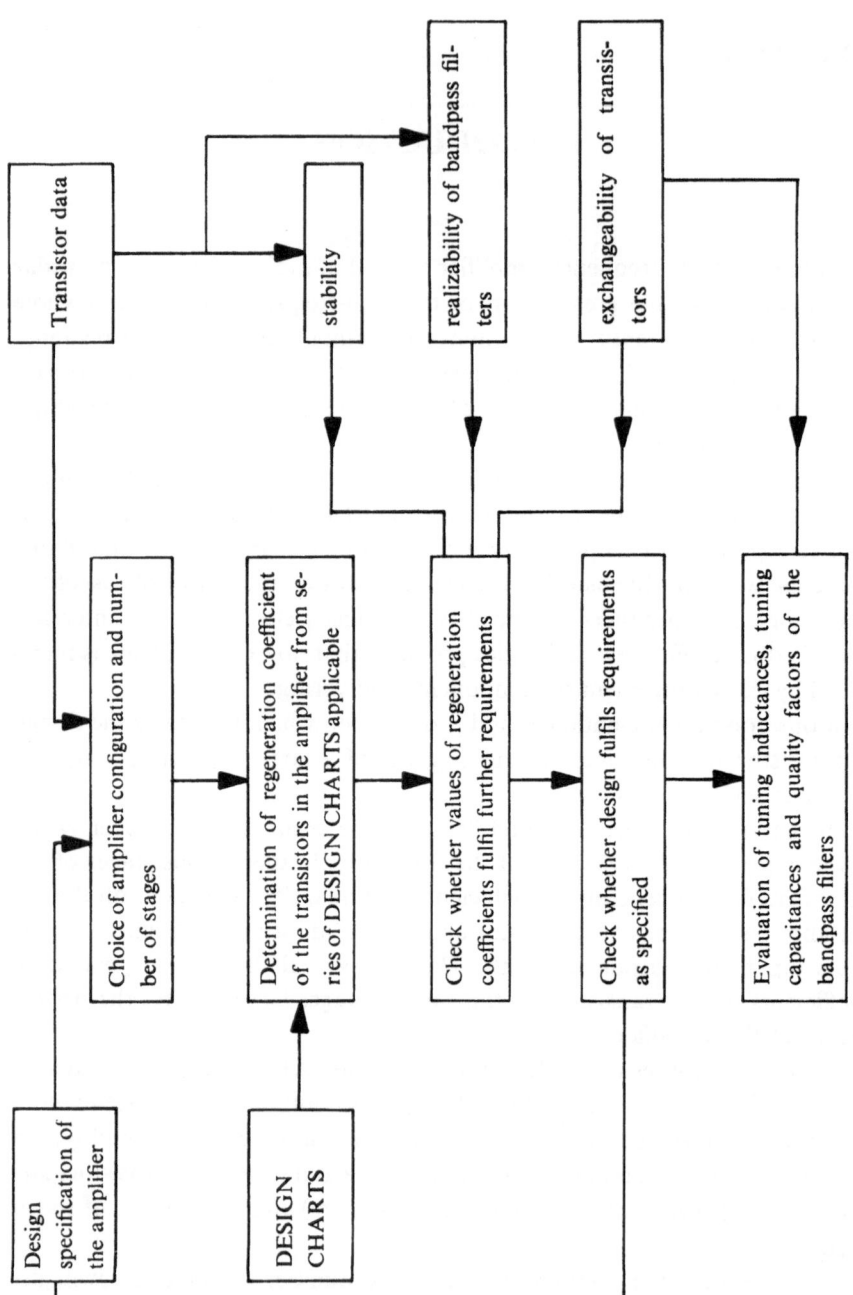

Fig. 1.1. Block diagram indicating various stages of the step-by-step method of designing
I.F. amplifiers.

can be selected, valid for a particular value of the regeneration coefficient in the amplifier.

The value of the regeneration coefficient obtained from the design charts fulfils most of the requirements for designing the amplifier. Further checks, however, must be made regarding stability, interchangeability of transistors of the type chosen, and realizability of the bandpass filters to be used. A value of regeneration coefficient should be chosen which meets all requirements.

When the final values of the regeneration coefficient of the various transistors of the amplifier are arrived at, the most important part of the design procedure will have been completed, provided all specified design requirements are fulfilled. Whether this is the case or not can now be checked in detail.

The last part of the design procedure consists of the practical design of the various bandpass filters of the amplifier.

The design procedure described has been fully set out in Chapter 13, which contains a step-by-step method for designing I.F. amplifiers. To facilitate design, the results of the various steps can be entered in a systematically arranged form, as shown on a fold-out page at the end of the book.

The step-by-step method shown includes in condensed form the considerations on various aspects of amplifier design dealt with in Chapters 2 to 8 inclusive.

In Chapter 2 design specifications of gain, amplitude response, and envelope delay characteristics for various types of I.F. amplifier are considered. It is argued that specifications should comprise not only nominal values of items but also acceptable spreads of values.

Spreads in performance that occur in a series of practical amplifiers arise from the use of active and passive components, the parameters of which are subject to production spreads.

The spreads in the admittance parameters of the transistors (or other active devices) as well as in the values of the components are usually specified by the manufacturer.

In Chapter 3 is discussed how the various spreads are taken into account to obtain a true picture of actual performance spreads.

In Chapter 4 are considered the small-signal admittance parameters of the transistors required and their dependence on biasing point, operating frequency, and ambient temperature.

The design of biasing circuits is not included in this book, since the subject is adequately treated in several works on transistor circuits.

The method of design of I.F. amplifiers presented here is based on the theoretical considerations set out in the book "Transistor Bandpass Amplifiers", which is in the present work referred to as Book I.*

In Chapter 5 (of the present work) a survey is given, for ease of reference and to facilitate its use, of the theory of I.F. amplifier design, which contains sundry references to relevant sections in the first-mentioned Book I.

In Chapter 6 a short account is given of neutralization of the internal feedback of transistors. In some instances neutralizing networks may considerably improve the performance of the amplifier.

In Chapter 7 are discussed problems of gain control in I.F. amplifiers.

Chapter 8 presents further considerations related to the development of the step-by-step method of designing practical I.F. amplifiers.

Chapters 9, 10, 11 and 12 give examples of the design of I.F. amplifiers.

In Chapter 9 examples of the design of I.F. amplifiers for radio receivers for A.M. and F.M. signals are presented.

In Chapter 10 examples of the design of I.F. amplifiers for television receivers are given.

In Chapter 11 the spreads in performance of the amplifiers designed in Chapter 10 are evaluated as they affect the choice of an A.G.C. system.

Chapter 12 presents an example of a theoretical method of calculating the spreads in performance that may occur in a vision I.F. amplifier due to spreads in the admittance parameters of the transistors.

Chapters 11 and 12 show how individual spreads of transistors and components should be evaluated and combined in order to obtain a realistic picture of the overall spread in the performance of an amplifier.

Chapter 13 shows a further example of the design of a vision I.F. amplifier.

Chapter 14 presents the step-by-step method of I.F. amplifier design and Chapter 15 contains the design charts referred to above.

---

*) W. Th. H. HETTERSCHEID, *Transistor Bandpass Amplifiers*, Philips Technical Library, Eindhoven, 1964.

CHAPTER 2

THE DESIGN SPECIFICATION OF I.F. AMPLIFIERS

Since an I.F. amplifier, as used in television or radio receivers, contains the stages linking the R.F. stages and the second detector, the requirements imposed on such an amplifier are closely related to the performance of these parts. The I.F. amplifier must, for example, be able to cope with and to supply signals of such a magnitude that proper functioning of the R.F. stages and detector circuit is ensured.

The undermentioned performance requirements are of special interest in the design procedure to be described in this book:

> gain,
> 3 dB bandwidth,
> 0 dB bandwidth, and
> adjacent channel selectivity.

Moreover, for some applications, special requirements must be met by the envelope delay characteristic.

These points are discussed in succession in the following sub-sections.

2.1 Gain

The gain required of an I.F. amplifier depends not only on the overall sensitivity requirements of the complete receiver, but also on the gain of the R.F. and mixer stages and on the type of second detector used. To specify the required gain of the I.F. amplifier for a given overall sensitivity of the complete receiver, it is therefore necessary to know both the magnitude of the signal required by the detector for proper operation and the gain of the R.F. and mixer stages. It should be recognized that the gain figures of the I.F. amplifier and of the R.F. and mixer stages should be expressed in such a way that they can conveniently be combined without the risk of errors due, for example, to different definitions being used.

Particularly when an amplifier is equipped with transistors, its gain is very often expressed in terms of transducer gain. This transducer gain is defined as the ratio of the power delivered to the load to the power available from the source. According to Fig. 2.1, we have:

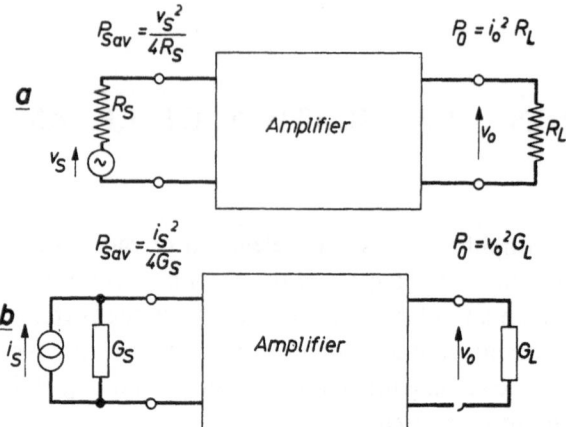

Fig. 2.1. Amplifier configurations for defining gain properties.

$$\Phi_t = P_0/P_{Sav} = 4(R_S/G_L) \cdot |i_0/v_S|^2$$
$$= 4(R_S/G_L)|Y_t|^2 \,,$$

or:
$$\Phi_t = 4G_S G_L \,|\, v_0/i_S|^2$$
$$= 4G_S G_L \,|\, Z_t|^2 \cdot$$

However, it is emphasized that the term "transducer gain" makes no sense unless the real parts of the source and load admittances of impedances have fixed values which cannot be affected by the design of the amplifier. If this condition is not fulfilled, any required transducer gain figure could be obtained for an amplifier with a given value of Z_t (or Y_t) by choosing G_S (or R_S) and G_L accordingly.

Very often the above requirements of the source and load terminations of an I.F. amplifier are not met. Then other means must be found, which allow an unambiguous specification of the gain properties. In the following subsection such a gain specification is considered for an I.F. amplifier of a television receiver. In Sections 2.1.2 and 2.1.3 these considerations are extended to cover I.F. amplifiers for radio receivers.

2.1.1 GAIN SPECIFICATIONS OF AN I.F. AMPLIFIER FOR A TELEVISION RECEIVER

For specifying the gain properties of tuners and I.F. amplifiers of television receivers, the following aspects must be taken into account:

a. The gain of the combination of tuner and I.F. amplifier is given by the required sensitivity figure of the receiver. When this combination is split into a "tuner" and an "I.F. amplifier", the gains of the individual parts should be specified in such a way that the gain of the combination can be determined in a straightforward manner.

b. The I.F. amplifier forms that part of the combination which mainly determines the amplitude and phase response curves of the complete receiver. Since the input (double-tuned) bandpass filter together with the adjacent channel traps has a major influence on these response curves, it must be considered as an integral part of this amplifier and, therefore, must be incorporated in the design of the I.F. part of the receiver. These considerations should not be affected by the fact that, for constructional reasons, the primary of the input double-tuned bandpass filter of the I.F. amplifier is situated in the tuner section.

When designing this primary, obviously, attention should be paid to requirements imposed on it from the view-point of tuner design.

As the input bandpass filter forms an integral part of the I.F. amplifier, the specification of the amplifying properties should include the characteristics of the input filter.

c. The above reasoning implies that the current delivered by the equivalent I.F. current source of the mixer transistor is the correct input reference for defining the gain properties of the I.F. amplifier.

The output signal of an I.F. amplifier is usually specified in terms of voltage (across a certain load impedance).

The combination of input current and output voltage suggests that the amplifying properties of the I.F. amplifier can best be expressed in terms of *transimpedance*.

d. A certain tuner may be used in combination with a variety of I.F. amplifier designs (with different input bandpass filters). It is therefore necessary to specify the gain properties of the tuner in such a way that they are independent of the I.F. load.

e. The reasoning in d. suggests that the current delivered by the equivalent I.F. current source of the mixer transistor is the correct output reference for specifying the tuner gain. The input signal of the tuner is usually specified in terms of antenna e.m.f.. This implies that the best characterization of the gain properties of the tuner can be given terms of *transadmittance*.

In addition, the output damping of the frequency changer transistor should be specified separately to facilitate the design of the input tuned circuit of the I.F. amplifier.

If the amplifying properties of the tuner and I.F. amplifier are expressed in this way, the transducer gain of the combination can be obtained from the expression:

$$\Phi_t = 4R_S \cdot G_L \cdot |Y_t|_{\text{tuner}} \cdot |Z_t|_{\text{I.F.}} \tag{2.1.3}$$

In this expression R_S denotes the source resistance of the tuner and G_L the load damping of the I.F. amplifier. Both R_S and G_L have well-defined values, so that the transducer gain thus determined gives a true indication of the gain performance.

Also the overall voltage gain*) of the combination of tuner and I.F. amplifier can be calculated as:

$$K_t = Y_t \text{ tuner} \cdot Z_t \cdot \text{I.F.} \tag{2.1.4}$$

Eqs. (2.1.3) and (2.1.4), show that the method of expressing the amplifying properties of the tuner in terms of transadmittance and those of the I.F. amplifier in terms of transimpedance leads to unambiguous results in calculating the overall gain of the combination.

2.1.2 GAIN SPECIFICATION OF AN I.F. AMPLIFIER FOR A RADIO RECEIVER.

2.1.2.1 *F.M. Receivers*

The method of specifying the gain of an I.F. amplifier of a television receiver in terms of transimpedance is also applicable to I.F. amplifiers for radio receivers for frequency modulated signals. The voltage level at the input terminals of the F.M. detector circuit should be of such magnitude that the amplitude limiting circuits of the amplifier are not operating.

The transimpedance of the amplifier is then the ratio of this detector input voltage and the I.F. current delivered by the mixer transistor.

The specification of the gain of an I.F. amplifier of an F.M. receiver in terms of transimpedance necessarily involves expressing the gain of the corresponding tuner in terms of transadmittance.

2.1.2.2 *A.M. Receivers*

In domestic radio receivers for amplitude modulated signals, the midband frequency of the I.F. amplifier is of the order of either 260 kc/s or 460 kc/s. At these frequencies the properties of present-day transistors are such that the gain per stage can easily be calculated in terms of voltage gain.

Since, moreover, an I.F. amplifier for an A.M. receiver consists of one or

---

*) For the voltage gain the symbol K_t is used here because this gain figure relates the output voltage of the I.F. amplifier to the e.m.f. of the voltage at the input of the tuner. The use of k_t is analogous to that of Z_t or Y_t.

two stages, there is no need to consider this amplifier as a single unit for specifying its gain.

The gain of an I.F. amplifier for A.M. receivers is therefore specified in terms of voltage gain (or power gain) per stage. For this specification, it is required that the transimpedances of the interstage coupling bandpass filters and the forward transadmittances of the transistors be known.

2.2 Amplitude Response Characteristic

2.2.1 RADIO RECEIVERS

It is customary to specify the amplitude response curve of an I.F. amplifier for a radio receiver by quoting the bandwidth between the -3 dB points and the attenuation of carriers at a frequency interval Δf above and below the carrier frequency of the desired signal. The latter characteristic is usually referred to as the adjacent channel selectivity, and Δf then equals the difference in carrier frequency between two channels. It is common practice to put $\Delta f = 9$ kc/s for I.F. amplifiers used in domestic A.M. receivers and $\Delta f = 300$ kc/s for I.F. amplifiers in F.M. receivers.

2.2.2 TELEVISION RECEIVERS

The amplitude response curve of an I.F. amplifier for the vision channel of a television receiver should fulfil a number of requirements regarding bandwidth and adjacent channel selectivity. For an I.F. amplifier of a receiver designed for reception of television signals according to the C.C.I.R. standard these requirements are the undermentioned:

— the I.F. vision carrier frequency has to be 38.9 Mc/s.
— at the frequency of the I.F. vision carrier the amplification has to be 6 dB lower than the maximum amplification in the frequency range from 35 to 38 Mc/s.
— the width of the amplitude response curve between the points at which the response is 6 dB down should be 5 Mc/s.
— the response at the I.F. sound carrier frequency (33.4 Mc/s) should be 20 dB below that at the I.F. vision carrier frequency.
— at the adjacent channel I.F. sound carrier frequency (40.4 Mc/s at V.H.F. reception and 41.4 Mc/s at U.H.F. reception) the response should be at least 36 dB below that at the own channel vision carrier frequency.
— at the adjacent channel I.F. vision carrier frequency (31.9 Mc/s at V.H.F. reception and 30.9 Mc/s at U.H.F. reception) the response of the I.F. amplifier should be at least 40 dB below that at own-channel vision carrier frequency.

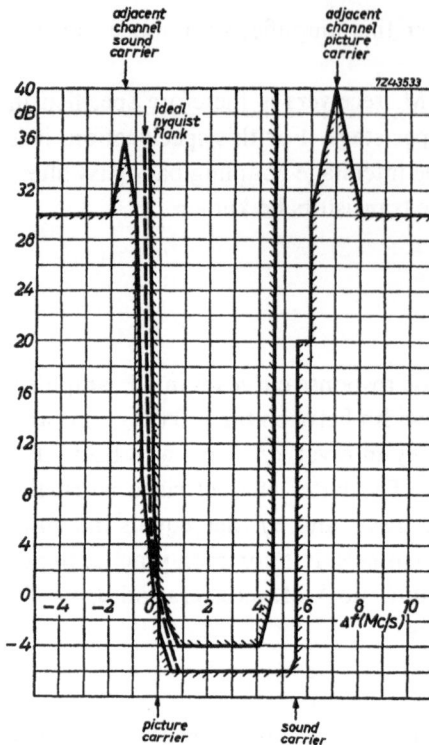

Fig. 2.2. Amplitude response curve, with tolerances, of a television I.F. amplifier for a receiver conforming with C.C.I.R. standards.

The amplitude response curve should, moreover, fulfil the following conditions:

— the amplification in the frequency range of 35 to 38 Mc/s should be constant within 2 dB.

— the shape of the response curve should be substantially independent of the gain reduction by the A.G.C. action.

In general, certain tolerances in the above requirement of the amplitude response curve of practical I.F. amplifiers will be accepted. As an example of what tolerances are considered to be acceptable, Fig. 2.2 represents the tolerance scheme of television receivers as recommended by the Postal Authorities of Western Germany*. In this tolerance scheme, which applies to the overall response curve of the complete receiver for V.H.F. reception, the relative frequency of the video signal is plotted along the horizontal axis.

---

* Tolerance scheme for "Fernsehempfänger für Dienstgebrauch". Fernmeldetechnisches Zentralamt der deutschen Bundespost, FTZ-Norm FTZ 142-TV6.

2.3 Envelope Delay Characteristic

For a faithful reproduction of the transmitted signal the envelope delay characteristic of I.F. amplifiers of television receivers and of radio receivers for frequency modulated signals should satisfy special requirements. In general, it should not exceed a certain variation within the frequency range of interest.

In specifying the permissible variation in I.F. amplifiers for the vision channel of a television receiver, the effects of wavetraps should also be taken into account.

CHAPTER 3

SPREADS, TOLERANCES AND VARIATIONS
IN DESIGNING I.F. AMPLIFIERS

When normal production type transistors, resistors, capacitors and inductors are used in the construction of an I.F. amplifier, allowance must be made in the design for the various manufacturing spreads and tolerances. Also, variations in operating conditions due to different environmental circumstances must be taken into account.

Only when all spreads, tolerances and variations are duly embodied in the design of the amplifier it can be expected that the desired performance will be obtained under all circumstances.

In the following sections the above mentioned aspects will be considered in some detail.

3.1 Definition of Various Groups of Deviations from Nominal Conditions

The deviations from nominal conditions encountered in the design of I.F. amplifiers can be divided into three groups, according to the character of the deviations. These groups are:

— spreads,
— tolerances, and
— variations.

3.1.1 SPREADS

Definition: The spread of a certain parameter of a device is the deviation of that parameter from the nominal value, which occurs during the manufacture of a large number of these devices.

Although attempts are made to keep spreads as small as possible, they are bound to occur.

Consider, for example, the manufacture of a high-frequency transistor. During the manufacturing process, and also subsequently to it, checks are carried out to see whether the transistor has the desired properties or not. Although certain values of the admittance parameters (specified at a certain signal frequency, d.c. biasing point and ambient temperature) are amongst the desired properties of such a transistor, these parameter values are usually not measured on each transistor during the manufacturing process. Instead, "good-or-faulty" types of measurement are carried out, which ensure that the average values of the various admittance parameters of all transistors of the type under consideration lie close to the specified values. The deviations from these specified values are called *spreads*.

3.1.2 TOLERANCES

Definition: The tolerance of a certain parameter of a device is the specified maximum deviation from the nominal value of that parameter which is allowed by the manufacturer or the user.

From the definition, a tolerance is a spread with limits to the maximum deviation from the nominal value.

For components used in electronic circuits the tolerance is usually expressed as a maximum allowable percentage deviation from the nominal value. The term "resistor, 100Ω, tolerance $\pm 10\%$", for example, means that, the actual resistance of this type of resistor lies between 90Ω and 110Ω.

3.1.3 VARIATIONS

Definition: A variation is a deviation from the nominal condition under which a device operates, brought about by causes external to the device.

An example of a variation is the change of the ambient temperature an electronic circuit is subjected to during its operation (unless special measures are taken to stabilize the ambient temperature). A variation, as defined here, is thus not affected by the choice of components of which the circuit is composed. In this respect the deviations from nominal conditions in the "variations" group have a completely different character from those in the "spreads" and "tolerances" groups.

3.2 Combination of Deviations from Nominal Conditions of Various Devices in Designing I.F. Amplifiers

Deviations from nominal values of the various active and passive devices used in the construction of the amplifier will cause performance to differ from that which may be expected of the strictly nominal case. When designing such an amplifier, it is therefore necessary to investigate the extent to which the deviation from the nominal value of each device affects overall performance of the amplifier. The deviations caused by each device separately must be combined to find out what deviation of amplifier performance may be expected due to the use of normal production type active and passive devices. Then it can be judged whether the design of the amplifier is acceptable or not in view of nominal performance and the deviation from it which will occur in practice.

3.2.1 METHOD OF COMBINING VARIOUS DEVIATIONS

The method of combining the separate deviations from the nominal performance greatly affects the final result. The method to be adopted for this purpose should ensure that nominal performance of nearly all amplifiers lies within the determined range of deviations. On the other hand it should not impose unnecessary strict requirements on building elements to meet a specified performance, since this would only lead to an expensive design.

It may be shown by means of the probability theory that a good method of combining the effects of individual deviations on the overall performance of the amplifier is the following:

— The effects on the overall performance are determined, of the *spreads* and *tolerances* of each individual device in the amplifier, with nominal values or conditions for all other devices. Let these effects, expressed as a percentage of the nominal performance figures a_1, a_2, a_3, etc., be denoted by Δa_1, Δa_2, Δa_3, etc. To find the total effect of the spreads and tolerances of all devices, the individual effects are added geometrically. When Δa is the total effect on the overall performance a:

$$\Delta a = \sqrt{\Delta a_1^2 + \Delta a_2^2 + \Delta a_3^2 + \text{etc.}} \qquad (3.2.1)$$

— The effects of *variations* on the overall performance of the amplifier are determined in the same way as above. Let the effects of the variations, when expressed as a percentage of the nominal performance figure b_1, b_2, b_3, etc., be denoted by Δb_1, Δb_2, Δb_3 etc. The total effect is found from a multiplication of the individual influences. Hence:

$$b + \Delta b = (b_1 + \Delta b_1)(b_2 + \Delta b_2)(b_3 + \Delta b_3). \text{ etc.,} \qquad (3.2.2)$$

or when the individual variations are relatively small:

$$\Delta b = \Delta b_1 + \Delta b_2 + \Delta b_3 + \text{etc.} \qquad (3.2.3)$$

— The effects of all deviations in the amplifier are obtained by combining linearly the geometrically added effects of spreads and tolerances and the linearly added effects of variations. If $\varDelta A$ denotes this total effect (as a percentage of the nominal performance figure):

$$\varDelta A = \varDelta a + \varDelta b. \tag{3.2.4}$$

The method of combining the various effects given above is based on the following considerations: An amplifier generally consists of a large number of active devices (transistors or electron tubes) and passive devices (resistors, capacitors and inductors). Although all devices are subject to spreads or tolerances it is very unlikely that extreme values of spreads and tolerances occur simultaneously in a single amplifier. A linear addition of the effects of the extreme values is therefore very unrealistic. The geometrical addition used in the method outlined above gives better results.

Variations, on the other hand, must be taken into account linearly because they are due to effects external to the amplifier. All the variations may therefore very well reach extreme values simultaneously.

3.3 Spreads, Tolerances and Variations Encountered in I.F. Amplifier Design

When designing I.F. amplifiers the following deviations from nominal values or conditions must be taken into account:

Spreads:

— Spreads in the admittance parameters of the transistors used in the amplifier at the nominal biasing point and nominal ambient temperature.
— Changes of the nominal values of the admittance parameters as a consequence of deviation of the biasing points from the nominal conditions. These changes can in some instances be included in the spreads of the admittance parameters.

Tolerances:

— Tolerances in the resistors determining the d.c. biasing points of the transistors in the amplifier.
— Tolerances in the adjustment of the supply voltage of the amplifier when fed from a stabilized power supply.
— Tolerances in capacitors, inductors and resistors in the high-frequency parts of the amplifier circuitry.
— Tolerances in the alignment procedure of the amplifier.

Variations:

— Variations of the ambient temperature at which the amplifier operates.

— Variations of the supply voltage at which the amplifier operates if the amplifier is not fed from a stabilized power supply. These supply voltage variations may be caused by fluctuations of the mains voltage, reduction during life of the terminal voltage of a dry battery or variations in the terminal voltage of a car-accumulator when charged or discharged.

In following chapters will be shown how the various spreads, tolerances and variations can be included in the design procedure of I.F. amplifiers.

CHAPTER 4

TRANSISTOR PARAMETERS

To facilitate designing I.F. amplifiers equipped with transistors, a method of expressing these active devices is required which allows a convenient combination of the their properties with those of the bandpass filters used in the amplifier. A possible way of doing this is to express the transistor properties in terms of admittance parameters or y-parameters.

4.1 Admittance Parameters

To arrive at a method of expressing the transistor properties which will be suitable for designing bandpass amplifiers, the transistor is considered as a black box with two terminal pairs as shown in Fig. 4.1. The voltages and currents at the terminal pairs will be considered to be of such magnitudes that the relations between them may to all intents and purposes be approximated by a set of linear functions. Such a set of functions is then given by:

$$\left. \begin{array}{l} i_1 = y_{11}\, v_1 + y_{12}\, v_2, \\ i_2 = y_{21}\, v_1 + y_{22}\, v_2, \end{array} \right\} \tag{4.1.1}$$

which is known as the admittance parameter relation.

As has been shown theoretically in Book I, Chapter 1, the "constants" or "parameters" y in Eq. (4.1.1) depend on the signal frequency, the d.c. biasing point and on the ambient temperature; this will also become apparent from the graphs given in Sections 4.4 to 4.8.

Considering the relations expressed by Eq.(4.1.1) it follows that a transistor may be represented by the equivalent four-terminal network shown in Fig. 4.2.

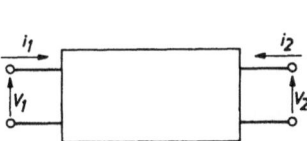

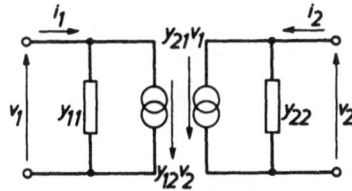

Fig. 4.1. Black-box representation of a transistor.

Fig. 4.2. Equivalent circuit diagram of a transistor based on the admittance parameter representation.

According to this figure and to Eq.(4.1.1), it follows that at a given d.c. biasing point of the transistor, a given ambient temperature, and a given signal frequency:

y_{11} is the small-signal input admittance of the transistor four-pole with the output terminals short-circuited (v_2 being zero);

y_{22} is the small-signal output admittance of the transistor four-pole with the input terminals short-circuited (v_1 being zero);

y_{12} is the small-signal reverse transfer admittance of the transistor four-pole, that is to say the ratio of the short-circuited input current to the output voltage (v_1 being zero);

y_{21} is the small-signal forward transfer admittance of the transistor four-pole, that is to say the ratio of the short-circuited output current to the input voltage (v_2 being zero).

In Table 4.1 these definitions are illustrated schematically.

TABLE 4.1.   Admittance Parameters

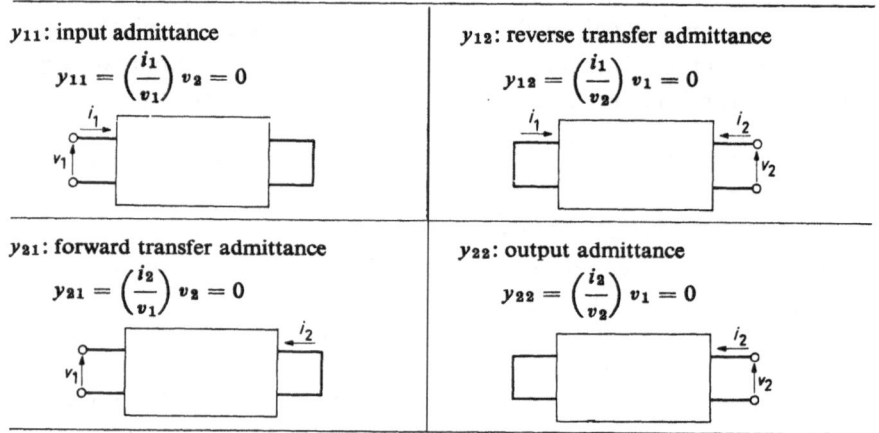

The admittances y_{11}, y_{12}, y_{21} and y_{22} are generally complex in character, so that each may be split up into a real and an imaginary part or written in a modulus and argument from:

$$y_{11} = \text{Re}\,(y_{11}) + j\text{Im}(y_{11}) = g_{11} + j\omega\,C_{11},$$
$$y_{12} = \qquad\qquad\qquad y_{12} \cdot e^{j\varphi_{12}},$$
$$y_{21} = \qquad\qquad\qquad y_{21} \cdot e^{j\varphi_{21}},$$
$$y_{22} = \text{Re}\,(y_{22}) + j\text{Im}(y_{22}) = g_{22} + j\omega\,C_{22}.$$

In bandpass amplifiers it is necessary to use a number of tuned circuits which form the coupling elements between the transistors. Since these tuned circuits are connected in parallel with the transistor terminals, it is convenient to express the properties both of these circuits and of the transistors in terms of admittance. The total admittance of the tuned circuit and transistor admittances connected in parallel can then be evaluated by simply adding the individual admittances. In practical amplifiers with purely parallel tuned circuits, the parameters y_{11} and y_{22} are always considered in connection with the tuned input and output circuits respectively. These tuned circuits are so designed that the imaginary parts of y_{11} and y_{22} are included in the tuning susceptances. The real parts g_{11} and g_{22} constitute the damping of the tuned circuits due to the transistor. This explains why y_{11} and y_{22} have been expressed in the form of $(g + jb)$ in the above expressions.

The admittances y_{12} and y_{21} are transfer properties of the four-pole, and can most conveniently be expressed in terms of modulus and argument because their product must be evaluated for analyzing the amplifier.

4.2 Transistor Parameter Nomenclature

At present it is customary to represent the four-pole parameters of transistors by the symbols recommened in I.E.E.E. standards. These symbols include an indication as to which of the three transistor terminals is common to both the input and output circuits. The common terminal may be either the base, the emitter or the collector. The indication is given by using the letter b, e or c respectively as the second suffix in the symbol denoting a given four-pole parameter. The first suffix of these symbols indicates which of the four-pole parameters is referred to; the input, reverse transfer, forward transfer and output parameters being denoted by the suffixes i, r, f and o respectively. The symbol y_{ie}, for example, denotes the input admittance parameter of a transistor in common emitter configuration; and so on.

TABLE 4.2. Nomenclature of Transistor Admittance Parameters

Admittance parameters	Common base	Common emitter	Common collector	General symbol
Input admittance	y_{ib}	y_{ie}	y_{ic}	y_{11}
Reverse transfer admittance	y_{rb}	y_{re}	y_{rc}	y_{12}
Forward transfer admittance	y_{fb}	y_{fe}	y_{fc}	y_{21}
Output admittance	y_{ob}	y_{oe}	y_{oc}	y_{22}

Table 4.2 summarises the notations using admittance parameters. These symbols will be used throughout this book but when a more general symbol is required (because the context is not necessarily restricted to a common base or common emitter connection), the symbols in the last column of Table 4.2 are employed.

In the following sections the various factors upon which transistor parameters depend are discussed for a particular type of transistor. The purpose is merely to show how the relevant parameter varies when one or more of the operating conditions is varied. This presentation therefore facilitates insight into the consequences of variation of operating conditions which are intentionally caused (e.g. choice of different biasing point) or which are due to environmental conditions (e.g. change of ambient temperature, variation of supply voltage).

4.3 Dependence of Transistor Admittance Parameters on Operational Conditions

In Book I, Chapter 1 it has been shown theoretically that the admittance parameters of a transistor are dependent on the d.c. biasing point, the signal frequency and the ambient temperature. In the following sections the various dependencies will be illustrated by means of a number of graphs. These graphs represent results of measurements on a batch of alloy-diffused germanium junction transistors of the type AF179 in the common emitter configuration. The average values of parameters measured on this batch of transistors are used for drawing the curves given in the various graphs.

It should be emphasized that it is not the intention to present in this chapter detailed data on the transistors AF179, but merely to show the tendencies in admittances parameter variations under various conditions of a high-frequency transistor.

4.4 Dependence of Transistor Admittance Parameters on Frequency

The dependence on frequency of the average values of the admittance parameters measured on a batch of high-frequency transistors (AF179) is shown in Figs. 4.3 to 4.14.

Three different biasing points are chosen; viz:

$$V_{CE} = -10 \text{ V}, I_E = 0.5 \text{ mA},$$
$$V_{CE} = -10 \text{ V}, I_E = 3 \text{ mA, and}$$
$$V_{CE} = -10 \text{ V}, I_E = 10 \text{ mA}.$$

In all cases the junction temperature of the transistor is 25 °C.

4.4.1 DEPENDENCE OF INPUT ADMITTANCE ON FREQUENCY

Fig. 4.3   p. 22

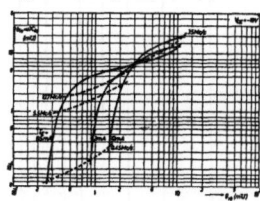

Complex plane representation of the common emitter input admittance y_{ie}.
$b_{ie} = f(g_{ie})$, $V_{CE} = c$, $I_E = c$, $f = c$, $T_j = 25$ °C.

Fig. 4.4
p. 22

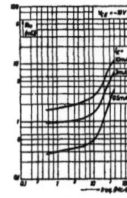

Input damping g_{ie} as a function of frequency.
$g_{ie} = f(\text{freq.})$, $V_{CE} = c$, $I_E = c$, $T_j = 25$ °C.

Fig. 4.5
p. 22

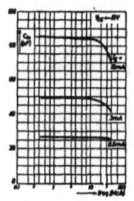

Input capacitance C_{ie} as a function of frequency.
$C_{ie} = f(\text{freq.})$, $V_{CE} = c$, $I_E = c$, $T_j = 25$ °C.

Considering Fig. 4.3 it is seen that the real as well as the imaginary part of the common-emitter input admittance of this transistor increases with frequency and with emitter current. The increase of the real part of y_{ie} may also be seen in Fig. 4.4.

At low values of emitter current the input capacitance C_{ie} is constant as a function of frequency. The input capacitance is larger at higher values of emitter current and then decreases with increasing frequency.

4.4.2 DEPENDENCE OF REVERSE TRANSFER ADMITTANCE ON FREQUENCY

Inspection of Figs. 4.6, 4.7 and 4.8 reveals that the modulus of the common emitter reverse transfer admittance of the transistor increases with frequency. At frequencies up to approximately 5.5 Mc/s the phase angle φ_{re} remains 270°. At higher frequencies φ_{re} becomes smaller, especially at higher values of emitter current. This means that y_{re} is purely capacitive

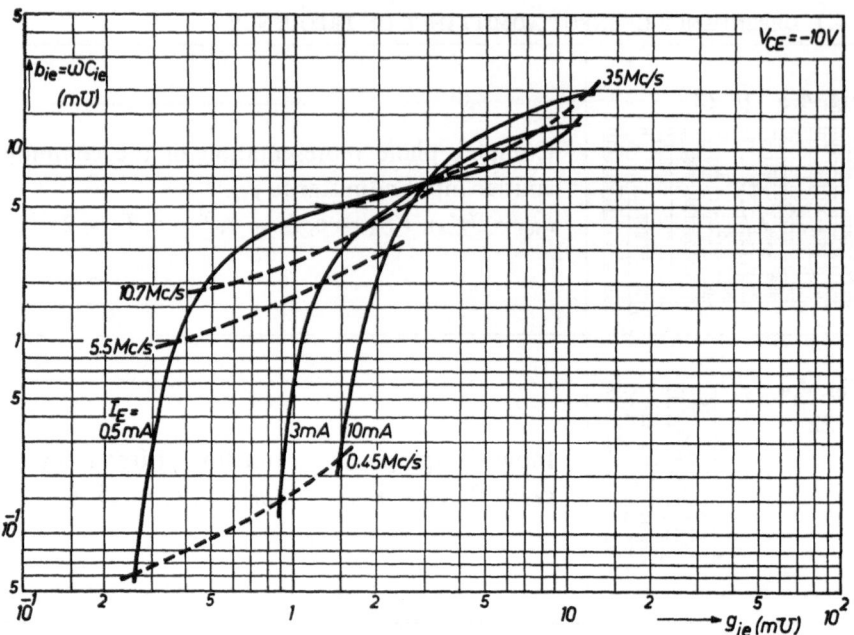

Fig. 4.3. Complex plane representation of the common-emitter input admittance y_{ie}.

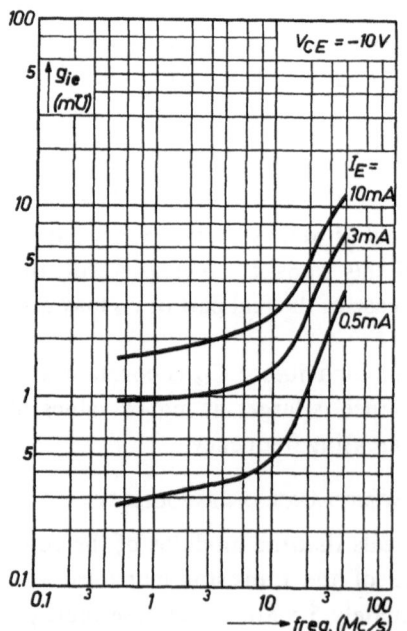

Fig. 4.4. Input damping g_{ie} versus frequency.

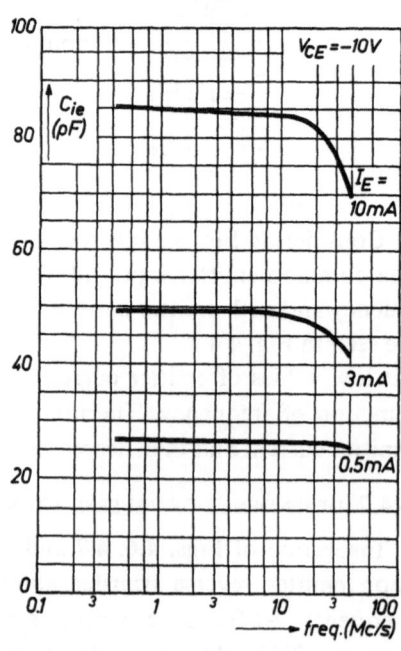

Fig. 4.5. Input capacitance C_{ie} versus frequency.

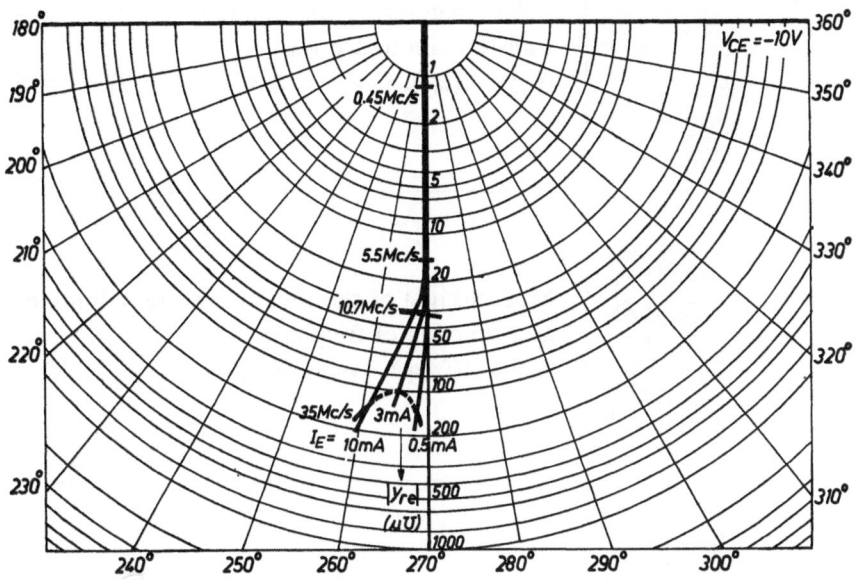

Fig. 4.6. Polar diagram of the common-emitter reverse transfer admittance y_{re}.

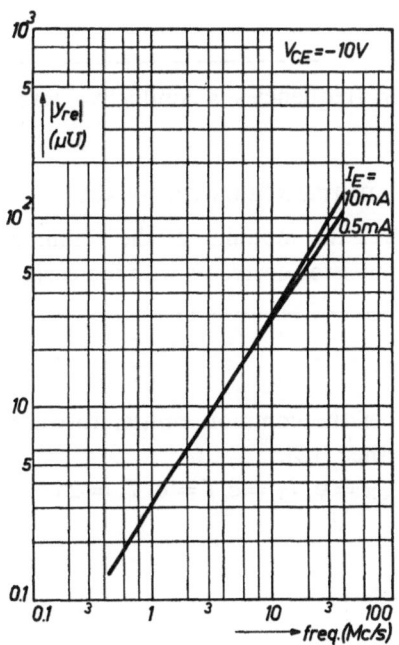

Fig. 4.7. Modulus $|y_{re}|$ of the reverse transfer admittance versus frequency.

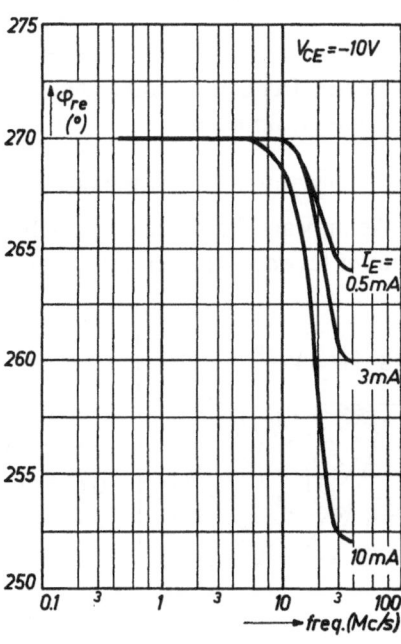

Fig. 4.8. Phase angle φ_{re} of the reverse transfer admittance versus frequency.

Fig. 4.6   p.23

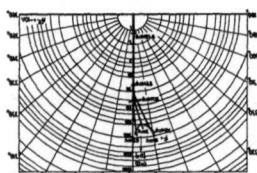

Polar diagram of the common emitter reverse transfer admittance y_{re}.

$|y_{re}| = f(\varphi_{re})$, $V_{CE} = c$, $I_E = c$, $f = c$, $T_j = 25$ °C.

Fig. 4.7
p.23

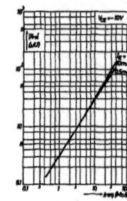

Modulus $|y_{re}|$ of the reverse transfer admittance as a function of frequency.

$|y_{re}| = f(freq.)$, $V_{CE} = c$, $I_E = c$, $T_j = 25$ °C.

Fig. 4.8
p.23

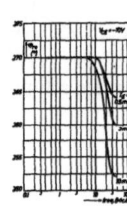

Phase angle φ_{re} of the reverse transfer admittance as a function of frequency.

$\varphi_{re} = f(freq.)$, $V_{CE} = c$, $I_E = c$, $T_j = 25$ °C.

at lower frequencies and that at higher frequencies a conductive term $|y_{re}| \cos \varphi_{re}$ appears.

4.4.3 DEPENDENCE OF FORWARD TRANSFER ADMITTANCE ON FREQUENCY

As appears from Figs. 4.9, 4.10 and 4.11 the modulus of the common emitter forward transfer admittance of a transistor decreases with increasing

Fig. 4.9   p.25

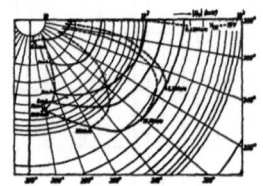

Polar diagram of the common emitter forward transfer admittance $|y_{fe}|$.

$|y_{fe}| = f(\varphi_{fe})$, $V_{CE} = c$, $I_E = c$, $f = c$, $T_j = 25$ °C.

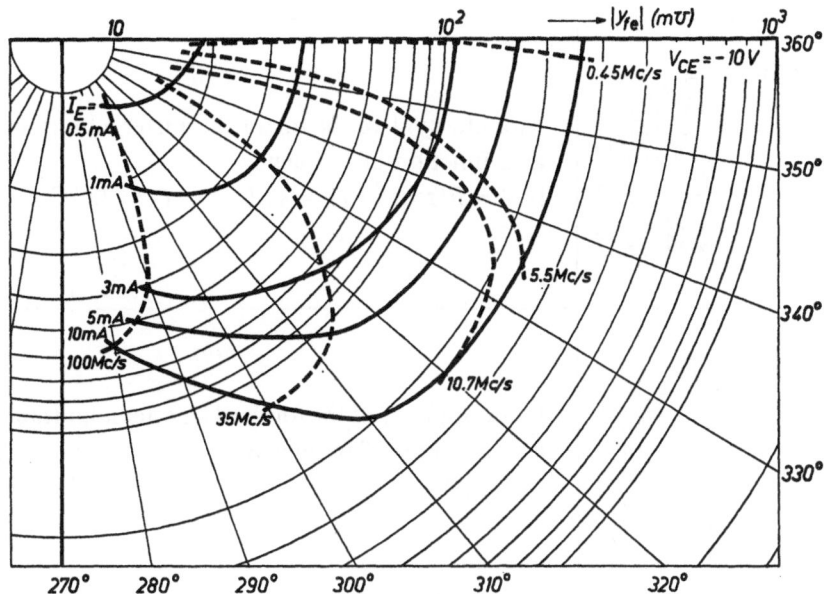

Fig. 4.9. Polar diagram of the common-emitter forward transfer admittance y_{fe}.

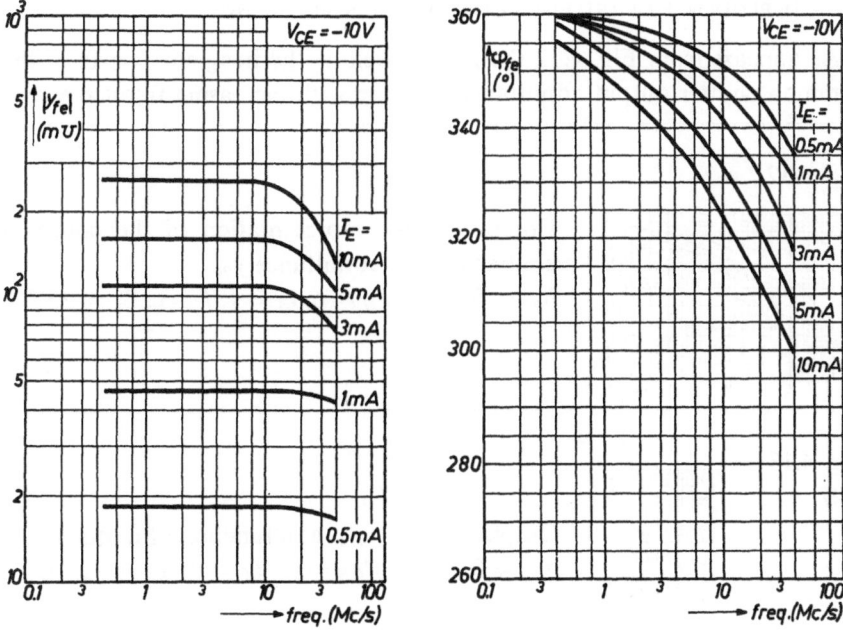

Fig. 4.10. Modulus $|y_{fe}|$ of the forward transfer admittance versus frequency.

Fig. 4.11. Phase angle φ_{fe} of the forward transfer admittance versus frequency.

Fig. 4.10
p.25

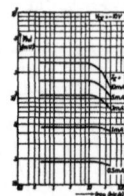

Modulus $|y_{fe}|$ of the forward transfer admittance
as a function of frequency.
$|y_{fe}| = \mathrm{f(freq.)}, V_{CE} = \mathrm{c}, I_E = \mathrm{c}, T_j = 25\ °\mathrm{C}$.

Fig. 4.11
p.25

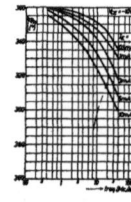

Phase angle φ_{fe} of the forward transfer admittance
as a function of frequency.
$\varphi_{fe} = \mathrm{f(freq.)}, V_{CE} = \mathrm{c}, I_E = \mathrm{c}, T_j = 25\ °\mathrm{C}$.

frequency and increases with increasing emitter current. The phase angle φ_{fe}
decreases with increasing frequency and with increasing emitter current.

4.4.4 DEPENDENCE OF OUTPUT ADMITTANCE ON FREQUENCY

As appears from Fig. 4.12 the real as well as the imaginary parts of the
common emitter output admittance increase with increasing frequency and

Fig. 4.12   p.27

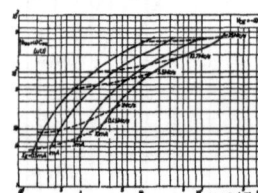

Complex plane representation of the common
emitter output admittance y_{oe}.
$b_{oe} = \mathrm{f}(g_{oe}), V_{CE} = \mathrm{c}, I_E = \mathrm{c}, f = \mathrm{c}, T_j = 25\ °\mathrm{C}$.

Fig. 4.13
p.27

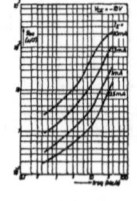

Output damping g_{oe} as a function of frequency.
$g_{oe} = \mathrm{f(freq.)}, V_{CE} = \mathrm{c}, I_E = \mathrm{c}, T_j = 25\ °\mathrm{C}$.

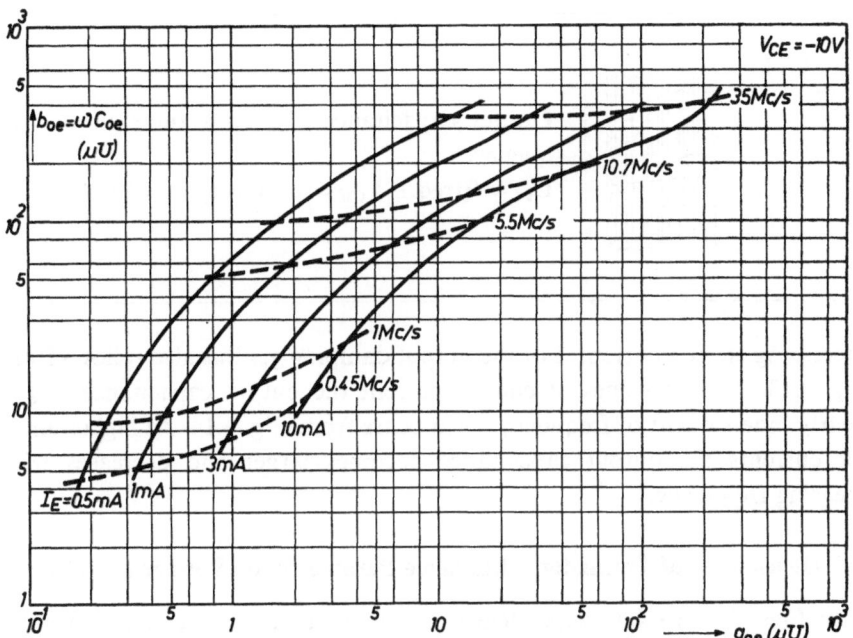

Fig. 4.12. Complex plane representation of the common-emitter output admittance y_{oe}.

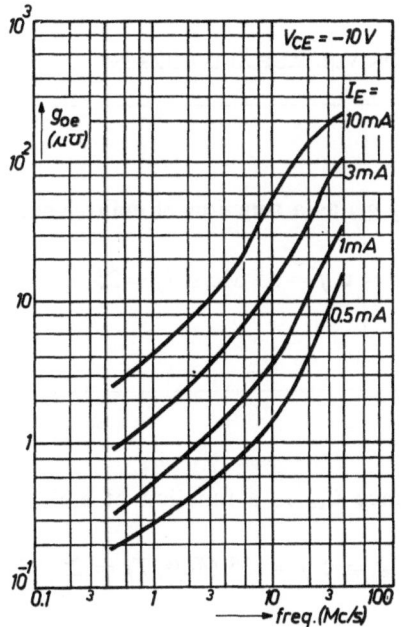

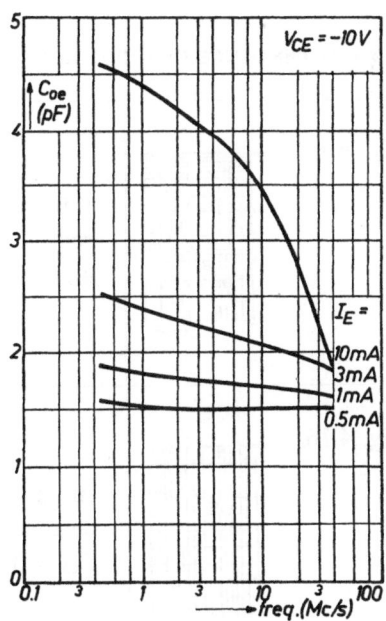

Fig. 4.13. Output damping g_{oe} versus frequency.

Fig. 4.14. Output capacitance C_{oe} versus frequency.

Fig. 4.14
p.27

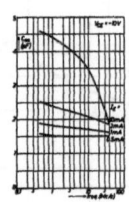

Output capacitance C_{oe} as a function of frequency.

$C_{oe} = f(\text{freq.})$, $V_{CE} = c$, $I_E = c$, $T_j = 25$ °C.

increasing emitter current. For the output damping g_{oe} this also follows from Fig. 4.13. At low values of emitter current the output capacitance C_{oe} is nearly independent of frequency, as appears from Fig. 4.14. At higher emitter currents the output capacitance becomes larger and decreases with creasing frequency.

4.5 Dependence of Transistor Admittance Parameters on Voltage and Current

In Figs. 4.15 to 4.33 dependence on voltage and current of the average values of the common emitter admittance parameters of a batch of transistors AF179 is shown, measured at a signal frequency of 35 Mc/s. The junction temperature of the transistor is 25 °C.

4.5.1 DEPENDENCE OF INPUT ADMITTANCE ON VOLTAGE AND CURRENT

Fig. 4.15
p.29

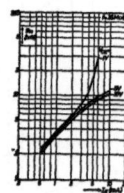

Input damping g_{ie} as a function of emitter current.

$g_{ie} = f(I_E)$, $V_{CE} = c$, $f = 35$ Mc/s, $T_j = 25$ °C.

Fig. 4.16
p.29

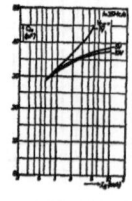

Input capacitance C_{ie} as a function of emitter current.

$C_{ie} = f(I_E)$, $V_{CE} = c$, $f = 35$ Mc/s, $T_j = 25$ °C.

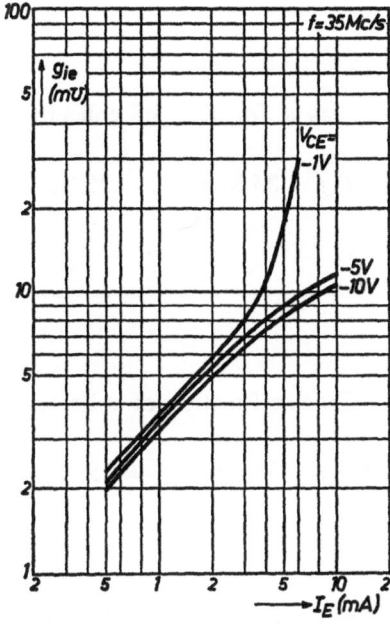

Fig. 4.15. Input damping g_{ie} versus emitter current.

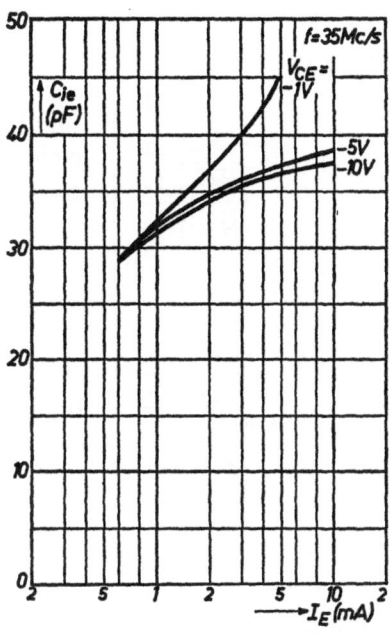

Fig. 4.16. Input capacitance C_{ie} versus emitter current.

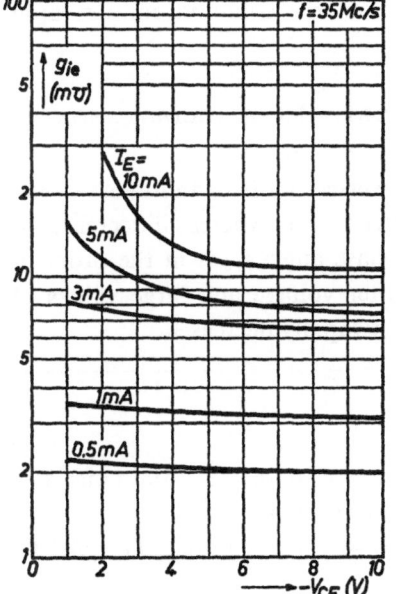

Fig. 4.17. Input damping g_{ie} versus collector-emitter voltage.

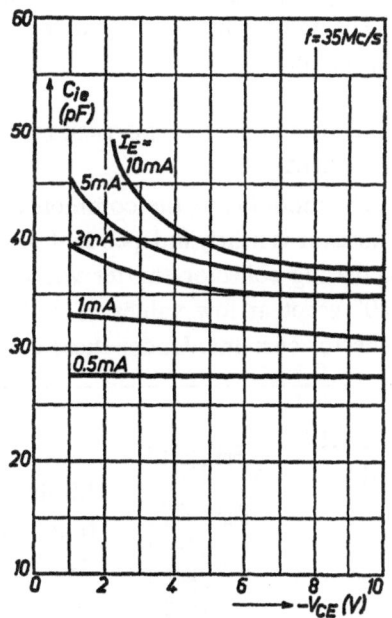

Fig. 4.18. Input capacitance C_{ie} versus collector-emitter voltage.

Fig. 4.17
p.29

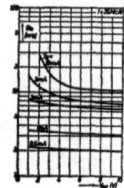

Input damping g_{ie} as a function of collector-emitter voltage.
$g_{ie} = f(V_{CE})$, $I_E = c$, $f = 35$ Mc/s, $T_j = 25$ °C.

Fig. 4.18
p.29

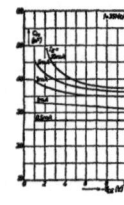

Input capacitance C_{ie} as a function of collector-emitter voltage.
$C_{ie} = f(V_{CE})$, $I_E = c$, $f = 35$ Mc/s, $T_j = 25$ °C.

It can be seen from these graphs that the common emitter input damping g_{ie} increases with increasing emitter current and is nearly independent of the collector-emitter voltage except at very low voltages. Then g_{ie} increases rapidly, especially at high values of emitter current.

The same conclusions can be drawn about the common emitter input capacitance C_{ie}.

4.5.2 DEPENDENCE OF REVERSE TRANSFER ADMITTANCE ON VOLTAGE AND CURRENT

The modulus of the common emitter reverse transfer admittance of the transistor is nearly independent of the emitter current whereas it decreases at increasing collector-emitter voltage. The phase angle φ_{re} is in the order of 260° ecept at low values of collector-emitter voltage and large values of collector currents. Under these conditions it rapidly decreases.

Fig. 4.19
p.31

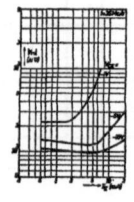

Modulus $|y_{re}|$ of reverse transfer admittance as a function of emitter current.
$|y_{re}| = f(I_E)$, $V_{CE} = c$, $f = 35$ Mc/s, $T_j = 25$° C.

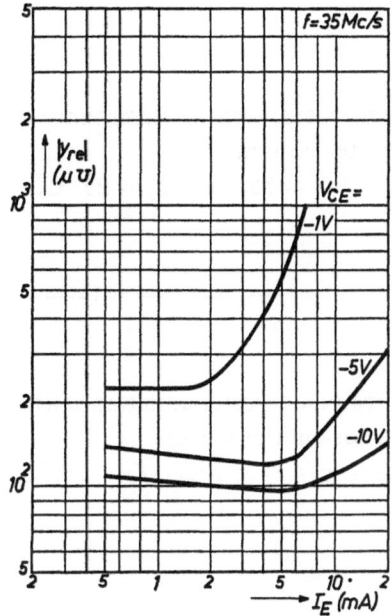

Fig. 4.19. Modulus $|y_{re}|$ of reverse transfer admittance versus emitter current.

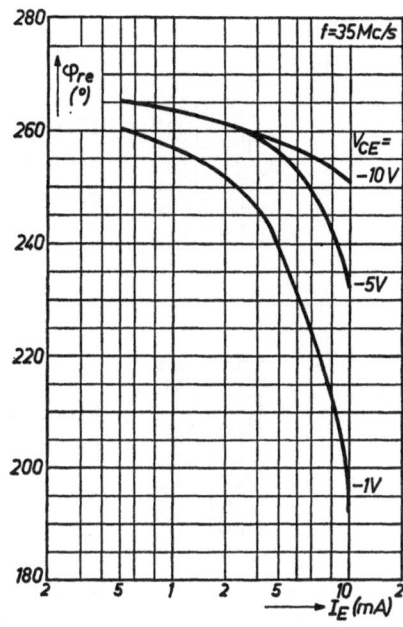

Fig. 4.20. Phase angle φ_{re} of reverse transfer admittance versus emitter current.

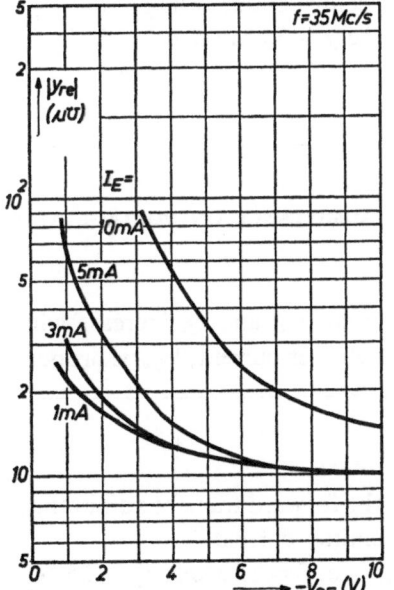

Fig. 4.21. Modulus $|y_{re}|$ of reverse transfer admittance versus collector-emitter voltage.

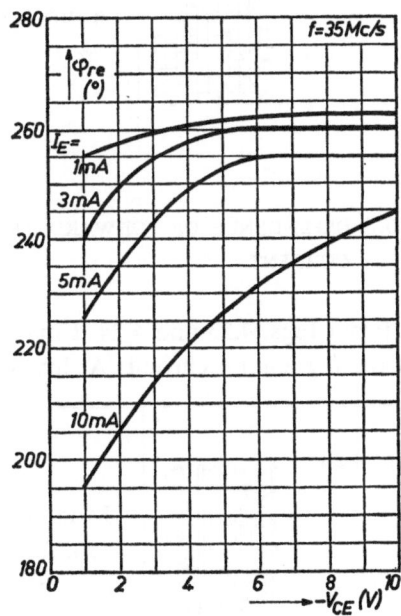

Fig. 4.22. Phase angle φ_{re} of reverse transfer admittance versus collector-emitter voltage.

Fig. 4.20
p.31

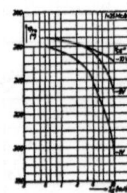

Phase angle φ_{re} of reverse transfer admittance as a function of emitter current.
$\varphi_{re} = f(I_E)$, $V_{CE} = c$, $f = 35$ Mc/s, $T_j = 25$ °C.

Fig. 4.21
p.31

Modulus $|y_{re}|$ of reverse transfer admittance as a function of collector-emitter voltage.
$|y_{re}| = f(V_{CE})$, $I_E = c$, $f = 35$ Mc/s, $T_j = 25$ °C.

Fig. 4.22
p.31

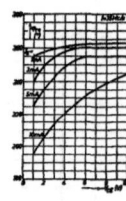

Phase angle φ_{re} of reverse transfer admittance as a function of collector emitter voltage.
$\varphi_{re} = f(V_{CE})$, $I_E = c$, $f = 35$ Mc/s, $T_j = 25$ °C.

4.5.3 DEPENDENCE OF FORWARD TRANSFER ADMITTANCE ON VOLTAGE AND CURRENT

From Figs. 4.23 and 4.24 it may be seen that $|y_{fe}|$ and φ_{fe} increase with increasing emitter current. At low values of emitter current, $|y_{fe}|$ is propor-

Fig. 4.23
p.33

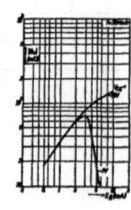

Modulus $|y_{fe}|$ of the forward transfer admittance as a function of emitter current.
$|y_{fe}| = f(I_E)$, $V_{CE} = c$, $f = 35$ Mc/s, $T_j = 25$ °C.

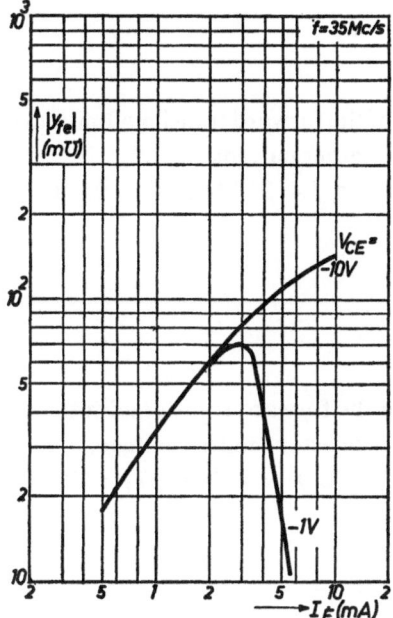

Fig. 4.23. Modulus $|y_{fe}|$ of the forward transfer admittance versus emitter current.

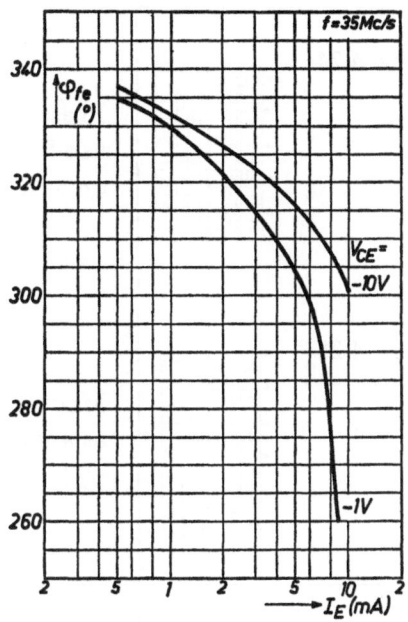

Fig. 4.24. Phase angle φ_{fe} of the forward transfer admittance versus emitter current.

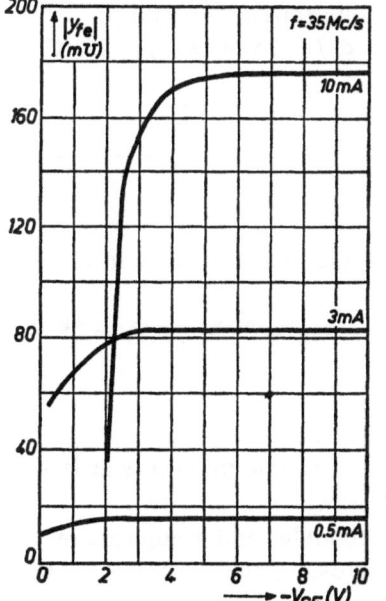

Fig. 4.25. Modulus $|y_{fe}|$ of the forward transfer admittance versus collector-emitter voltage.

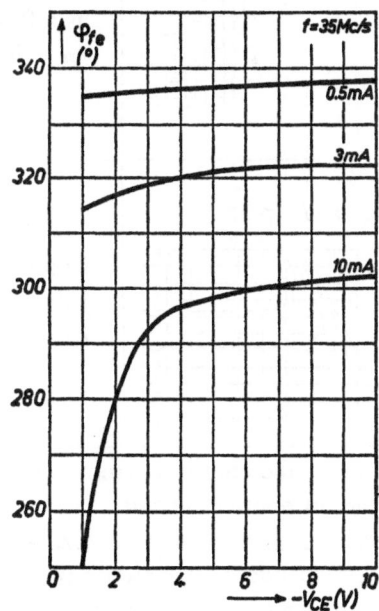

Fig. 4.26. Phase angle φ_{fe} of the forward transfer admittance versus collector-emitter voltage.

Fig. 4.24
p.33

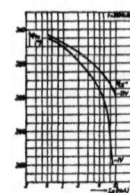

Phase angle φ_{fe} of the forward transfer admittance as a function of emitter current.

$\varphi_{fe} = f(I_E)$, $V_{CE} = c$, $f = 35$ Mc/s, $T_j = 25$ °C.

Fig. 4.25
p.33

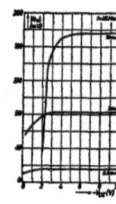

Modulus $|y_{fe}|$ of the forward transfer admittance as a function of collector-emitter voltage.

$|y_{fe}| = f(V_{CE})$, $I_E = c$, $f = 35$ Mc/s, $T_j = 25$ °C.

Fig. 4.26
p.33

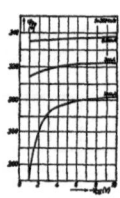

Phase angle φ_{fe} of the forward transfer admittance as a function of collector-emitter voltage.

$\varphi_{fe} = f(V_{CE})$, $I_E = c$, $f = 35$ Mc/s, $T_j = 25$ °C.

Fig. 4.27   p.35

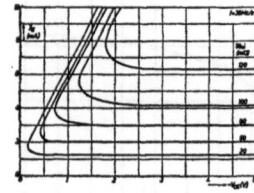

Representation of the modulus of y_{fe} in the I_C, V_{CE} plane.

$I_C = f(V_{CE})$, $|y_{fe}| = c$, $f = 35$ Mc/s, $T_j = 25$ °C.

tional to the emitter current, whereas a kind of saturation occurs at larger values of emitter current.

As is seen in Figs. 4.25, 4.26 and 4.27 the modulus and arguments of y_{fe} are independent of collector-emitter voltage, provided this voltage is not too low. A kind of "knee effect" occurs at low collector-emitter voltages. This "knee region" in which the high-frequency properties of the transistor rapidly deteriorate, lies well above the statical knee (as measured d.c.-wise).

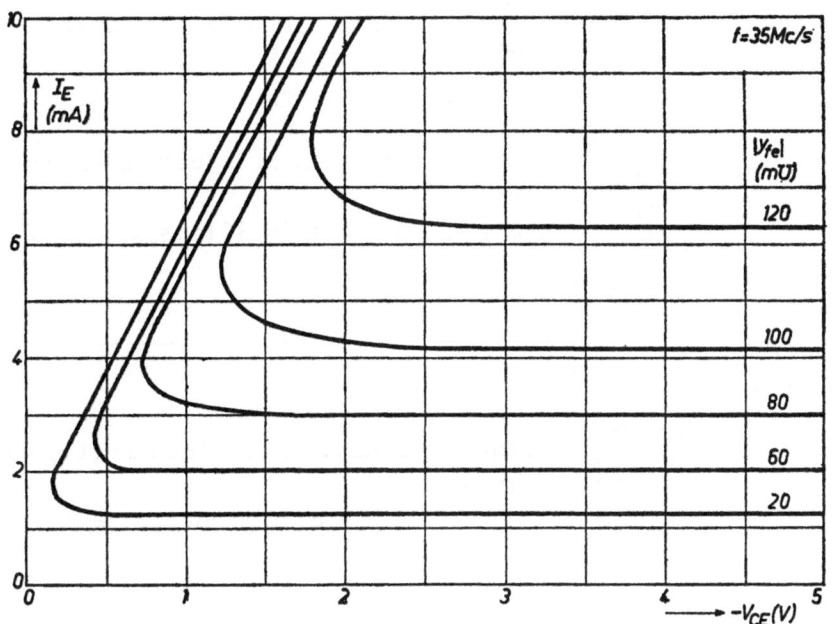

Fig. 4.27. Representation of the modulus of y_{fe} in the $I_E - V_{CE}$ plane.

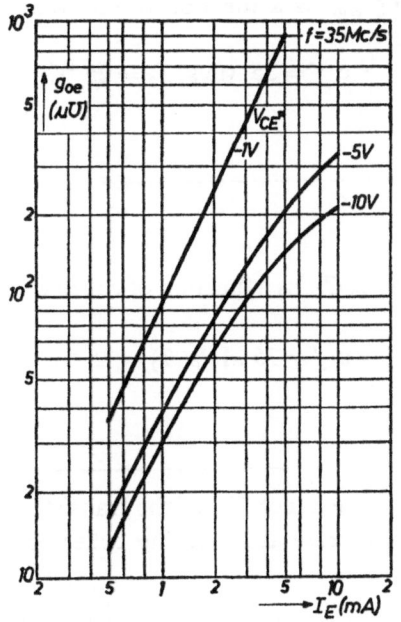

Fig. 4.28. Output damping g_{oe} versus emitter current.

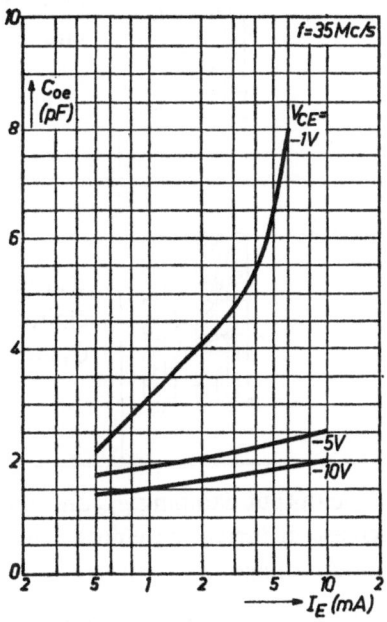

Fig. 4.29. Output capacitance C_{oe} versus emitter current.

4.5.4 DEPENDENCE OF OUTPUT ADMITTANCE ON VOLTAGE AND CURRENT

Fig. 4.28 p.35	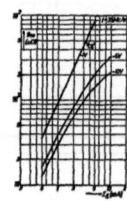	Output damping g_{oe} as a function of emitter current. $g_{oe} = \mathrm{f}(I_E)$, $V_{CE} = \mathrm{c}$, $f = 35$ Mc/s, $T_j = 25°$ C.
Fig. 4.29 p.35	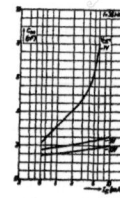	Output capacitance C_{oe} as a function of emitter current. $C_{oe} = \mathrm{f}(I_E)$, $V_{CE} = \mathrm{c}$, $f = 35$ Mc/s, $T_j = 25$ °C.
Fig. 4.30 p.37	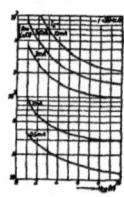	Output damping g_{oe} as a function of collector-emitter voltage. $g_{oe} = \mathrm{f}(V_{CE})$, $I_E = \mathrm{c}$, $f = 35$ Mc/s, $T_j = 25$ °C.
Fig. 4.31 p.37	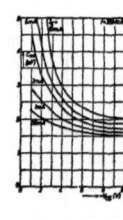	Output capacitance C_{oe} as a function of collector-emitter voltage. $C_{oe} = \mathrm{f}(V_{CE})$, $I_E = \mathrm{c}$, $f = 35$ Mc/s, $T_j = 25$ °C.

The output damping g_{oe} and the output capacitance C_{oe} of the transistor increase at increasing emitter current and decrease at increasing collector-emitter voltage.

The $I_C - V_{CE}$ plane representation of g_{oe} and C_{oe} shown in Figs. 4.32 and 4.33 respectively, facilitates the determination of the variation in output

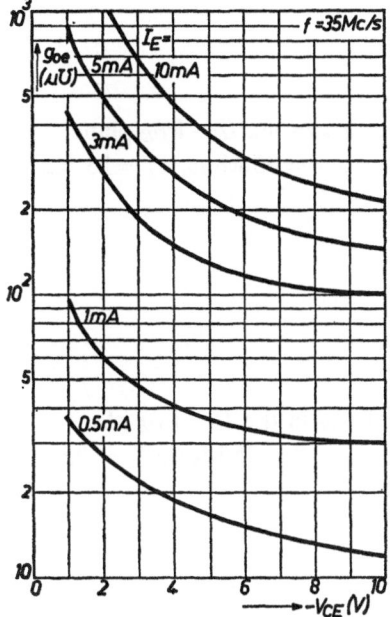

Fig. 4.30. Output damping g_{oe} versus collector-emitter voltage.

Fig. 4.31. Output capacitance C_{oe} versus collector-emitter voltage.

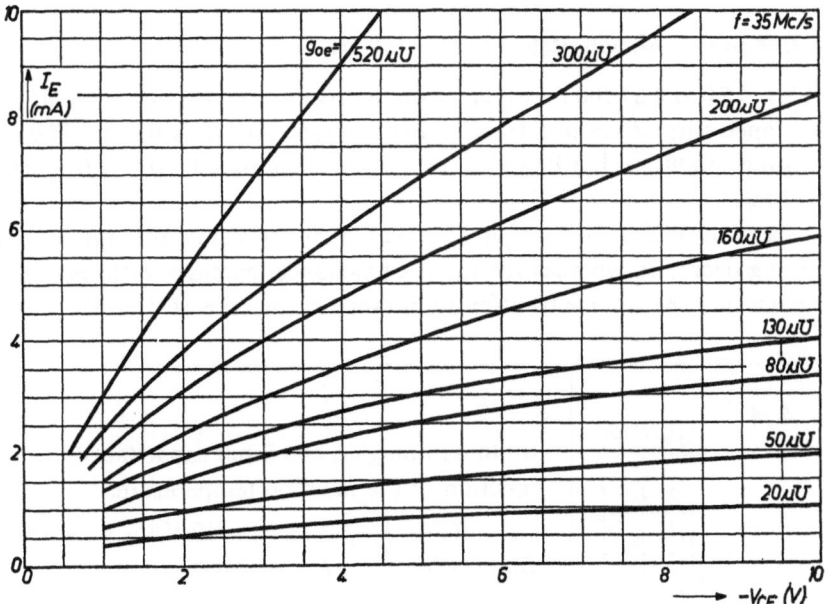

Fig. 4.32. Representation of the output damping g_{oe} in the $I_E - V_{CE}$ plane.

Fig. 4.32   p.37

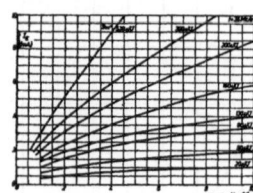

Representation of g_{oe} in the I_C-V_{CE} plane.
$I_C = \mathrm{f}(V_{CE})$, $g_{oe} = \mathrm{c}$, $f = 35$ Mc/s, $T_j = 25$ °C.

Fig. 4.33   p.39

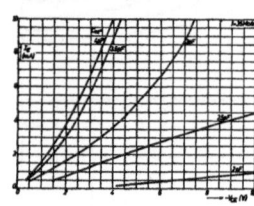

Representation of C_{oe} in the I_C-V_{CE} plane.
$I_C = \mathrm{f}(V_{CE})$, $C_{oe} = \mathrm{c}$, $f = 35$ Mc/s, $T_j = 25$° C.

damping and output capacitance that occurs when the operating point of the transistor moves along a certain load line during large signal excursion.

4.6 Dependence of Admittance Parameters on Temperature

In Figs. 4.34 to 4.37 the dependence on temperature of the common emitter admittance parameters of a transistor of the type AF179 is shown. The d.c. operating conditions are $V_{CE} = -10$V and $I_E = 3$mA. The signal frequency is 35 Mc/s.

Inspection of Figs. 4.34 to 4.37 reveals that the common emitter input damping decreases whereas the input capacitance increases with increasing temperature. The reverse transfer admittance of the type of transistor under

Fig. 4.34
p.39

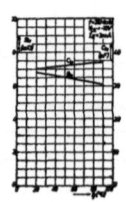

Input damping g_{ie} and input capacitance C_{ie} as a function of the junction temperature.

$g_{ie} = \mathrm{f}(T_j)$, $\quad V_{CE} = -10$ V, $I_E = 3$ mA,
$C_{ie} = \mathrm{f}(T_j)$, $\quad f = 35$ Mc/s.

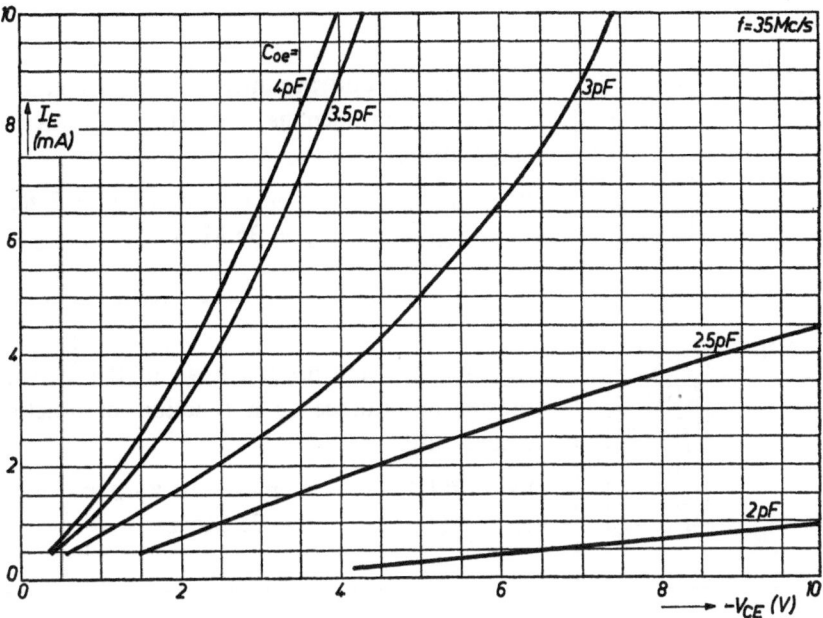

Fig. 4.33. Representation of the output capacitance C_{oe} in the $I_E - V_{CE}$ plane.

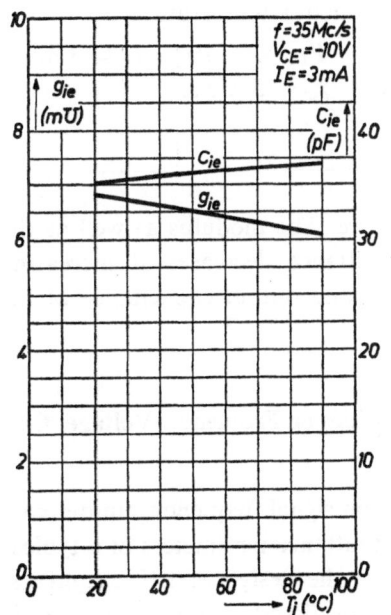

Fig. 4.34. Input damping g_{ie} and input capacitance C_{ie} versus junction temperature.

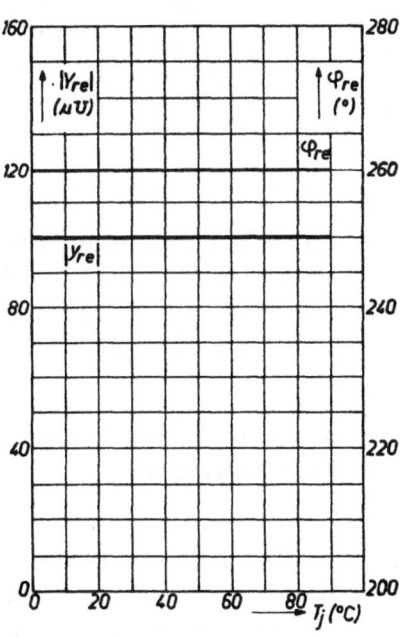

Fig. 4.35. Modulus $|y_{re}|$ and argument φ_{re} of the reverse transfer admittance versus junction temperature.

Fig. 4.35 p. 39	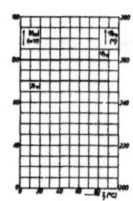	Modulus $\lvert y_{fe}\rvert$ and argument φ_{fe} of the reverse transfer admittance as a function of junction temperature. $\lvert y_{re}\rvert = \mathrm{f}(T_j),$ ⟩ $\quad V_{CE} = -10\ \mathrm{V},\ I_E = 3\ \mathrm{mA},$ $\varphi_{re}\ = \mathrm{f}(T_j),$ ⟨ $\quad f = 35\ \mathrm{Mc/s}.$
Fig. 4.36 p. 42	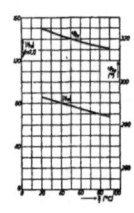	Modulus $\lvert y_{fe}\rvert$ and argument φ_{fe} of the forward transfer admittance as a function of junction temperature. $\lvert y_{fe}\rvert = \mathrm{f}(T_j),$ ⟩ $\quad V_{CE} = -10\ \mathrm{V},\ I_E = 3\ \mathrm{mA},$ $\varphi_{fe}\ = \mathrm{f}(T_j),$ ⟨ $\quad f = 35\ \mathrm{Mc/s}.$
Fig. 4.37 p. 42	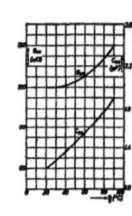	Output damping g_{oe} and output capacitance C_{oe} as a function of junction temperature. $g_{oe}\ = \mathrm{f}(T_j),$ ⟩ $\quad V_{CE} = -10\ \mathrm{V},\ I_E = 3\ \mathrm{mA},$ $C_{oe} = \mathrm{f}(T_j),$ ⟨ $\quad f = 35\ \mathrm{Mc/s}.$

consideration is independent of temperature. The modulus as well as the phase angle of the forward transfer admittance decreases at increasing junction temperature. The output damping and output capacitance of the transistor increase with temperature.

4.7 Dependence of Derived Transistor Quantities on Frequency, Voltage, Current and Temperature

In the preceding sections the dependence on signal frequency, emitter current and collector-emitter voltage of the admittance parameters of a typical high-frequency transistor are considered. In the design of I.F. amplifiers, however, these admittances themselves are not directly employed but use is made of quantities derived from these parameters. These quantities are:

— the unilateralized power gain Φ_{uM},

— the regeneration phase angle Θ,

— the transfer admittance ratio N, and

— the intrinsic regeneration coefficient t.

These quantities are defined according to the expressions given in Eqs. (4.7.1) to (4.7.4) below. (Detailed discussion of these quantities is found in relevant sections in Book I)

$$\Phi_{uM} = \frac{|y_{fe}|^2}{4 \, g_{ie} \, g_{oe}}. \tag{4.7.1}$$

$$\Theta = \varphi_{re} + \varphi_{fe}, \tag{4.7.2}$$

$$N = \frac{|y_{fe}|}{|y_{re}|}, \tag{4.7.3}$$

and

$$t = \frac{|y_{re} \, y_{fe}|}{g_{ie} \, g_{oe}}. \tag{4.7.4}$$

In Figs. 4.38 to 4.42 the ways in which Φ_{uM}, Θ, N and t depend on signal frequency, d.c. biasing conditions and junction temperature are plotted. The various curves are derived from the relationships considered in Sections 4.4, 4.5 and 4.6.

Inspection of Fig. 4.38 reveals that both Φ_{uM} and Θ decrease at increasing frequencies. This is also true of the quantities N and t, see Fig. 4.39.

Fig. 4.38   p.42

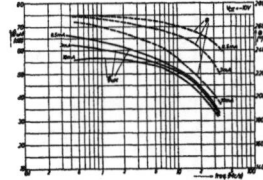

Maximum unilateralized power gain Φ_{uM} and regeneration phase angle Θ as a functions of frequency.

$\Phi_{u\,M} = $ f(freq.) $\Big\}$ $I_E = $ c, $V_{CE} = -10$ V,

$\Theta = $ f(freq.) $T_j = 25$ °C.

Fig. 4.39   p.43

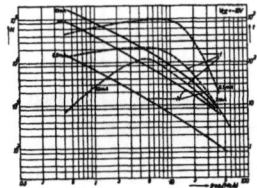

Transfer admittance ratio N and intrinsic regenration coefficient t as functions of frequency.

$N = $ f(freq.), $I_E = $ c, $V_{CE} = -10$ V,

$t = $ f(freq.), $T_j = 25$ °C.

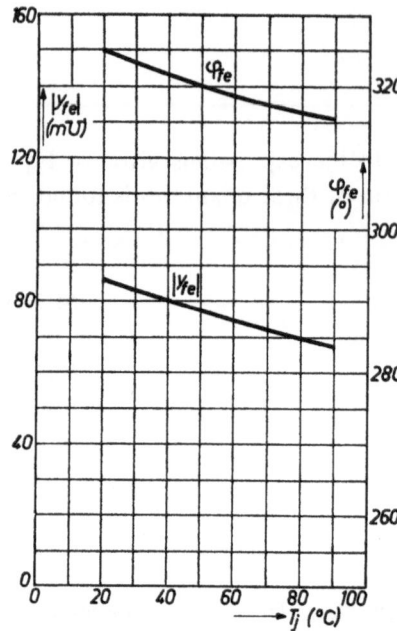

Fig. 4.36. Modulus $|y_{fe}|$ and argument φ_{fe} of the forward transfer admittance versus junction temperature.

Fig. 4.37. Output damping g_{oe} and output capacitance C_{oe} versus junction temperature.

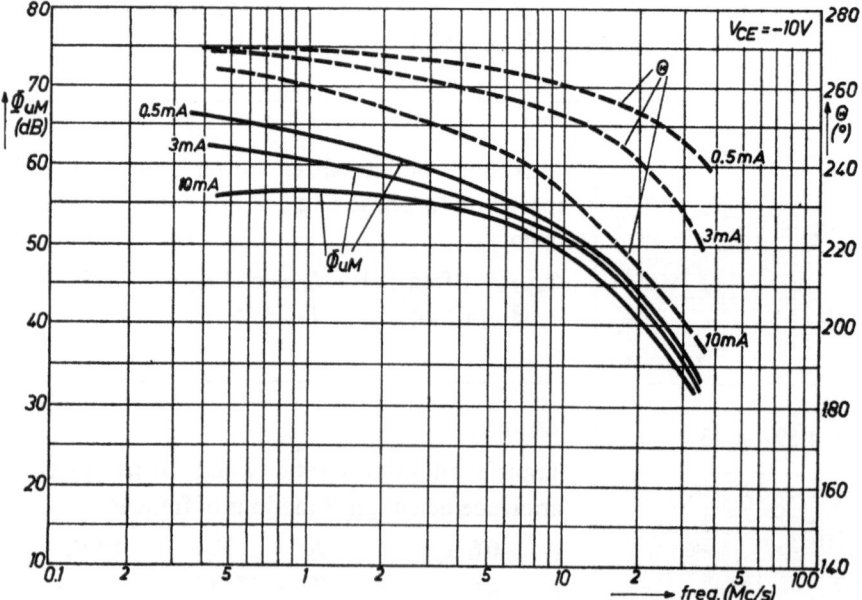

Fig. 4.38. Maximum unilateralized powergain Φ_{uM} and regeneration phase angle Θ versus frequency.

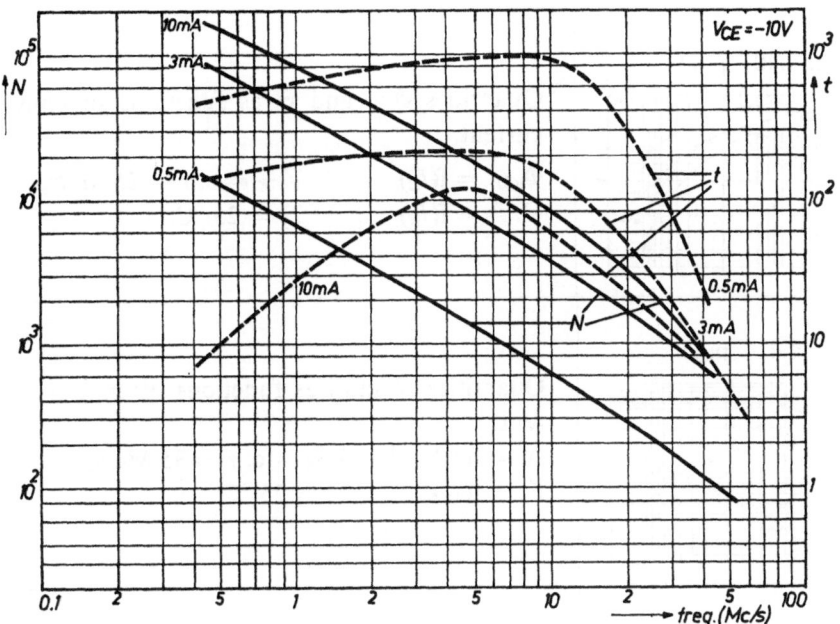

Fig. 4.39. Transfer admittance ratio N and intrinsic regeneration coefficient t versus frequency.

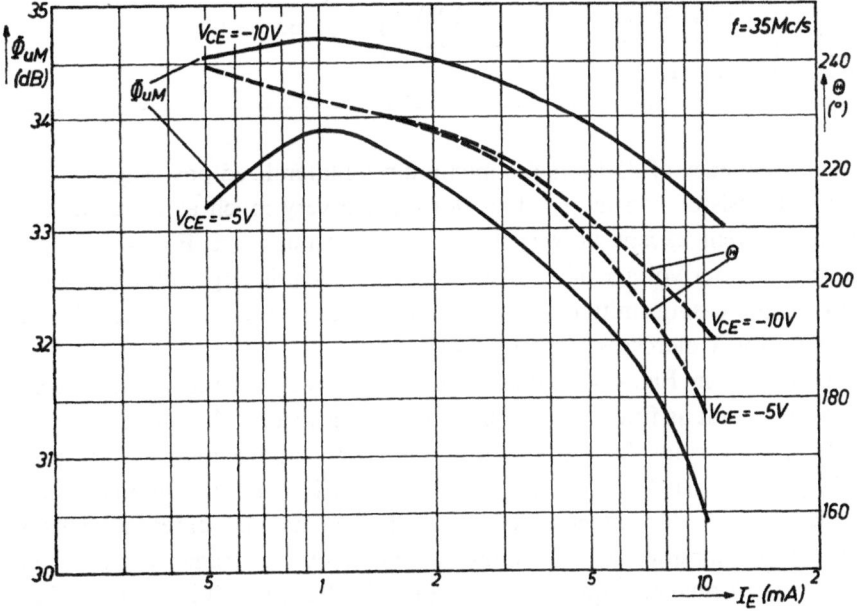

Fig. 4.40. Quantities Φ_{uM} and Θ versus emitter current.

Fig. 4.40   p.43

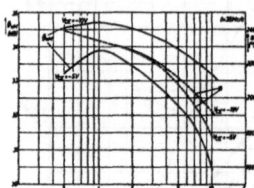

Quantities Φ_{uM} and Θ as functions of emitter current.

$\Phi_{uM} = \mathrm{f}(I_E),$ $V_{CE} = \mathrm{c}, f = 35 \text{ Mc/s},$
$\Theta \quad = \mathrm{f}(I_E),$ $T_j = 25 \text{ °C}.$

Fig. 4.41   p.45

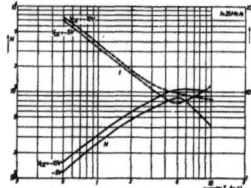

Quantities N and t as functions of emitter current.

$N = \mathrm{f}(I_E),$ $V_{CE} = \mathrm{c}, f = 35 \text{ Mc/s},$
$t = \mathrm{f}(I_E),$ $T_j = 25 \text{ °C}.$

Fig. 4.42   p.45

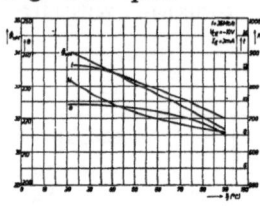

Quantities Φ_{uM}, Θ, N and t as functions of junction temperature.

$\Phi_{uM} = \mathrm{f}(T_j),$
$\Theta \quad = \mathrm{f}(T_j),$ $V_{CE} = -10 \text{ V}, I_E = 3 \text{ mA},$
$N \quad = \mathrm{f}(T_j),$ $f = 35 \text{ Mc/s}.$
$t \quad = \mathrm{f}(T_j),$

At a frequency of 35 Mc/s the maximum unilateralized power gain of the type of transistor considered does not change very much as a function of emitter current except at very large current values, see Fig. 4.40. At a collector-emitter voltage of $-$ 10V the value of Φ_{uM} is larger than at $V_{CE} = -5$V. At increasing emitter current the quantities Θ and t decrease whereas the quantity N increases, see Fig. 4.41.

The curves in Fig. 4.42 indicate that the various derived quantities are almost constant as functions of junction temperature. Although the dependence of the various derived quantities considered above gives an indication of the gain that may be expected from a transistor in a given amplifier circuit, the actual gain dependencies of the amplifier may be different. This is because the quantities Φ_{uM}, Θ, N and t are based only on transistor properties and any performance requirement the amplifier has to fulfil is not taken into account.

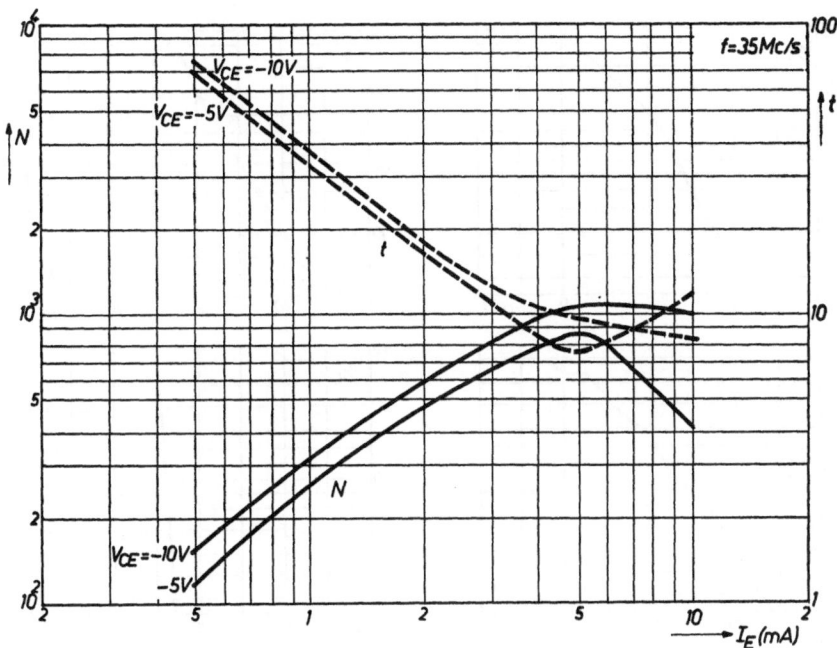

Fig. 4.41. Quantities N and t versus emitter current.

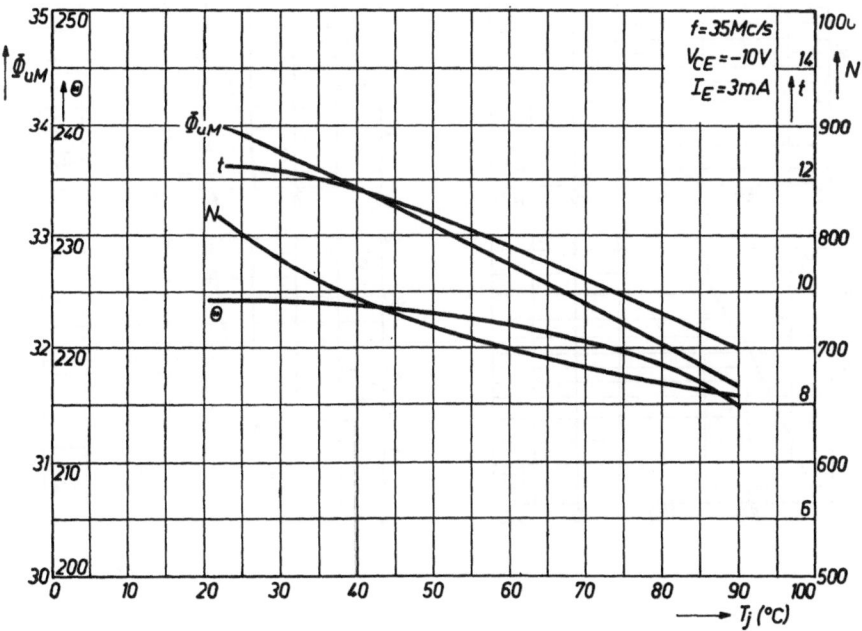

Fig. 4.42. Quantities Φ_{uM}, Θ, N and t versus junction temperature.

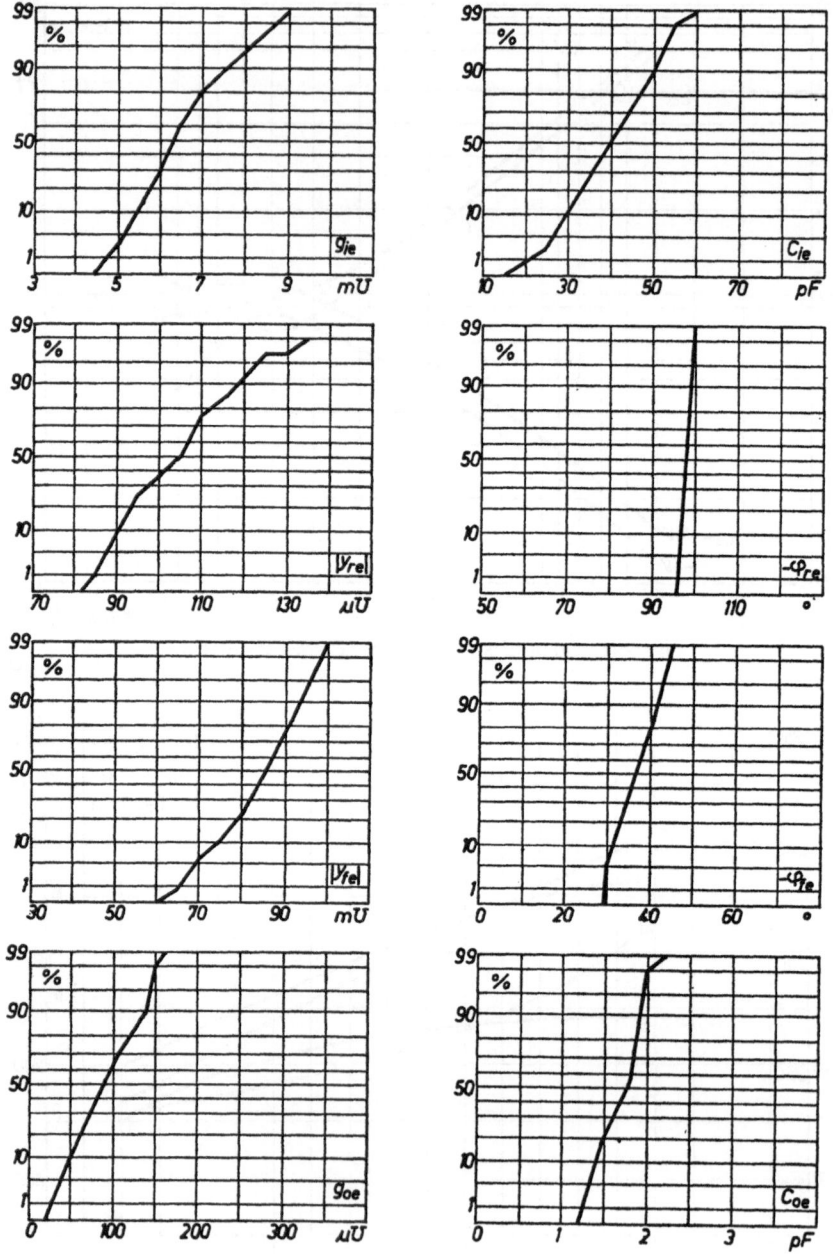

Fig. 4.43. Spreads in admittance parameters measured on a batch of transistors AF 179.

4.8 Spreads in Transistor Admittance Parameters

In the batch of high-frequency transistors of the type AF179 measured for illustrative purposes in this chapter, the parameters are found to spread around an average value as shown in the graphs given in Fig. 4.43. The spread graphs shown in this figure are derived from measurements at a signal frequency of 35 Mc/s, with a transistor biasing point of $V_{CE} = -10\text{V}$ and $I_E = 3$ mA and a transistor junction temperature of 25 °C.

The spread graphs of Fig. 4.43, obviously, do not indicate any correlation in the spreads of the admittance parameters of the type of high-frequency transistor considered.

CHAPTER 5

SURVEY OF I.F. AMPLIFIER DESIGN THEORY

In this chapter a general survey will be given of the theory of designing transistorized I.F. amplifiers as presented in Book I. This survey constitutes the basis for the practical design procedure which will be developed in the following chapters.

The various aspects of amplifier design that need be considered are:

— stability
   this term refers to the property of the amplifier to be protected to a certain extent against self-regeneration.

— alignment
   this term covers the property of the amplifier of being alignable in a straightforward manner.

— amplification
   this usually denotes the gain of the amplifier in the centre or in the flat portion of the amplitude response curve: for transistorized I.F. amplifiers this amplification is, in most instances, specified in terms of transimpedance or transducer gain.

— response curve
   generally the amplitude response as well as the envelope delay of the amplifier as functions of the frequency, in most cases, normalized with respect to the mid-band frequency and magnitude.

The various design aspects mentioned are considered in some detail in the following sections. We will confine ourselves to amplifiers consisting of one or more stages in which the interstage coupling networks are either single-tuned or double-tuned bandpass filters.

Throughout this chapter references will be made to relevant sections of Book I.

5.1 Stability

The prime design requirement of an I.F. amplifier is that it be adequately stable in all conditions that may occur in practice. It is therefore necessary to investigate the conditions in which the amplifier reaches the verge of self-oscillation. A certain safety margin must then be included in the design, to ensure that the amplifier remains stable in all practical conditions. In Book I boundaries of stability of various amplifier arrangements were determined. In the following sub-sections these stability considerations will be summarized.

5.1.1. SINGLE-STAGE AMPLIFIER WITH TWO SINGLE-TUNED BANDPASS FILTERS

The simplest arrangement is a single-stage amplifier with two single-tuned circuits. In Fig. 5.1 a schematic circuit diagram is shown, which can be simplified to that shown in Fig. 5.2 by combining the various admittances at input and output terminals of the transistor.

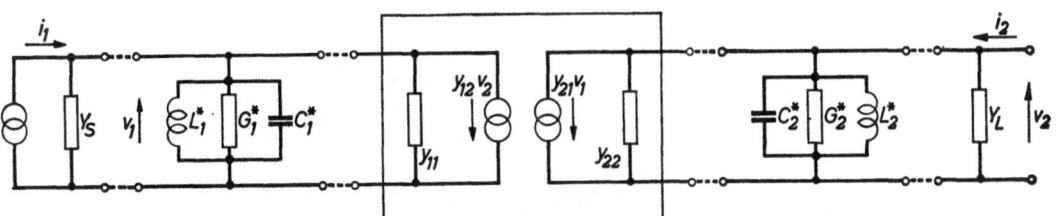

Fig. 5.1. Schematic diagram of a single-stage amplifier with single-tuned circuits at the input and output terminals. The active fourpole represents the transistor, Y_S denotes the admittance of the current source which drives the amplifier, and Y_L the load admittance of the amplifier.

Fig. 5.2. Simplified diagram of the circuit presented in Fig. 5.1. The admittances Y_S, Y_1^* (of the input tuned circuit) and y_{11} have been combined in a single admittance $Y_1 = Y_S + Y_1^* + y_{11}$. Also $Y_2 = y_{22} + Y_2^* + Y_L$.

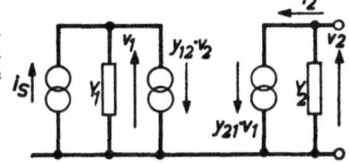

The admittances Y in Fig. 5.2 represent single-tuned circuits for which we define:

$$Y = G\,(1 + jx) \tag{5.1.1}$$

$$x = \beta Q \tag{5.1.2}$$

$$\beta = \frac{\omega}{\omega_0} - \frac{\omega_0}{\omega} \tag{5.1.3}$$

and $$Q = \frac{\omega_0 C}{G}$$ (5.1.4)

In these expressions:

> G is the total damping of the tuned circuit
> x is the normalized detuning
> β is the relative detuning, and
> Q is the quality factor.

These concepts are extensively dealt with in Book I, Appendix II.

Furthermore we introduce a *regeneration coefficient T* and an associated *regeneration phase angle* Θ for the transistor, which are defined as follows:

$$T = \frac{|y_{12}\, y_{21}|}{G_1 G_2}$$ (5.1.5)

and $$\Theta = \varphi_{12} + \varphi_{21}.$$ (5.1.6)

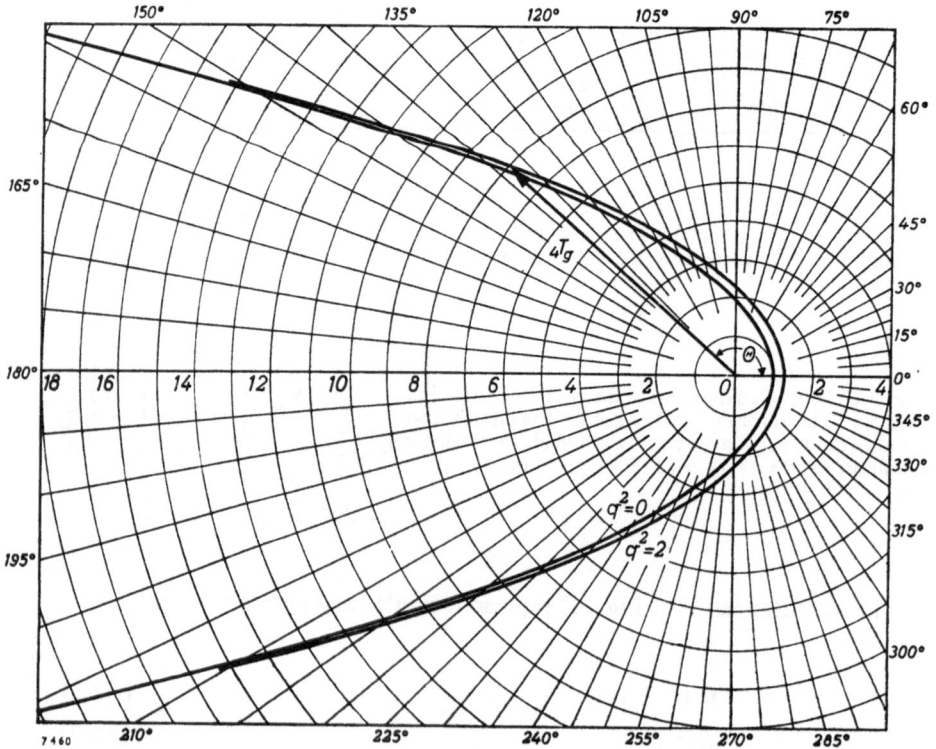

Fig. 5.3. Boundary of stability of a single-stage amplifier with two identical single-tuned bandpass filters.

It can then be calculated that the single-stage amplifier is at the boundary of stability if:

$$(1 + jx_1)(1 + jx_2) - T_g \exp(j\Theta) = 0. \tag{5.1.7}$$

In this expression T_g denotes the value of T at the boundary of stability. For identical tuned circuits at the input and output terminals of the amplifier, it follows that:

$$T_g = \frac{2}{1 + \cos \Theta}. \tag{5.1.8}$$

In Fig. 5.3 this boundary of stability has been plotted as a function of Θ. Stability of the amplifier is ensured when:

$$T < T_g. \tag{5.1.9}$$

To prevent the amplifier from becoming unstable due to spreads in transistor parameters or to change of operational or environmental conditions, T is usually made smaller than T_g by a certain factor s, the *stability factor*. Then:

$$T = \frac{T_g}{s}. \tag{5.1.10}$$

A detailed discussion of the boundary of stability of the single-stage amplifier is given in Book I, Section 2.2, especially sub-sections 2.2.2 and 2.2.4.

5.1.2 MULTI-STAGE AMPLIFIER WITH SINGLE-TUNED BANDPASS FILTERS

In Fig. 5.4 a schematic diagram is shown of a multi-stage amplifier with single-tuned circuits as interstage coupling networks. An equivalent circuit diagram is presented in Fig. 5.5. In such an amplifier the stability of the various stages is less than that of an isolated stage due to the fact that the feedback of each stage affects the operation of all other stages.

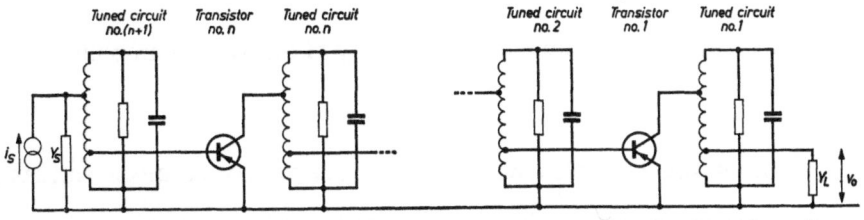

Fig. 5.4. Schematic diagram of a multi-stage amplifier with single-tuned bandpass filters.

Fig. 5.5. Equivalent circuit diagram of the amplifier presented in Fig. 5.4.

The boundary of stability of the multi-stage amplifier can be obtained from:

$$T_g = \frac{2}{1 + \cos \Theta} \cdot {}_n u_g \qquad (5.1.11)$$

in which the constant ${}_n u_g$ takes account of the reduction in stability due to the cascading of the stages.

The value of ${}_n u_g$ is given in Table 5.1.

TABLE 5.1.

Stability reduction factor

n	${}_n u_g$
1	1.00
2	0.50
3	0.38
4	0.33
5	0.31
6	0.29
7	0.28
8	0.28
9	0.27
10	0.27
∞	0.25

For a practical amplifier, including a stability factor s, we find:

$$T = \frac{2 {}_n u_g}{s(1 + \cos \Theta)} \qquad (5.1.12)$$

For further details on the boundary of stability of this type of amplifier, see Book I, Chapter 6.

5.1.3 SINGLE-STAGE AMPLIFIER WITH TWO DOUBLE-TUNED BANDPASS FILTERS

A single-stage amplifier with double-tuned bandpass filters as shown

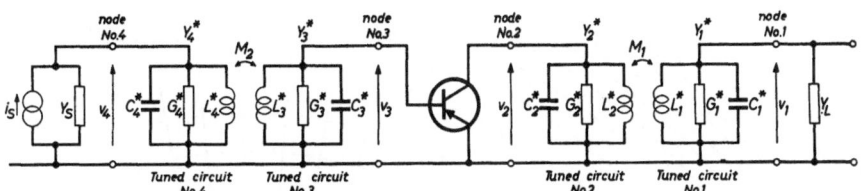

Fig. 5.6. Schematic circuit diagram of a single-stage amplifier with two double-tuned bandpass filters.

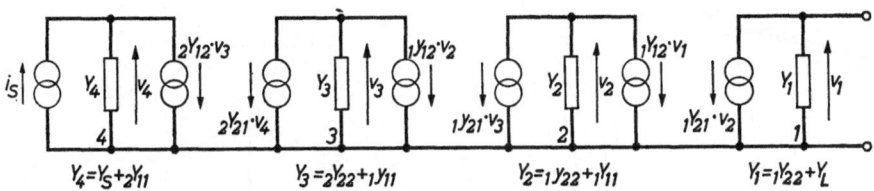

Fig. 5.7. Simplified equivalent circuit of the amplifier represented in Fig. 5.6.

schematically in Fig. 5.6 can be represented by the simplified diagram of Fig. 5.7.

The same notations are used for the single-tuned circuits of which the double-tuned bandpass filters are composed as in sub-section 5.1.1. Furthermore, a coupling coefficient q^2 is introduced representing the amount of coupling between primary and secondary of the double-tuned bandpass filters. This coupling coefficient is defined as:

$$q = kQ, \qquad\qquad\qquad (5.1.13)$$

in which $\quad Q = \sqrt{Q_p Q_s} \qquad\qquad\qquad (5.1.14)$

and k represents the (normalized) coupling factor of primary and secondary. Further details can be found in Appendix III of Book I, which is devoted to defining various concepts regarding double-tuned bandpass filters.

In Fig. 5.8 the boundaries of stability of this type of amplifier are shown for different values of the coupling coefficent q^2. Also the boundary of stability for the single-stage amplifier with two single-tuned bandpass filters ($q^2 = 0$) is shown.

These boundaries are expressed in the complex T-plane in which a logarithmic scale has been used for T. In this way the same relative accuracy is achieved for different values of T_g.

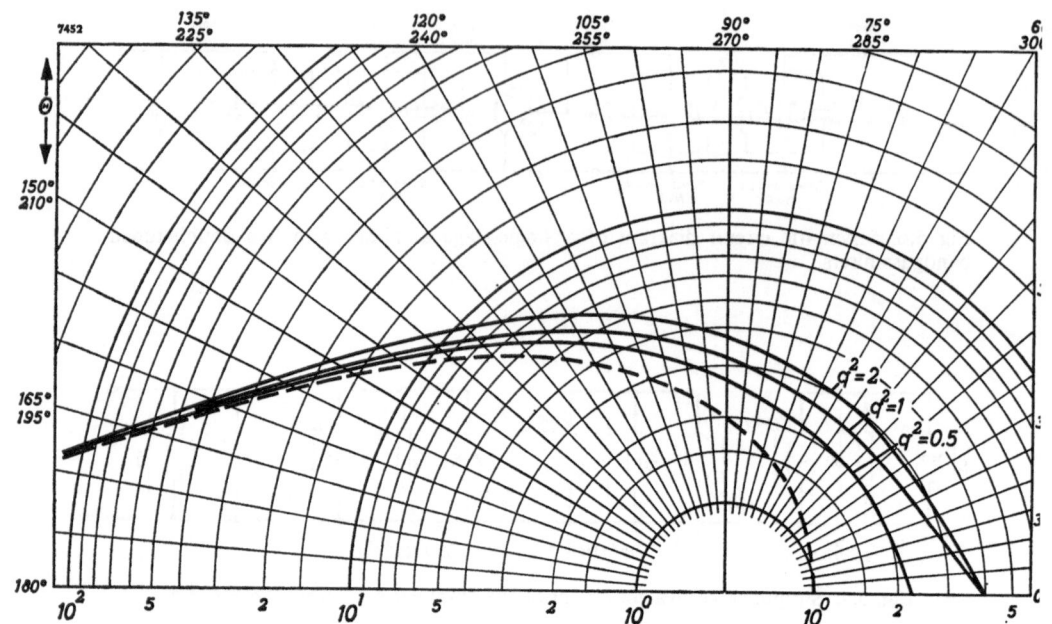

Fig. 5.8. Boundaries of stability of a single-stage amplifier with two double-tuned band-pass filters for various values of the coupling coefficient q^2. The bandpass filters are assumed to be identical. For comparison, the curve valid for an amplifier with two single-tuned bandpass filters ($q^2 = 0$) is also shown (dashed curve).

Because the various boundaries of stability, when plotted in the complex T-plane, are symmetrical with respect to the axis $\Theta = 0$, only the upper halves of the curves are drawn.

The stability of the single-stage amplifier with double-tuned bandpass filters is further discussed in Book I, Section 5.6.

5.1.4 MULTI-STAGE AMPLIFIERS WITH DOUBLE-TUNED BANDPASS FILTERS

In Fig. 5.9 a schematic diagram is shown of a multi-stage amplifier with double-tuned bandpass filters as interstage coupling networks. The admittances of the tuned circuits of which the bandpass filters are composed and the transistor self-admittances are combined in the same way as in Figs. 5.6 and 5.7.

The boundary of stability of these types of amplifier is considered in detail in Book I, Section 7.3.

In Figs. 5.10, 5.11 and 5.12, boundaries of stability of two, three and four-stage amplifiers are shown for different values of q^2. These boundaries are

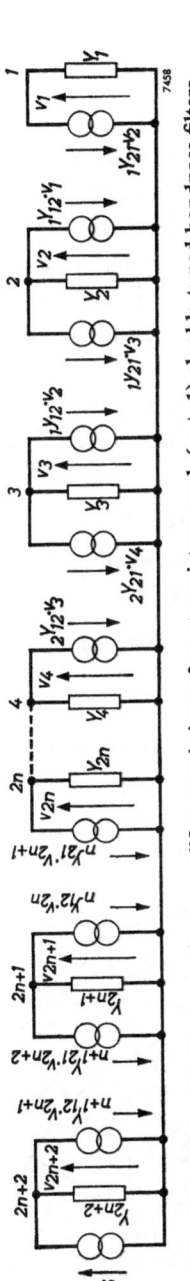

Fig. 5.9. Schematic diagram of an amplifier consisting of n stransistors and $(n + 1)$ double-tuned bandpass filters.

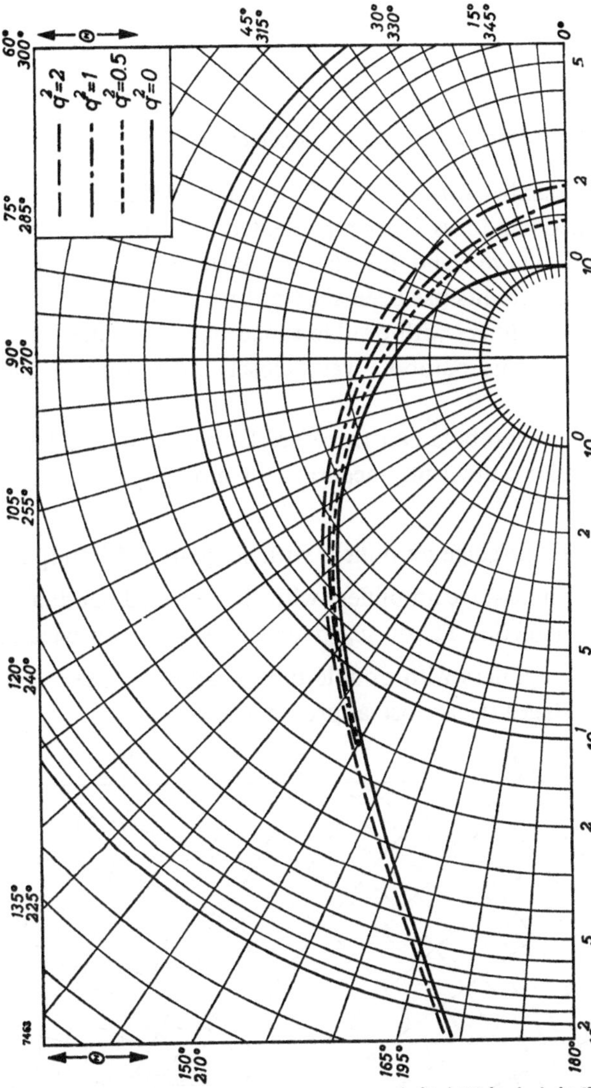

Fig. 5.10. Boundaries of stability of a two-stage amplifier with three identical double-tuned bandpass filters with equal primaries and secondaries for different values of q^2. For comparison, the boundary of stability of a single-stage amplifier with two single-tuned circuits has also, been plotted.

also represented in the complex T-plane in which a logarithmic scale has been used for T.

5.1.5 BASIC BOUNDARY OF STABILITY

Together with the actual boundaries of stability for amplifiers with double-tuned bandpass filters represented in Figs. 5.8, 5.10, 5.11, and 5.12, the parabolic curve valid for the single-stage amplifier with two identical single-tuned circuits has been shown. From Figs. 5.10, 5.11 and 5.12 it follows that for amplifiers with double-tuned bandpass filters consisting of two or more stages, this parabola very closely approximates the actual boundaries of stability.

The parabola can therefore be regarded as a basic boundary of stability which is sufficiently accurate for most types of amplifiers. It can be applied to:

— single-stage amplifiers with single-tuned bandpass filters (exact).

— multi-stage amplifiers with single-tuned bandpass filters by introducing a reduction factor $_nu_g$ given in Table 5.1 on page 52 (exact).

— amplifiers with double-tuned bandpass filters consisting of two or more stages (to a close approximation).

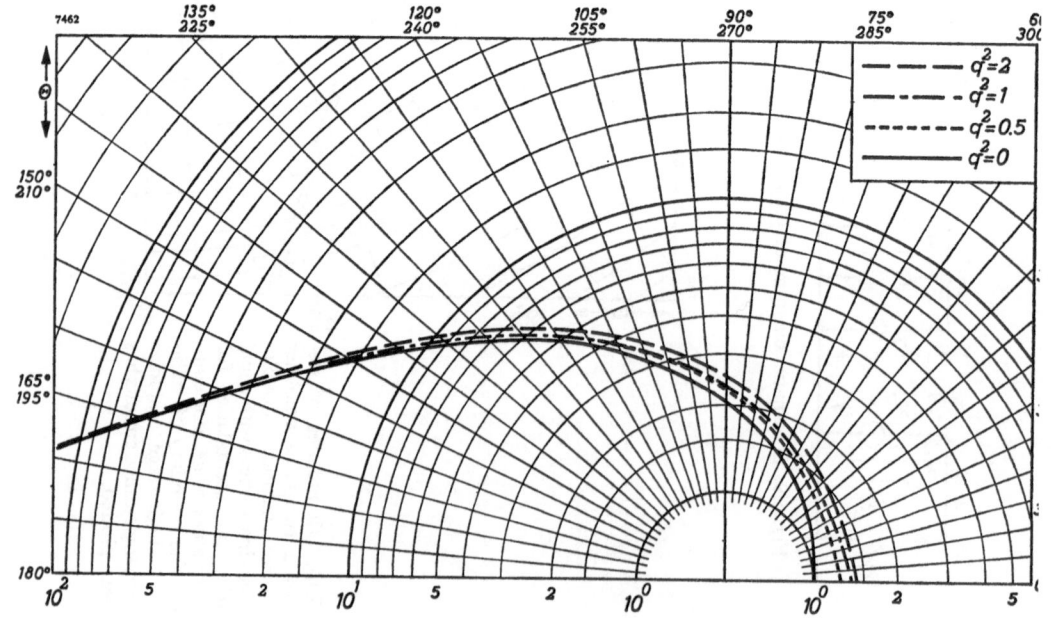

Fig. 5.11. As Fig. 5.10, but for a three-stage amplifier.

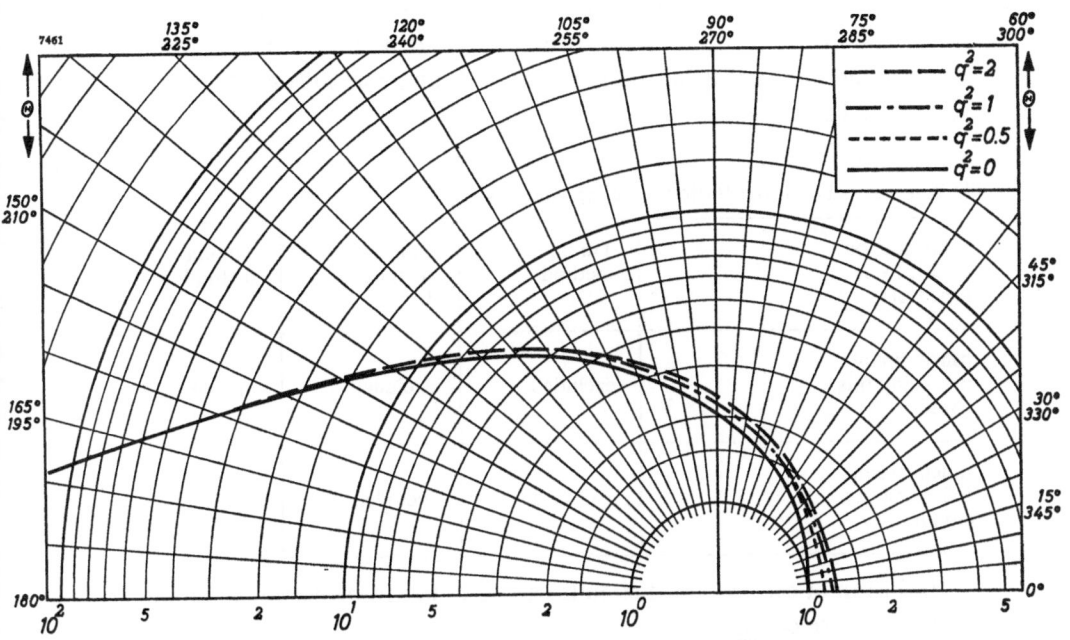

Fig. 5.12. As Fig. 5.10, but for a four-stage amplifier.

For single-stage amplifiers with double-tuned bandpass filters the approximation is not so good as for multi-stage amplifiers. In many instances, however, the accuracy will be sufficient for practical design procedures. It may therefore be concluded that in nearly all practical cases the boundary of stability of a stage of an amplifier can be calculated by means of the simple expression (5.1.8) given on page 51.

5.2 Alignment

The alignment of an amplifier equipped with transistors often poses difficulties due to interaction between the various tuned circuits. These interactions are due to feedback in the transistors, which causes the input admittance of a certain transistor in the amplifier to depend on the admittance of the tuned circuit at its output side and, hence, on the tuning of this circuit. The reverse also occurs.

In amplifiers which are not designed in such a way that the effects of feedback in the transistors become negligibly small, good results are achieved only by employing systematic tuning procedures. In Book I, Sections 2.3 and 5.7 three such tuning procedures are developed. These procedures systematically exclude and include, during the tuning process, effects of other tuned circuits in the amplifier on the circuit to be tuned. The differences between the three procedures, referred to as tuning methods A, B and C, are set out in Table 5.2.

TABLE 5.2.

Differences between the Various Tuning Methods

Influences on circuit to be tuned of:	Tuning method A	Tuning method B	Tuning method C
tuned circuits preceding this circuit	excluded	excluded	included
tuned circuits following this circuit	excluded	included	excluded

Exclusion of certain tuned circuits can be obtained by heavily damping or detuning these circuits.

In Table 5.3 the practical methods of carrying out the various tuning procedures have been indicated. The tuning can be carried out either by means of an oscilloscope and a wobbulator or by means of a valve voltmeter and a normal signal generator. In Table 5.3 the oscilloscope or the valve voltmeter is termed the "indicator".

From the three tuning methods considered, method B has established itself as the most practical one. This being due to the preference of most designers to start aligning an amplifier with the output stage. Then, immediately, an indication is obtained whether this stage operates properly. The alignment then proceeds, stage by stage, to the input side of the amplifier. During the alignment of each stage, a check is obtained on the performance of this particular stage and the part of the amplifier following this stage.

The gain and response curve calculations to be carried out in this book are therefore confined to tuning method B.

Such a restriction is, moreover, necessary in order to arrive at a practical amplifier design procedure. Otherwise, the number of gain expressions and computed design charts would become unwieldily large.

5.3 Amplification

According to Chapter 2, Section 2.1 the amplification of an I.F. amplifier equipped with transistors can best be expressed in terms of transducer gain

TABLE 5.3   Practical Tuning Procedures

1	2	3	4	5
Method of tuning	Generator connected to:	Indicator connected to:	Large damping or detuning required for:	Operation required for tuning each circuit of the amplifier
A (Sections 2.3.2 and 5.7.2 of Book I)	tuned circuit preceding the circuit to be aligned: must have a low impedance, except for tuning the input circuit.	tuned circuit following the circuit to be aligned: must have a low impedance, except for tuning the output circuit.	tuned circuits immediately preceding and following the circuit to be aligned; is automatically provided for by the generator and the indicator.	(1) Connect generator to the circuit preceding the circuit to be tuned. (2) Connect indicator to the circuit following the circuit to be tuned. (3) Tune the circuit.
B (Sections 2.3.3 and 5.7.3 of Book I)	tuned circuit preceding the circuit to be aligned: must have a low impedance, except for tuning the input circuit.	output terminals of the amplifier: must have a high impedance.	tuned circuit preceding the circuit to be aligned: is automatically provided for by the generator.	Connect the indicator to the output terminals of the amplifier. Subsequently: (1) Connect the generator. (2) Tune the circuit.
C (Sections 2.3.4 and 5.7.4 of Book I)	input terminals of amplifier: must present a high impedance to the input terminals of the amplifier.	tuned circuit following the circuit to be aligned: must have a low input impedance, except for tuning the output circuit.	tuned circuit following the circuit to be aligned: is automatically provided for by the indicator.	Connect the signal generator to the input terminals of the amplifier. Subsequently: (1) Connect the indicator. (2) Tune the circuit.

or transimpedance. In chapter 8 and also in the design chart section of this book expressions are given from which transducer gain and transimpedance of various types of amplifiers can be determined, assuming tuning method B to be employed. Gain expressions for amplifiers according to methods A or C can be found in the relevant sections of Book I.

5.4 Response Curve

The amplitude response curves as well as the envelope delay curves of various types of I.F. amplifier have been determined for a wide range of parameters by evaluating the amplifier determinant by means of an electronic computer. Again tuning method B is assumed to be employed, see Section 5.2. The computed curves are included in the design charts in the second part of this book. (Chapter 15)

5.5 Resumé

In this chapter we have surveyed the most important points of the theory of I.F. amplifier design presented in Book I. Together with the gain considerations in Chapter 2, Section 2.1 this chapter establishes a link between the theory of I.F. amplifier design and the practical design procedures to be described in the following chapters.

CHAPTER 6

NEUTRALIZATION

Feedback in transistors may easily lead to instability in I.F. amplifiers unless special precautions are taken. These precautions consist either of choosing such a damping for the tuned circuits connected to the transistor terminals that stability is ensured or of employing a technique known as neutralization. A combination of both methods is also possible.

Neutralization is a process of cancelling all or part of the feedback of the transistor by means of circuits external to the transistor, and can be achieved in many different ways as described in Book I, Chapter 3. In this book we will confine ourselves to Y-neutralizing networks in which fixed elements are employed. Moreover, parasitic effects due to non-ideal components used in practical neutralizing network will be disregarded.

The principle of neutralization will first be briefly discussed. Suppose that the amplifying stage includes a transistor with a given value of reverse transfer admittance, y_{12}. The transistor is said to be perfectly neutralized when the neutralizing network in parallel with the transistor fourpole has such a transfer admittance that the resulting reverse transfer admittance of the parallel combination is zero. In other words, neutralization is perfect when:

$$y_{12} + Y_{12N} = 0, \tag{6.1.1}$$

in which Y_{12N} denotes the reverse transfer admittance of the neutralizing network. Fig. 6.1 shows the circuit of a practical neutralized amplifier; Y_N represents the Y-neutralizing network of which Y_{12N} must satisfy Eq.(6.1.1). The network Y_N usually consists of a series connection of R and C, see Fig. 6.2. It can be shown that in order to obtain perfect neutralization, the values of C and R should be:

$$C = \frac{|y_{12}|}{n \, \omega \, \sin \varphi_{12}} \tag{6.1.2}$$

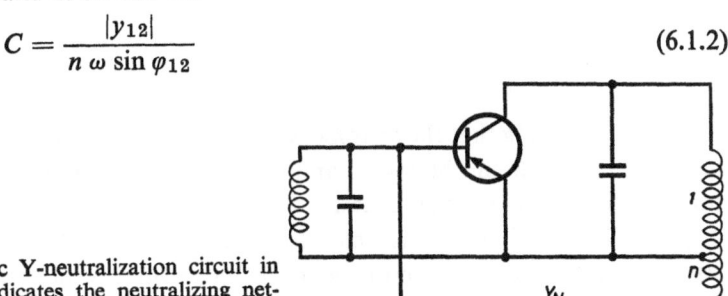

Fig. 6.1. Basic Y-neutralization circuit in which Y_N indicates the neutralizing network.

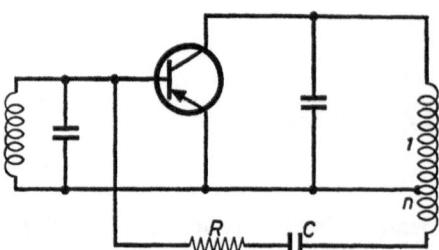

Fig. 6.2. Basic Y-neutralization circuit. If C and R are fixed elements, they should be so chosen that satisfactory results are obtained with all transistors of a given type.

and

$$R = n \cos \varphi_{12}/|y_{12}|, \qquad (6.1.3)$$

in which n denotes the transformer ratio of the output circuit. For transistors of a given type, the values of $|y_{12}|$ and φ_{12} will show some spread above and below quoted average values. Perfect neutralization could obviously be obtained for all transistors by making C and R adjustable. In this context, however, neutralization by means of fixed elements will be considered. The question then arises how C and R should be chosen to ensure that neutralization is most effective for all transistors of a given type.

As regards stability, the best choice of C and R is that at which the stability factor of the amplifier equipped with a transistor having a higher-limit value of $|y_{12}|$ is equal to the stability factor obtained with a transistor having a lower-limit value of $|y_{12}|$.

According to the theory outlined in Book I, Sections 3.7 and 11.4 it can be shown that the best values for C and R are then those given by Eqs (6.1.2) and (6.1.3) respectively, substituting

$$|Y_{12N}| = \frac{2M_a + \Delta M \cos \Theta}{2|y_{21a}| + \Delta|y_{21}| \cos \Theta} \qquad (6\;..4)$$

and

$$\varphi_{12N} = \varphi_{12}, \qquad (6.1.5)$$

where

$$M = |y_{12}| \cdot |y_{21}|, \qquad (6.1.6)$$

the index a indicating that the arithmetical mean of the extreme values of the quantity involved should be taken, and the symbol Δ standing for the difference between these extreme values.

If $\varphi_{12} = 270°$ and $|y_{21}|$ is assumed to be the same for all transistors, the resistor R can be omitted from the neutralizing network, which then consists of the capacitor C_N of value:

$$C_N = \frac{1}{n} \cdot (C_{12a} + \tfrac{1}{2} \varDelta C_{12} \cdot \sin \varPhi_{21}). \qquad (6.1.7)$$

In this formula, Eq. (6.1.2) has already been taken into account.

An extensive treatment of the subject of neutralization is given in Book I, Chapter 3. Spreads in transistor parameters and parameters of the neutralizing network are considered in Sections 3.7 and 11.4 of the mentioned book.

CHAPTER 7

AUTOMATIC GAIN CONTROL OF TRANSISTOR
I.F. AMPLIFIERS

In I.F. amplifiers intended for use in radio receivers for A.M. signals and in television receivers it is desirable that the output signal be constant or nearly constant for a wide range of input signal strength variations at the aerial terminals of the receiver, see Fig. 7.1. This can be achieved by employing a system that automatically provides the desired reduction of gain of the R.F. and I.F. sections of such receivers when the input signal strength increases. Such a system is referred to as *automatic gain control* (abbreviated to AGC). Gain control should be applied to a receiver in such a way that no

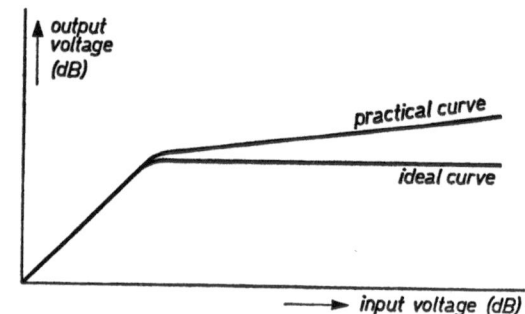

Fig. 7.1. A.G.C. curve.

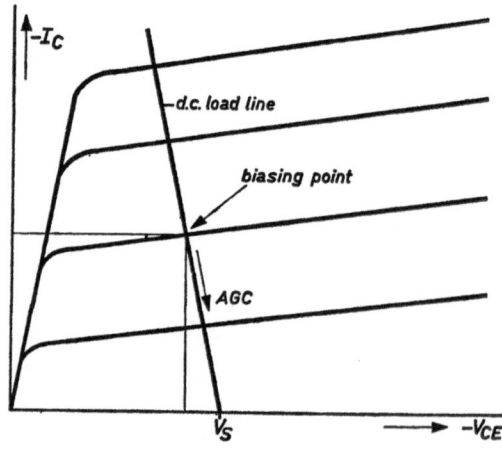

Fig. 7.2. Biasing point variation of a transistor during application of reverse AGC.

overloading occurs on any of the stages in the R.F. and I.F. sections for the whole range of aerial signals handled. This means that AGC must be applied to the input stages of the I.F. and the R.F. sections and that these input stages must have adequate signal handling capability in controlled and non-controlled conditions.

In this chapter we will restrict ourselves to basic AGC systems as used in transistor I.F. amplifiers for A.M. radio receivers and television receivers.

In transistor I.F. amplifiers gain control can be achieved by varying the biasing point of the transistor in the input stage. (In some cases the second stage also is controlled.) This change of biasing point may be obtained in two different ways which are referred to as *reverse gain control* and *forward gain control* respectively.

7.1 Reverse Gain Control

In the reverse gain control system the biasing point of a transistor is varied in such a way that its collector current is reduced to very small values with very little change in collector-emitter voltage. This has been illustrated in Fig. 7.2 for a *pnp* transistor. The resistances used in the emitter and collector circuits of an I.F. transistor are usually comparatively small so that the d.c. load line is rather steep. The collector-emitter voltage then changes only slightly when the collector current is varied down to zero. As will be apparent form Chapter 4 the actual gain control mechanism of the transistor is the decrease of the forward transfer admittance with decreasing emitter (collector) current. In Fig. 7.3 a typical reverse gain control characteristic as a function of emitter current has been depicted. At lower values of emitter current there is a linear relationship between gain and emitter current. There are two causes of departure from this linear relationship at larger emitter currents. The first is that at these values of emitter current the forward transadmittance

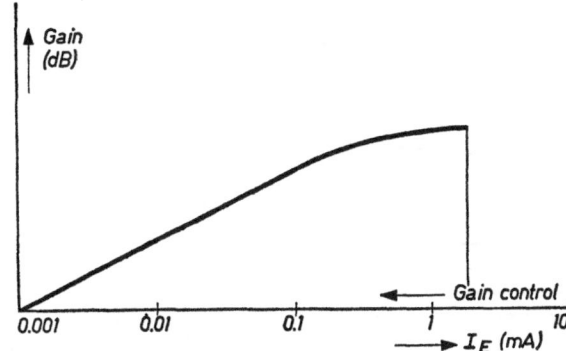

Fig. 7.3. Typical gain-emitter current relationship for a reverse controlled transistor.

of the transistor varies less than proportionally to the emitter current, due to the presence of an internal base resistance.

The second is the variation of input and output admittances of the transistor. At the normal biasing point the real parts of these admittances are not negligible with respect to the damping of the tuned circuits connected to the transistor terminals. When the emitter current is decreased, input and output damping (g_{11} and g_{22}) decrease and, hence, counteract the gain reduction caused by the variation of the forward transfer admittance. At lower values of emitter current g_{11} and g_{22} become negligible with respect to the tuned circuit damping and, therefore, do not affect the gain reduction.

The same comments can be made regarding the input capacitance C_{11}, when the tuned circuit at the input terminals of the transistor is tapped capacitively.

The method of reverse gain control can be applied to all existing I.F. transistors suitable for use at the operating frequency of the amplifier.

7.2 Forward Gain Control

The gain control method which is referred to as forward gain control basically consists of *increasing* the emitter current of the transistor to such levels that its amplifying properties start to deteriorate.

In Fig. 7.4 the variation of the biasing point of a forward gain controlled transistor is illustrated. The manner in which the d.c. load line traverses the d.c. collector characteristics depends on the type of transistor.

The mechanism inside the transistor bringing about forward gain control is a so-called "high-frequency knee" region. In transistors specifically designed for this type of gain control this region is made extra large. This has

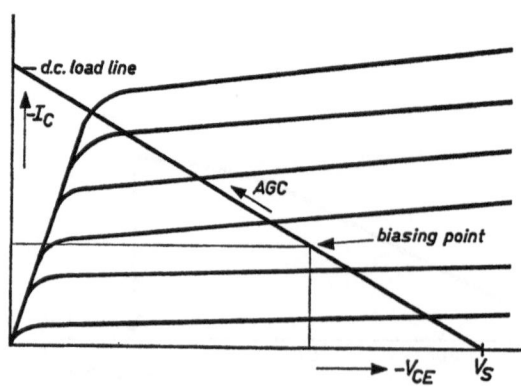

Fig. 7.4. Biasing point variation of a transistor during application of forward AGC.

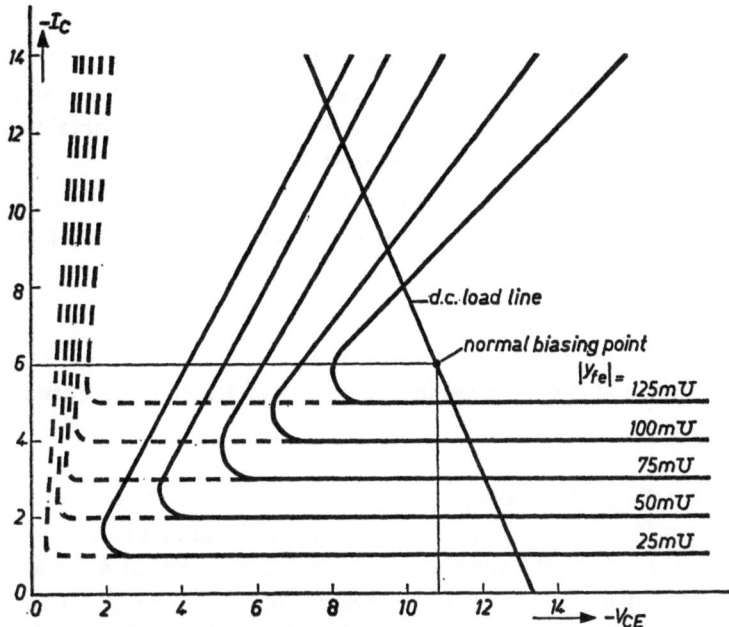

Fig. 7.5. Lines of constant forward transfer admittance in the $I_C - V_{CE}$ plane of a transistor to illustrate the differences in "high frequency knee" region for a typical forward gain control transistor (solid curves) and for a transistor not suitable for this type of gain control (dashed curves).
The curves have been drawn for the purpose of illustrating this difference. They do not apply to actual transistors.

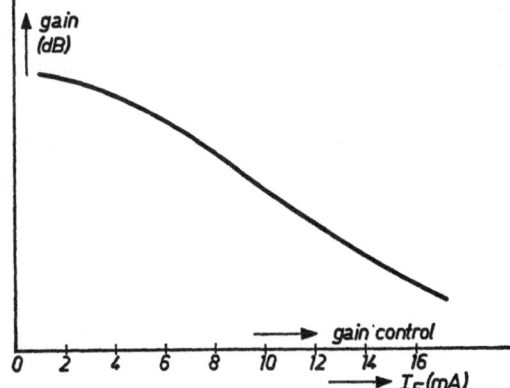

Fig. 7.6. Typical gain-emitter current relationship for a forward gain controlled transistor.

been illustrated in Fig. 7.5 in which lines of constant $|y_{fe}|$ are shown in the $I_C - V_{CE}$ plane. The solid curves apply to a transistor which is specially designed for forward gain control and the dashed curve to a transistor which is not suitable for forward gain control. The d.c. load line has also been drawn. It may be seen that the forward transadmittance of the transistor gradually decreases when the working point moves upwards along the load line. In Fig. 7.6 the gain control curve of a transistor designed for forward gain control has been depicted (solid curve). The dashed curve shows the gain reduction that would be achieved if it was attempted to obtain forward gain control with a transistor not designed for this purpose.

Because forward gain control makes use of deterioration of the high frequency properties of transistors, this type of gain control is only possible at relatively high frequencies.

7.3 Comparison of Signal Handling Capability of Reverse and Forward Gain Control

The term "signal handling capability" when used in connection with transistors refers to the magnitude of the input signal that can be applied to the transistor before the output signal reaches a specified amount of distortion. This distortion may be expressed and specified as either "cross modulation" or "modulation distortion"[1]). The problem of how much signal can be handled by a transistor is very complex due to the many factors involved. A detailed study is therefore considered to be beyond the scope of this book. Only a simplified aspect of the signal handling problem will be discussed, from which the differences between reverse and forward gain control may be appreciated.

In Fig. 7.7 a simplified equivalent circuit diagram of the input side of a transistor in the common emitter connection is shown. The emitter-base

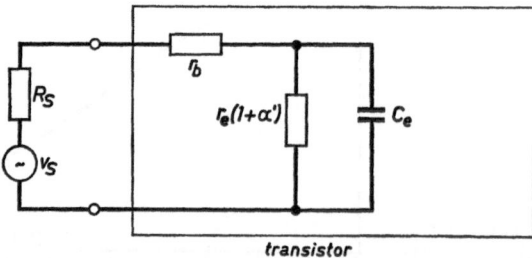

transistor

Fig. 7.7 Simplified equivalent circuit diagram of the input side of a transistor in common emitter connection. The base resistance is represented by r_b and the base-emitter junction by the resistance $r_e (1 + a')$ and the capacitance C_e. The circuit connected externally to the input terminals of the transistor is represented by a voltage source with e.m.f.v_s and source resistance R_s.

---

[1]) A. H. J. NIEVEEN VAN DIJKUM and J. J. SIPS, *Cross Modulation and Modulation Distortion in A.M. Receivers equipped with Transistors*, Electronic Applications, Vol. 20, no. 3 p. 107

junction is represented by the combination of the resistance $r_e (1 + a')$ and the capacitance C_e. The quantity r_e is the forward differential resistance of the junction defined as:

$$r_e = \frac{kT}{q} \cdot \frac{1}{I_E}.$$ (7.3.1)

In this expression:
— q is the charge of an electron,
— k is Bolzman's constant,
— T is the absolute temperature in degrees Kelvin, and
— I_E is the forward emitter current of the transistor.

Furthermore:

$$C_e \approx \frac{1}{2\pi f_1 r_e}$$ (7.3.2)

in which:

— f_1 is the transitional frequency of the transistor, i.e. that frequency at which the common emitter current gain a' fulfils the condition $|a'| = 1$.

The impedance between the base terminal of the transistor and the actual base-emitter junction is represented by the resistance r_b.

The circuitry external to the transistor is simplified to a voltage source of e.m.f. v_s and source resistance R_s.

The specified amount of distortion in the output signal of the transistor referred to above will occur at a certain value of the signal voltage across the actual base-emitter junction. For the purpose of this chapter, this signal voltage level may be assumed to be independent of the forward emitter current I_E.

The relation between the source e.m.f. v_s and the voltage across the actual base emitter junction v_j depends on the relation:

$$\frac{\text{junction impedance}}{\text{junction impedance} + r_b + R_s} = \frac{v_j}{v_s}.$$ (7.3.3)

As the junction impedance is dependent on emitter current, see Eqs. (7.1.1) and (7.3.2), the value of the source e.m.f. at which the specified amount of distortion is reached also depends on the emitter current.

Let the allowable source e.m.f. be equal to a value v_1 at a certain value of I_E, the value with no AGC applied. When the emitter current decreases (reverse AGC), the junction impedance increases and the ratio given by Eq.

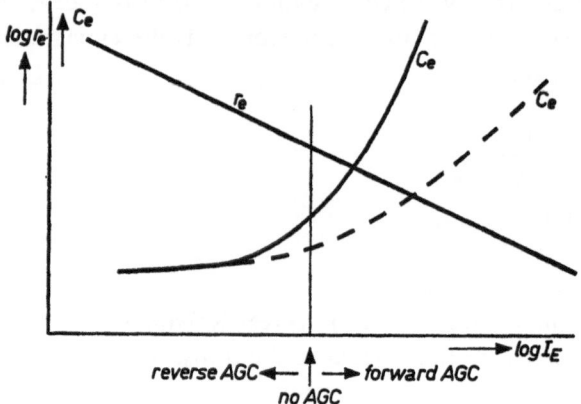

Fig. 7.8. Variation of the forward differential resistance r_e and capacitance C_e of the base-emitter junction of a transistor as a function of the emitter current. The dashed curve for C_e applies to a normal transistor and the solid curve to a transistor designed for forward gain control.

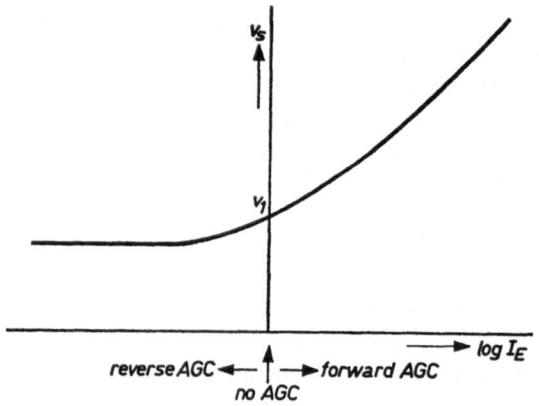

Fig. 7.9. Allowable value of the source e.m.f. (see Fig. 7.7) at which a certain distortion level in the output signal of the transistor is reached as a function of the emitter current.

(7.3.3) becomes larger. At very small values of I_E the value of $r_b + R_s$ can be neglected with respect to the junction impedance, so that the source e.m.f. equals the tolerable junction voltage. When the emitter current is increased from the value with no AGC (forward AGC) the ratio expressed by Eq. (7.3.3) decreases because of the decreasing junction impedance. Then the allowable source e.m.f. at which the distortion limit is reached increases.

The variations of the junction impedance and the allowable value of the source e.m.f., both as a function of emitter current, have been depicted in Figs. 7.8 and 7.9 respectively. As appears from Fig. 7.8 the signal handling capability of a typical forward gain control transistor is greatly enlarged by the rapid increase of C_e at increasing I_E.

From the above considerations it follows that by means of a system of forward AGC in an amplifier, much larger signals can be handled without exceeding distortion criteria than with reverse AGC. This conclusion, however, is only true when the signal levels at other places in the amplifier as well as in the gain-controlled transistor do not exceed tolerable values due to the high levels allowable at the controlled stage.

7.4 Effects of Input and Output Admittance Variations on the Response Curve

When gain control is applied to a transistor, large variations will occur in the input and output admittances y_{11} and y_{22}, see Chapter 4. These variations will affect the tuning as well as the quality factors of the tuned circuits connected to input and output terminals of this transistor. To avoid large variation in the response curve of the amplifier during gain control, the design of the relevant tuned circuits must be such that the variations of the transistor admittances are swamped by the tuned circuit admittances. This point will be dealt with further in the next chapter.

CHAPTER 8

PRACTICAL DESIGN METHOD FOR
I.F. AMPLIFIERS

In the preceding chapters and also in Book I, various aspects of the design of transistor I.F. amplifiers were considered. In this chapter these aspects will be reviewed in the light of a practical design procedure. Also limitations which might be imposed on the design due to the construction of the band-pass filters are considered. These considerations lead to a step-by-step design procedure to be presented in Chapter 14.

The following method of designing I.F. amplifiers mainly consists in determining the proper value of the regeneration coefficient of each stage of the amplifier. When this has been done, further design merely amounts to dimensioning the tuned circuits of the amplifier and determining their tapping ratios.

If neutralization is applied, the method of design will be slightly different from that for amplifiers in which no neutralization is used. These two cases will therefore be dealt with separately where necessary.

8.1 Transistor Parameters Required for the Design

The design of an I.F. amplifier for a given application will usually be based on the use of transistors which are recommended by the manufacturer for this particular purpose. Information on these transistors should comprise the admittance parameters at the frequency of interest for a range of d.c. operating points. These parameters are: g_{11}, C_{11}; $|y_{12}|$, φ_{12}; $|y_{21}|$, φ_{21} and g_{22}, C_{22}.

If these admittance parameters are not given for a particular type of transistor, they must be determined by measurement. These measurements, which should be carried out on a sufficiently large batch of transistors and at high frequencies, require specialized equipment.

The parameters mentioned above allow the following quantities to be calculated directly:

(a) the maximum unilateralized power gain:

$$\Phi_{uM} = |y_{21}|^2/(4g_{11}g_{22});$$ (8.1.1)

(b) the intrinsic regeneration coefficient of the transistor and the regeneration phase angle:

$$t = |y_{12}y_{21}|/(g_{11}g_{22}),$$ (8.1.2)

and

$$\Theta = \varphi_{12} + \varphi_{21};$$ (8.1.3)

(c) the approximate boundary of stability (without neutralization):

$$T_g = 2/(1 + \cos \Theta).$$ (8.1.4)

8.2 The Regeneration Coefficient

It was shown in Chapter 5, and more extensively, in Book I, that the regeneration coefficient T of each stage of the amplifier has considerable influence on its performance (stability, gain, amplitude response and envelope delay). A high value of T is desirable to obtain high gain, but for several reasons an upper limit is imposed on its value:

(a) The amplifier should be sufficiently stable with "nominal" transistors, and it should still remain reasonably stable even when "upper-limit" transistors are used.

(b) The requirements imposed on the amplitude response and envelope delay curves should be met with a "nominal" transistor and the properties of the bandpass filters should not be appreciably affected by spreads in transistor parameters.

With each of the conditions mentioned above corresponds a certain maximum value of T. The amplifier should be so designed that the lowest upper limit of T thus imposed is not exceeded. As a rule this value will turn out to be relatively low, so that the practicable gain of the amplifier will be considerably less than the maximum value which would be calculated on the basis of the transistor data. The specification of the above requirements should be unambiguous so that the usable gain is not unduly reduced.

8.3 Limit Imposed on the Regeneration Coefficient by the Stability Requirement

It has been shown that the amplifier is on the boundary of stability when the regeneration coefficient becomes equal to T_g and that this coefficient should therefore not exceed a value T_g/s, where s — termed the stability factor — is at least unity.

Since the recommended value of s differs for amplifiers without and with neutralizing elements, distinction will be made between these two cases when discussing the design for stable operation.

8.3.1 AMPLIFIER WITHOUT NEUTRALIZATION

An amplifier without neutralization should be so designed that it remains reasonably stable for all transistors of the type chosen. This means that if the amplifier is designed for the nominal transistor with a stability factor s, this factor should still be larger than unity when the amplifier is equipped with upper or lower limit transistors of this type (see Chapter 11, Book I). In practice the stability requirement is met by choosing $s \simeq 4$ for the nominal transistor and it is recommended that designers make sure whether this is indeed so for every design.

For the nominal transistor the value of T_g can be evaluated from the graphs in Section 5.1 or from Eq.(8.1.4). The regeneration coefficient T becomes:

$$T = T_g/s. \tag{8.3.1}$$

For all types of amplifier except multi-stage amplifiers with single-tuned bandpass filters (see Section 5.1), Eq. (8.3.1) becomes:

$$T = \frac{2}{s(1 + \cos \Theta)}. \tag{8.3.2}$$

For multi-stage amplifiers with single-tuned bandpass filters, Eq.(8.3.1) becomes:

$$T = \frac{2_n u_g}{s(1 + \cos \Theta)} \tag{8.3.3}$$

in which $_n u_g$ follows from Table 5.1 on page 52.

It is considered beyond the scope of this chapter to carry out calculations on the exchangeability criterion. This problem will be dealt with separately in Chapter 12. The value of T according to Eqs.(8.3.2) or (8.3.3) will therefore be considered as the value of T *determined by stability*.

8.3.2 AMPLIFIER WITH NEUTRALIZATION

Once the proper dimensioning of the neutralizing network has been ascertained, the regeneration coefficient which may occur under extreme conditions can be calculated from the expression:

$$T_{MN} = |y_{21M}| \cdot \{|y_{12M}| - |Y_{12N}|\}/G_1 G_2. \tag{8.3.4}$$

In this expression the suffix M denotes the maximum value of the quantity concerned and the suffix N refers to a neutralized or neutralizing quantity.

Since the amplifier must remain stable when equipped with transistors having extreme values of transfer admittance, the following condition must be satisfied:

$$T_{MN} \leqslant T_g/s', \tag{8.3.5}$$

in which T_g is given by Eq.(8.3.2) and s' denotes the lowest permissible stability factor for a transistor with extreme parameters. As a rule it is considered sufficient to choose the lowest value of s' roughly equal to 2.

8.4 Limit imposed on the Regeneration Coefficient by the Response Curve

It was shown in Book I that the values of T and q^2 have great influence on the shape of the amplitude response and envelope delay curves of amplifiers with double-tuned bandpass filters. Both quantities must therefore be chosen with much care so that the response curves[1]) have the required form.

In general, the quantity T gives rise to some skewing of the response curves, whereas the quantity q^2 affects the centre part of these curves symmetrically (see Section 5.9 of Book I). The combined effect of T and q^2, however, is very complicated. For this reason a large number of computed response curves are given in this book, which greatly facilitates the correct choice of T. If other amplifier parameters are given (such as the amplifier configuration, the number of stages and the values of Θ and q^2), these curves offer the possibility of ascertaining rapidly the particular value of T at which the imposed requirements are satisfied as closely as possible.

For amplifiers with single-tuned bandpass filters obviously the value of q^2 need not be determined.

The value of T thus found meets the requirements for the nominal transistor. However, spread in transistor parameters may have an adverse effect on the response curves, so that the amplifier should be so designed that the effect of the spreads is acceptable. It depends obviously on the extent of these spreads whether the requirements imposed on the value of T are more or less stringent than those imposed by a nominal transistor.

The transistor parameters y_{11} and y_{22} directly influence the properties of the bandpass filters.

Distinction can be made between the effect of spreads of the resistive component and that of spreads of the capacitive component of these parameters.

---

[1]) The term "response curve" will be used to denote both the amplitude response and the envelope delay curves.

To minimize the influences of these spreads on the response curves, the total damping of the resonant circuit to which either the input or output terminals of the transistor are connected should be large compared with the spreads of the resistive components of either y_{11} and/or y_{22}, whilst, similarly, the total tuning capacitance of these circuits should be large compared with the spread of the capacitive components of either y_{11} and/or y_{22}.

It will now be useful to investigate the various spreads more closely. For this investigation distinction will be made between amplifiers with single-tuned and those with double-tuned bandpass filters. The latter type of inter-stage coupling will be considered firstly.

8.4.1 AMPLIFIERS WITH DOUBLE-TUNED BANDPASS FILTERS

8.4.1.1 *Spreads in g_{22}*

The effect of spreads in g_{22} will be investigated with reference to Fig. 8.1 showing part of a transistor I.F. amplifier. It is seen that the output terminals of the transistor are connected across the primary of a double-tuned bandpass filter.

The total damping of the primary is obviously:

$$G_p = G_p{}^* + g_{22}, \qquad (8.4.1)$$

in which g_{22} is subjected to spread around the nominal value. The larger $G_p{}^*$ with respect to g_{22}, the less will be the effect of this spread on G_p and hence on Q_p, but the greater will be the losses due to $G_p{}^*$. This can be seen by putting:

$$G_p{}^*/G_p = w_p, \qquad (8.4.2)$$

whence:

$$G_p = g_{22}/(1 - w_p) = g_{22}/\Phi_p. \qquad (8.4.3)$$

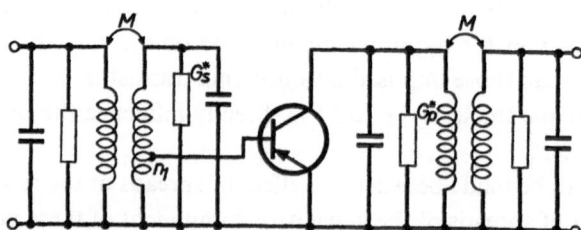

Fig. 8.1. Schematic circuit of part of an I.F. amplifier, showing the connection of the transistor to the double-tuned bandpass filters. (The transistor is connected, by way of example, in common emitter configuration).

If Φ_p is specified for the nominal transistor, its value is a measure of the dependence of G_p on the spread in g_{22}: the lower the value of Φ_p (or the more negative this value expressed in dB), the smaller will be this dependence.

8.4.1.2 *Spreads in g_{11}*

Fig. 8.1 shows that the input terminals of the transistor are connected to a tapping n_1 of the secondary of a double-tuned bandpass filter. Analogous to the argument used for g_{22}, it may be written:

$$G_s = G_s{}^* + n_1{}^2 g_{11}. \tag{8.4.4}$$

Putting

$$G_s{}^*/G_s = w_s, \tag{8.4.5}$$

and

$$\Phi_s = 1 - w_s, \tag{8.4.6}$$

gives:

$$G_s = g_{11}/\Phi_s. \tag{8.4.7}$$

Hence the larger w_s, that is the smaller Φ_s, the smaller will be the effect of spreads in g_{11} on G_s, assuming again that Φ_s is specified for the nominal transistor.

8.4.1.3 *Combined Effect of Spreads in g_{22} and g_{11}*

By combining Eqs.(8.4.3) and (8.4.7) an upper limit for T is again obtained, namely:

$$T = \frac{|y_{12}\,y_{21}|}{G_{p2}\,G_{s1}} = \frac{|y_{12}\,y_{21}|}{g_{11}\,g_{22}} \cdot \Phi_{p2}\,\Phi_{s1},$$

whence, substituting Eq. (8.1.2):

$$T = t \cdot \Phi_{p2}\Phi_{s1}. \tag{8.4.8}$$

The suffixes 1 and 2 refer to the bandpass filters at the input and output terminals of the transistor respectively.

8.4.1.4 *Spreads in C_{22} and C_{11}*

In the total tuning capacitances of the tuned circuits to which the transistor input and output terminals are connected, C_{11} and C_{22} respectively are included[1]). The total tuning capacitance of these circuits need not be large with respect to C_{11} and C_{22} because it is generally accepted that an amplifier must be re-aligned after replacement of a transistor. This does not, however, mean that these tuning capacitances may be made arbitrarily small. In fact, a lower limit is set to their values by the following circumstances:

(a) In modern I.F. amplifiers the bandpass filters are aligned by adjusting the position of a ferroxcube or dust iron core in the tuning coil. The tuning range thus obtained is restricted, and the possible spreads in total tuning capacitance must obviously remain within this range. With miniature bandpass filters the possible inductance variation is of the order of only $\pm 10\%$.

(b) In receivers fed by dry batteries or accumulators the supply voltage may drop well below its nominal value. This changes the biasing points of the transistors which leads in turn, amongst other things, to variations of C_{11} and C_{22}.

   To ensure that the response curve of the amplifier is not seriously affected by this drop in supply voltage, the tuning capacitances of the tuned circuits should have certain minimum values. Similar effects may be experienced in mains-fed receivers with a non-stabilized power supply due to line-voltage variations.

(c) Even if in view of the spreads and other variations of C_{11} and C_{22} an extremely small tuning capacitance were permissible, it should be kept in mind that physical reasons set a lower limit to this capacitance. This is due to the self-capacitance of the coil, the wiring and stray capacitances, and so forth.

The lowest practicable value of the tuning capacitance is, however, of little importance unless it can be combined with a low value of G. And since $Q = \omega_0 C/G$, the ratio C/G is fixed for a given value of Q. If, for example, for reasons of stability, a higher value of G is required, the value of C must be increased accordingly.

If it is indeed possible to give the capacitance the minimum value C_{min}, the damping of the primary of the double-tuned bandpass filter will be:

$$G_p = \omega_0 C_{p\,min}/Q_p.$$

and that of the secondary, referred to the tapping (input terminals of the transistor)[2]:

---

[1]) This holds for tuning method A. For tuning method B an extra capacitance appears in parallel with C_{11} due to the presence of transistor feedback. For tuning method C such a capacitance is in parallel with C_{22}. See Book I, Chapters 2 and 5.

[2]) If a capacitive tap is used, spreads in C_{11} result in spreads in n. In that case these spreads should also be taken into account.

$$G_s = \omega_0 C_{s\,min}/n^2 Q_s.$$

These expressions again lead to an upper limit of T for the nominal transistor, because:

$$T = \frac{|y_{12}\,y_{21}|}{G_p\,G_s} = \frac{|y_{12}\,y_{21}|}{\dfrac{\omega_0 C_{p2\,min}}{Q_{p2}} \cdot \dfrac{\omega_0 C_{p1\,min}}{n^2\,Q_{s1}}} \qquad (8.4.9)$$

8.4.2 AMPLIFIERS WITH SINGLE-TUNED BANDPASS FILTERS

8.4.2.1 *Influences of Spreads in g_{11} and g_{22}*

In Fig. 8.2 part of an I.F. amplifier with single-tuned band-pass filters is shown. It can be seen that each bandpass filter is loaded by the output damping g_{22} of one transistor and the input damping g_{11} of the following transistor.

The total damping of the bandpass filter equals:

$$G = g_{22} + G^* + n_1^2\,g_{11}. \qquad (8.4.10)$$

When the single-tuned bandpass filter forms the coupling element between two successive stages the design must be made such that in the nominal case:

$$g_{22} = n_1^2\,g_{11} \qquad (8.4.11)$$

Then no mismatch losses occur across this bandpass filter and, moreover, the stability of one stage is not increased at the expense of a decrease in stability of the other stage (see Book I, Chapter 6).

The insertion losses of a single-tuned bandpass filter are defined as (see Book I Appendix II):

$$\Phi_i = \left\{\frac{g_{22} + n_1^2\,g_{11}}{G}\right\}^2 \qquad (8.4.12)$$

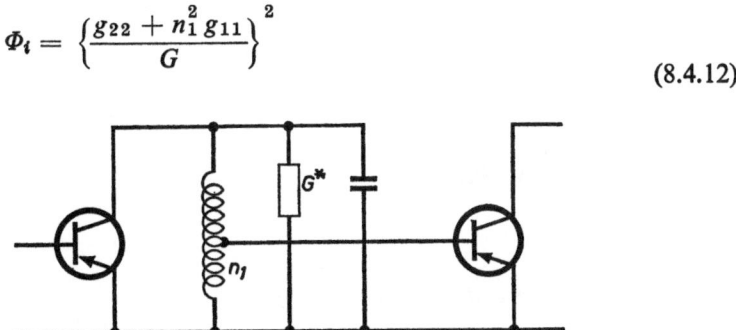

Fig. 8.2. Schematic circuit of part of an I.F. amplifier with single-tuned bandpass filters, showing the connection of the transistors to the bandpass filter.

This equation may also be written as:

$$G = \frac{g_{22} + \overset{2}{n_1} g_{11}}{\sqrt{\Phi_i}} \qquad (8.4.13)$$

The last expression clearly indicates that for a certain nominal value of G, spreads in g_{22}, n, and g_{11} have less effect on the total bandpass filter damping G when the insertion losses Φ_i have a larger value.

A condition for the regeneration coefficient T can now be derived as follows:

Assume that the various single-tuned bandpass filters in a multistage amplifier are designed so that the matching condition expressed by Eq. (8.4.11) is fulfilled. Then the total damping at the input terminals of a particular transistor in this amplifier amounts to:

$$G_1 = \frac{2g_{11}}{\sqrt{\Phi_{i1}}}. \qquad (8.4.14)$$

This damping includes the damping g_{11} of the transistor. The quantity Φ_{i1} denotes the insertion losses of the bandpass filter preceding the transistor under consideration. For the output side of the transistor an analogous expression can be derived; viz:

$$G_2 = \frac{2g_{22}}{\sqrt{\Phi_{i2}}}. \qquad (8.4.15)$$

With Eqs. (8.4.14) and (8.4.15) the regeneration coefficient of this transistor becomes:

$$T = \frac{|y_{12}\, y_{21}|}{4\, g_{11}\, g_{22}} \cdot \sqrt{\Phi_{i1}\, \Phi_{i2}}, \qquad (8.4.16)$$

or with Eq. (8.1.2):

$$T = \frac{t}{4} \sqrt{\Phi_{i1}\, \Phi_{i2}}. \qquad (8.4.17)$$

8.4.2.2 Effects of Spreads in C_{11} and C_{22}

As regards the effects of spreads in the capacitances C_{11} and C_{22} of the transistor in amplifiers with single-tuned bandpass filters, the same remarks are applicable as in the case of amplifiers with double-tuned bandpass filters (see sub-section 8.4.1.2).

If $C_{1\,min}$ and $C_{2\,min}$ denote the minimum tuning capacitances of the bandpass filters at the transistor input and output terminals respectively and Q_1 and Q_2 denote the respective quality factors, we obtain an upper limit for T as:

$$T = \frac{|y_{12}\,y_{21}|}{\dfrac{\omega_0\,C_{1\,min}}{n^2\,Q_1} \cdot \dfrac{\omega_0\,C_{2\,min}}{Q_2}} \qquad (8.4.18)$$

In this expression n denotes the tap on the bandpass filter at the input side of the transistor.

8.4.3 SUMMARY OF REMARKS ON THE REGENERATION COEFFICIENT

In the preceding sub-sections the various points which set a limit to the regeneration coefficient T were discussed. As already emphasized, the actual amplifier design must be based on the lowest value of T.

In Table 8.1 the various considerations of regeneration coefficient of the preceding sections are summarized.

Once T has been fixed, and in amplifiers with double-tuned bandpass filters, q^2 has also been chosen, the amplitude response and envelope delay curves can be found from the relevant set of calculated response curves in the Design Chart section of this book.

8.5 The Value of q^2

In amplifiers in which double-tuned bandpass filters are used as interstage coupling networks, the value of q^2 has also to be determined.

For simplicity the determination of T with a view to response curve requirements was explained by assuming that the value of q^2 was already known. In practice, however, this will not be so and both T and q^2 are chosen simultaneously in such a way that at the selected combination of these quantities the response curves are in good agreement with the requirements. Therefore, in the design chart section referred to above, graphs are given for several values of q^2, with T as parameter.

8.6 The Gain

In this book the gain of an I.F. amplifier is expressed in terms of either transimpedance or transducer gain of the complete amplifier. If source and

TABLE 8.1. Summarizing

Determination of parameter	Case	Type of interstage bandpass filters	Important equations	number of stages	Relevant graphs in this book: Fig.no.	page
Regeneration coefficient determined by stability	without neutralization	single-tuned	$T = \dfrac{T_g}{s}$ $T_g = \dfrac{2n^u g}{1 + \cos\Theta}$	1 n	5. 3	50
		double-tuned	$T = \dfrac{T_g}{s}$ $T_g = \dfrac{2}{1 + \cos\Theta}$	1 2 3 4	5. 8 5.10 5.11 5.12	54 55 56 57
	with fixed neutralizing elements with regard to spreads	single-tuned or double-tuned	$T_{MN} = \dfrac{|y_{21M}| \cdot \{|y_{12M}| - |y_{12N}|\}}{G_1 G_2} = \dfrac{T_g}{s'}$ $s' = 2$ $R_N = n_N \cos\varphi_{12} / |Y_{12N}|$ $C_N = \dfrac{1}{n_N} \cdot \dfrac{1}{\omega} \cdot \dfrac{|y_{12N}|}{\sin\varphi_{12}}$	n		
		single-tuned or double-tuned		n		
Regeneration coefficient determined by response curve requirements	for the average transistor	single-tuned	T: value of T chosen from the appropriate Design Chart.	n		
		double-tuned		n		
	on account of spreads in g_{11} and g_{22}	single-tuned	$T = \dfrac{t}{4} \cdot \sqrt{\Phi_{i1}\,\Phi_{i2}}$	n		
		double-tuned	$T = t \cdot \Phi_{s1} \cdot \Phi_{p2}$	n		
	on account of spreads in C_{11} and C_{22} or of the lowest physically realizable capacitances	single-tuned	$T = |y_{12}\,y_{21}| \cdot \dfrac{Q_2 Q_1}{\dfrac{\omega_0 C_{1m}}{n^2}} \cdot \omega_0 C_{2m}$	n		
		double-tuned	$T = |y_{12}\,y_{21}| \cdot \dfrac{Q_{p2} Q_{s1}}{\dfrac{\omega_0 C_{s1m}}{n^2}} \cdot \omega_0 C_{p2m}$	n		

The suffixes 1 and 2 in the various symbols used in this table refer to quantities relevant to

Remarks on Regeneration Coefficient

Design aspects dealt with				Remarks
in this book		in Book I		
section	page	section	page	
5.1.1 5.1.2 8.3.1	49 51 74	2.2 6.3	25 161	T = value of regeneration coefficient for average transistor s = stability factor, usually $s \simeq 4$ for reasons of interchangeability
5.1.3 5.1.4 5.1.5 8.3.1	52 54 56 74	5.6 7.3	128 190	$_nu_g$ = reduction factor for stability of multistage amplifier with single-tuned bandpass filters. For values: see Table 5.1 on page 52.
6.0.0 8.3.2	61 74	3.3 3.7	75 90	T_{MN} = value of the regeneration coefficient for the extreme transistor occurring in the amplifier neutralized for the average transistor. A stability factor $s' = 2$ is considered to be sufficient in this case.
		3.7 11.2 11.4 11.5	90 227 236 244	Only necessary if the assumptions of $s = 4$ in the case without neutralization or $s^I = 2$ in the case with neutralization are considered inadequate.
		2.5 6.7	53 184	$\sqrt{\Phi_{i1}\ \Phi_{i2}} = \dfrac{4T}{t}$
		5.9	150	$\Phi_{s1}\ \Phi_{p2} = \dfrac{T}{t}$
8.4.2	79			Φ_{i1}, Φ_{i2} and Φ_{s1}, Φ_{s2} respectively are given such values that, in the opinion of the designer, the variations in response curve remain sufficiently small.
8.4.1	76			
8.4.2	79			n = tapping on the bandpass filter to which the input terminals of the transistor are connected.
8.4.1	76			

the bandpass filters connected to input and output terminals of the transistor respectively.

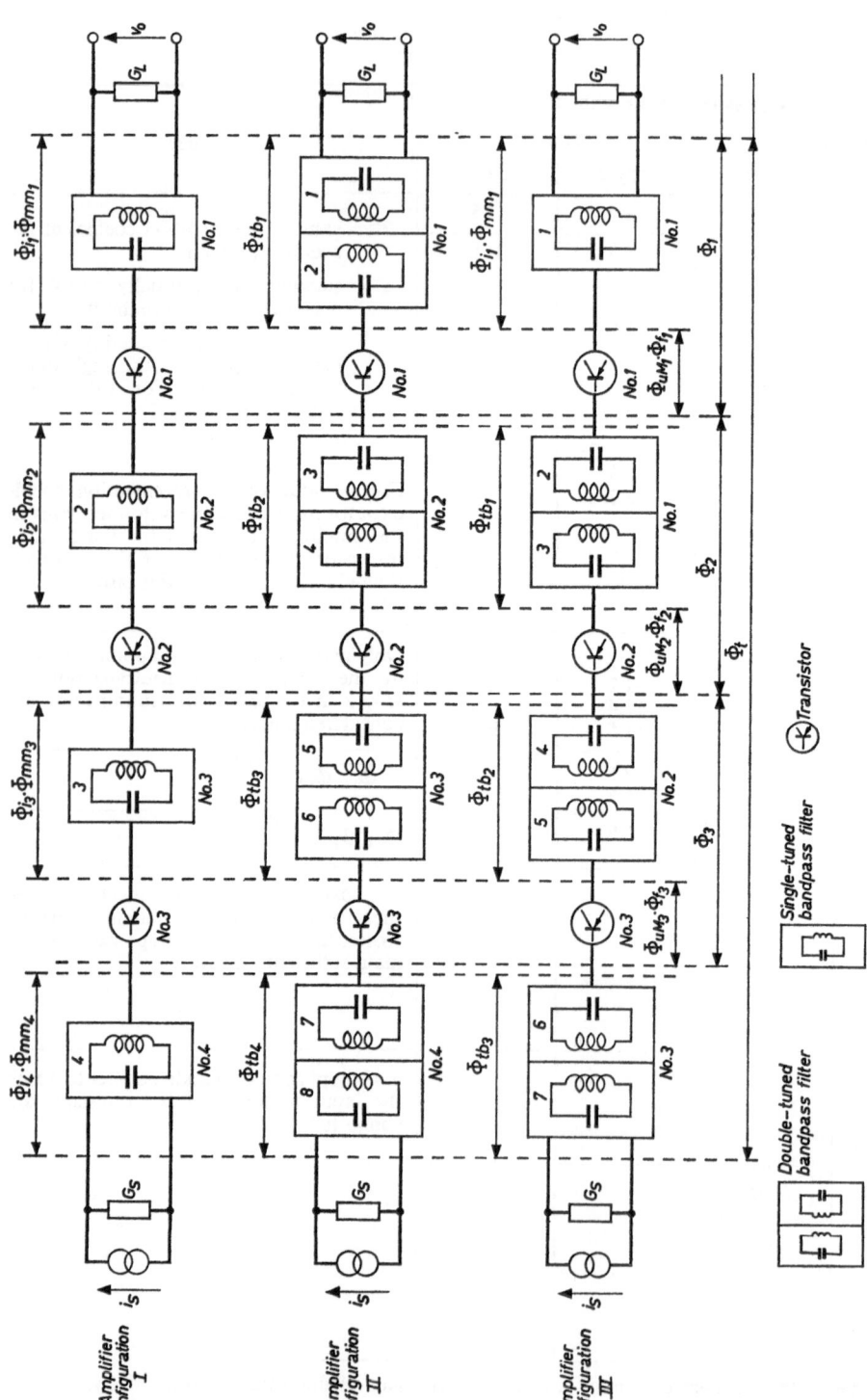

Fig. 8.3. Schematic representation of the three amplifier configurations considered. By way of example, three-stage amplifiers are shown. The various gain figures that can be defined are indicated.

load terminations are known, the transimpedance or transadmittance of the amplifier can easily be derived from the transducer gain figure or vice versa.

The gain is also expressed in terms of power gain and voltage gain per stage of the amplifier. The latter method of expressing gain is very convenient for checking the performance of each individual stage of the amplifier.

Three different amplifier configurations will be considered;
— configuration   I: an n-stage amplifier with $(n + 1)$ single-tuned bandpass filters,
— configuration II: an n-stage amplifier with $(n + 1)$ double-tuned bandpass filters, and
— configuration III: an n-stage amplifier with n double-tuned bandpass filters and one single-tuned bandpass filter at the output terminals of the amplifier.

These three amplifier configurations are schematically represented in Fig. 8.3, it being assumed, by way of example, that they consist of three stages each.

8.6.1 THE TRANSDUCER GAIN OF THE COMPLETE AMPLIFIER

General formulae for the transducer gain of amplifiers according to the configurations I, II and III have been derived in Book I, Chapters 6, 7 and 8 respectively. It will be shown that these formulae are greatly simplified, so that they can readily be employed if the amplifiers are composed of identical transistors and identical bandpass filters. However, when this is not so, recourse must be had to the general formulae. Expressions for the transducer gain of amplifiers according to the three configurations and consisting of identical stages are entered in Table 8.2.

The different factors of which the transducer gain expressions are composed are indicated in the schematic representation of the amplifier configurations in Fig. 8.3.

The transducer losses which occur in single-tuned bandpass filters are indicated as $\Phi_i \Phi_{mm}$. These losses are composed of the insertion losses (damping ratio) Φ_i and the mismatch losses Φ_{mm} as illustrated in Fig. 8.4. (see also Book I, Appendix II).

The transducer losses Φ_{tb} occurring in a double-tuned bandpass filter are composed of the factors:

Φ_p; primary damping ratio,

Φ_s; secondary damping ratio, and

Φ_q; coupling losses.

TABLE 8.2 Transducer Gain Expressions for n-Stage Amplifiers with Identical Stages

Configuration I	Configuration II	Configuration III
n transistors $(n + 1)$ single-tuned bandpass filters	n transistors $(n + 1)$ double-tuned bandpass filters	n transistors n double-tuned bandpass filters 1 single-tuned bandpass filter
$$_n\Phi_t = 4\frac{G_L}{G_n+1}\cdot\frac{G_S}{G_1}\cdot T^n\cdot N^n\cdot\frac{1}{\lvert _n\delta_0\rvert^2}$$ or: $$_n\Phi_t = \Phi_{uM}^n\cdot\Phi_t^{(n+1)}\cdot\Phi_{mm}^{(n+1)}\cdot{}_n\Phi_f$$ $$_n\Phi_f = \frac{1}{\lvert _n\delta_0\rvert^2}$$	$$_n\Phi_t = 4\frac{G_S}{G_{2n+2}}\cdot\frac{G_L}{G_1}\cdot T^n\cdot N^n\cdot q^{(2n+1)}\frac{1}{\lvert _n\delta_0\rvert^2}$$ or: $$_n\Phi_t = \Phi_{uM}^n\,(\Phi_p\,\Phi_s\,\Phi_Q)^{(n+1)}\cdot{}_n\Phi_f$$ $$_n\Phi_f = \frac{1}{\lvert _n\delta_0\rvert^2}\cdot(1+q^2)^{(2n+1)}$$	$$_n\Phi_t = 4\frac{G_S}{G_{2n+1}}\cdot\frac{G_L}{G_1}\cdot T^n\cdot N^n\cdot q^{2n}\,\frac{1}{\lvert _n\delta_0\rvert^2}$$ or: $$_n\Phi_t = (\Phi_{uM}\,\Phi_p\cdot\Phi_s\cdot\Phi_Q)^n\cdot\Phi_t\cdot\Phi_{mm}\cdot{}_n\Phi_f$$ $$_n\Phi_t = \frac{1}{\lvert _n\delta_0\rvert^2}(1+q^2)^{2n}$$
$$\Phi_t = \left\{\frac{g_{22}+n^2g_{11}}{G}\right\}^2$$ $$\Phi_{mm} = \frac{4\,g_{22}\cdot n^2g_{11}}{(g_{22}+n^2g_{11})^2}$$	$$\Phi_p = \frac{g_{22}}{G_p}$$ $$\Phi_s = \frac{g_{11}}{G_s}$$ $$\Phi_q = \left\{\frac{2q}{1+q^2}\right\}^2$$	Φ_t and Φ_{mm} refer to the single-tuned band-pass filter at the output of the amplifier.

$$\Phi_{uM} = \frac{\lvert y_{21}\rvert}{4g_{11}g_{22}}$$

The quantities $_n\delta_0$ and $_n\Phi_f$ are dependent on the amplifier configuration and the tuning method, see Tables 8.3 and 8.4.

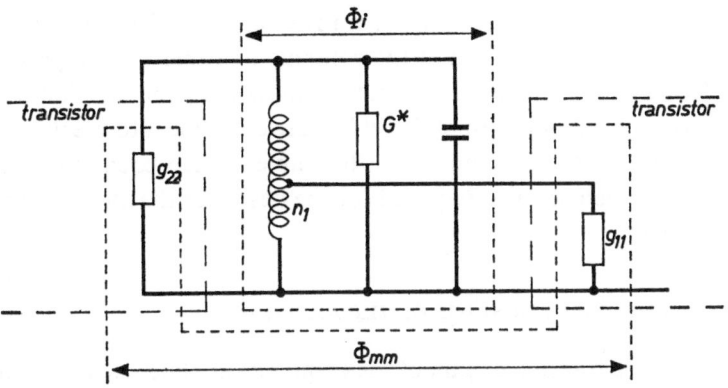

Fig. 8.4. Schematic diagram illustrating the definition of the insertion losses Φ_i and the mismatch losses Φ_{mm} of a single-tuned bandpass filter.

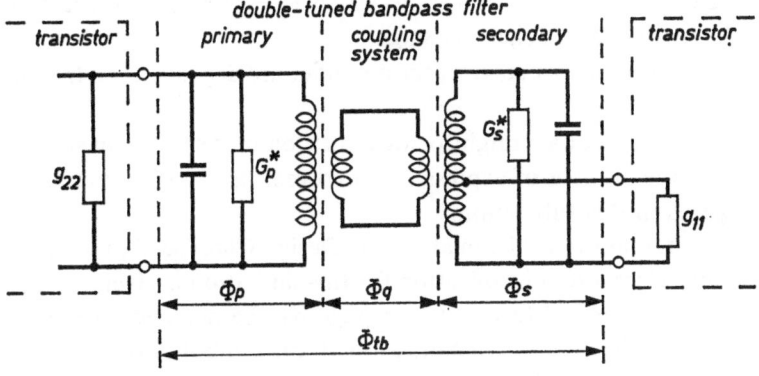

Fig. 8.5. Schematic diagram illustrating the definition of the transducer losses, and the factors of which it is composed, of a double-tuned bandpass filter.

The different factors have been indicated schematically in Fig. 8.5.

In the various transducer gain expressions in Table 8.2 the quantities $|_n\delta_o|$ and $_n\Phi_f$ are dependent on the amplifier configuration and the tuning method employed.

In the Design Chart section of this book curves are given from which $|_n\delta_o|$ and $_n\Phi_f$ can be determined graphically for the proper amplifier configuration and parameters. These curves, however, are valid only when tuning method B is employed.

The value of the quantity $_n\Phi_f$ can also be computed from the expressions given in Table 8.2 when the value of the quantity $|_n\delta_o|$ has been ascertained, which can be done by means of the method outlined in the following subsection.

When it is necessary to ascertain the transducer gain of the complete amplifier, the first equation in each of the columns of Table 8.2 will preferably be used. However, if it is desirable also to know the gain contributions or losses of various parts of the amplifier, it is more convenient to use the second equation in the different columns of this table. These equations contain several factors which are also required for the design of the individual filters of the amplifier.

8.6.2 THE POWER GAIN PER STAGE OF THE AMPLIFIER

In Table 8.2 the transducer gain of the complete amplifier is given in terms of the maximum unilateral power gain Φ_{uM} of each transistor, the transducer losses of each bandpass filter (single or double-tuned), and a factor $_n\Phi_f$ representing the losses due to real feedback. To ascertain the power gain of each stage of the amplifier it is also necessary to express $_n\Phi_f$ in factors which give the losses due to feedback per stage.

The feedback losses per stage of the amplifier are dependent on the method of tuning.

In order to arrive at a straightforward method of determining stage gains, and to limit the number of expressions involved, only tuning method B will be considered in this subsection.

The same limitation will be made in the Design Chart section (Chapter 15) of this book. Here also a justification for this limitation is given.

The losses due to real feedback per stage will be denoted by the symbol Φ_{fr} (*followed* by the index r), in contrast with the symbol $_n\Phi_f$ (*preceded* by an index n), which denotes these losses for the whole n-stage amplifier.

Expressions for the feedback losses per stage are given in Table 8.3 for each of the three amplifier configurations considered. The quantities P_M in these expressions, which are different for each of the three configurations, are given in Table 8.4 for amplifiers containing up to four stages.

TABLE 8.3   Losses per Stage Due to Real Feedback (Tuning Method B)

Amplifier configuration I	$\Phi_{fr} = \left\{ \dfrac{P_{rM}}{P_{(r+1)M}} \right\}^2$
Amplifier configuration II	$\Phi_{fr} = \left\{ 1 + q^2{}_{(r+1)} \right\}^2 \cdot \left\{ \dfrac{P_{2rM)}}{P_{(2r+2)M}} \right\}^2$
Amplifier configuration III	$\Phi_{fr} = (1 + q_r{}^2)^2 \left\{ \dfrac{P_{(2r-1)M}}{P_{(2r+1)M}} \right\}^2$

TABLE 8.4 Values for P_M for Amplifiers with up to Four Stages (Tuning Method B)

Amplifier (Fig. 8.3)	Configuration I	Configuration II	Configuration III	stage number	corresponding $|_n\delta_0|$
Single-stage	$P_{1M} = 1$ $P_{2M} = 1 - T_1\cos\Theta_1$	$P_{1M} = 1$ $P_{2M} = 1 + q_1^2$ $P_{3M} = P_{2M} - T_1\cos\Theta_1 . P_{1M}$ $P_{4M} = P_{3M} + q_2^2 P_{2M}$	$P_{1M} = 1$ $P_{2M} = 1 - T_1\cos\Theta_1$ $P_{3M} = P_{2M} + q_1^2 P_{1M}$	1	$|_1\delta_0|$
Two-stage	$P_{3M} = P_{2M} - T_2\cos\Theta_2$	$P_{5M} = P_{4M} - T_2\cos\Theta_2 . P_{3M}$ $P_{6M} = P_{5M} + q_3^2 P_{4M}$	$P_{4M} = P_{3M} - T_2\cos\Theta_2 P_{2M}$ $P_{5M} = P_{4M} + q_2^2 P_{3M}$	2	$|_2\delta_0|$
Three-stage	$P_{4M} = P_{3M} - T_3\cos\Theta_3 P_{2M}$	$P_{7M} = P_{6M} - T_3\cos\Theta_3 P_{5M}$ $P_{8M} = P_{7M} + q_4^2 P_{6M}$	$P_{6M} = P_{5M} - T_3\cos\Theta_3 P_{4M}$ $P_{7M} = P_{6M} + q_3^2 P_{5M}$	3	$|_3\delta_0|$
Four-stage	$P_{5M} = P_{4M} - T_4\cos\Theta_4 P_{3M}$	$P_{9M} = P_{8M} - T_4\cos\Theta_4 P_{7M}$ $P_{10M} = P_{9M} + q_5^2 P_{8M}$	$P_{8M} = P_{7M} - T_4\cos\Theta_4 P_{6M}$ $P_{9M} = P_{8M} + q_4^2 P_{7M}$	4	$|_4\delta_0|$

TABLE 8.5 Values for Z_t for Amplifiers with up to Four Stages (Tuning Method B)

Amplifier	Configuration I	Configuration II	Configuration III	Stage number
Single-stage	$Z_{t1} = \dfrac{1}{G_1} \cdot \dfrac{1}{P_{1M}}$	$Z_{t1} = \dfrac{q_1}{\sqrt{G_2 G_1}} \cdot \dfrac{1}{P_{2M}}$	$Z_{t1} = \dfrac{1}{G_1} \cdot \dfrac{1}{P_{1M}}$	1
Two-stage	$Z_{t2} = Z_{t1} \cdot \dfrac{_1y_{21}}{G_2} \cdot \dfrac{P_{1M}}{P_{2M}}$	$Z_{t2} = Z_{t1} \cdot \dfrac{|_1y_{21}| \cdot q_2}{\sqrt{G_4 G_3}} \cdot \dfrac{P_{2M}}{P_{4M}}$	$Z_{t2} = Z_{t1} \cdot \dfrac{|_1y_{21}| \cdot q_1}{\sqrt{G_3 G_2}} \cdot \dfrac{P_{1M}}{P_{3M}}$	2
Three-stage	$Z_{t3} = Z_{t2} \cdot \dfrac{_2y_{21}}{G_3} \cdot \dfrac{P_{2M}}{P_{3M}}$	$Z_{t3} = Z_{t2} \cdot \dfrac{|_2y_{21}| \cdot q_3}{\sqrt{G_6 G_5}} \cdot \dfrac{P_{4M}}{P_{6M}}$	$Z_{t3} = Z_{t2} \cdot \dfrac{|_2y_{21}| \cdot q_2}{\sqrt{G_5 G_4}} \cdot \dfrac{P_{3M}}{P_{5M}}$	3
Four-stage	$Z_{t4} = Z_{t3} \cdot \dfrac{_3y_{21}}{G_4} \cdot \dfrac{P_{3M}}{P_{4M}}$	$Z_{t4} = Z_{t3} \cdot \dfrac{|_3y_{21}| \cdot q_4}{\sqrt{G_8 G_7}} \cdot \dfrac{P_{6M}}{P_{8M}}$	$Z_{t4} = Z_{t3} \cdot \dfrac{|_3y_{21}| \cdot q_3}{\sqrt{G_7 G_6}} \cdot \dfrac{P_{5M}}{P_{7M}}$	4

The values of P_M are not given explicitly in this table because this would lead to very cumbersome expressions. In the Design Chart section of this book, however, these complete expressions are included for each of the amplifier arrangements considered, see Chapter 15. Very often it will be preferable to evaluate P_{1M}, etc. successively, the more so as these quantities are required in any case to determine the power gain Φ per stage.

If the value of Φ_{uM} required for the calculation of the stage gain is not published by the manufacturer of the transistors, it can be evaluated by means of Eq. (4.7.1) on page 41.

The value of Φ_{fr} follows from Tables 8.3 and 8.4.

The way in which the transducer losses Φ_i Φ_{mm} of a single-tuned bandpass filter or the transducer losses Φ_{tb} of a double-tuned bandpass filter depend on T is explained in Section 8.4.

The losses due to real feedback of the transistors are defined in this book as the losses which occur in the amplifier in excess of the losses in the single or double-tuned bandpass filters in case there was no feedback. This method has the advantage that the effect of the feedback of the transistors on the gain of the amplifier clearly stands out. A drawback of this method, however, is that it does not allow a simple addition of partial gains or losses when calculating gain figures per stage of the amplifier. This will be illustrated by the following example:

Suppose that the input terminals of a transistor are connected to the secondary of a double-tuned bandpass filter (see Fig. 8.6). Let the secondary tuned circuit damping $G_s{}^*$ be equal to the input damping g_{ie} of the transistor. Then the damping ratio of the secondary, which forms a part of the transducer losses of the complete bandpass filter, is:

$$\Phi_s = \frac{g_{ie}}{g_{ie} + G_s{}^*} = 0.5, \quad \text{or} \quad \Phi_s = -3 \text{ dB} \tag{8.6.1}$$

This damping ratio, and hence the transducer losses, are defined without taking into account the extra damping which appears at the transistor input terminals due to the real component of the feedback.

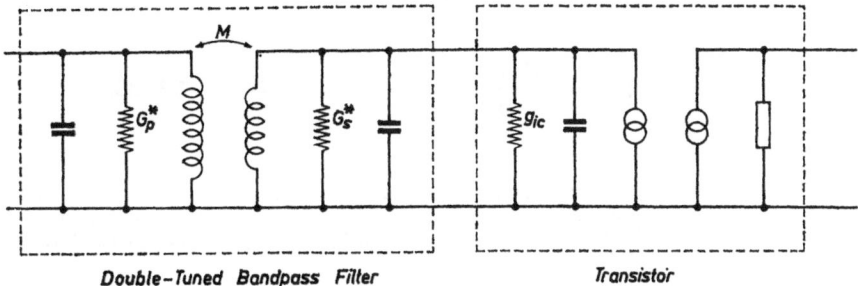

Double-Tuned Bandpass Filter   Transistor

Fig. 8.6 Part of an I.F. amplifier circuit used for illustrating the term "physically correct losses".

Let this feedback damping be positive and equal to g_{ie}. Then the "physically correct" value of the damping ratio would be:

$$\frac{2g_{ie}}{2g_{ie} + G_s^*} = 0.67, \quad \text{or} \quad -1.77 \text{ dB.}$$

The coupling losses of a double tuned bandpass filter are defined as:

$$\Phi_q = \left(\frac{2q}{1 + q^2}\right)^2, \tag{8.6.2}$$

with:

$$q^2 = k^2 Q_p Q_s .$$

In these expressions the quantities Q_p and Q_s are defined without incorporating feedback effects. When we assume: $q^2 = 2$, we obtain:

$$\Phi_q = 0.89, \quad \text{or} \quad \Phi_q = -0.51 \text{ dB.}$$

To determine the "physically correct" value of the coupling losses the secondary quality factor Q_0 must be reduced by a factor 2/3 because of the real feedback damping assumed. Then q^2 reduces to 1.33 and the coupling losses become:

$$\frac{4 \cdot 1.33}{(1 + 1.33)^2} = 0.97, \quad \text{or} \quad -0.09 \text{ dB.}$$

According to the definition for feedback losses adopted in this book, they are expressed as (see Book I):

$$\Phi_{fr} = \left(\frac{(g_{ie} + G_s^*)(1 + q^2)}{(g_{ie} + G_s^*)(1 + q^2) + q_{in\ feedback}}\right)^2 \tag{8.6.3}$$

In this expression q^2 is defined without regard to the non-unilateral properties of the transistor, see above.

With the various values assumed, Φ_{fr} becomes:

$$\Phi_{fr} = \left(\frac{2 q_{ie} (1 + 2)}{2 q_{ie} (1 + 2) + q_{ie}}\right)^2 = 0.73, \quad \text{or} \quad \Phi_{fr} = -1.34 \text{ dB.}$$

Because $g_{in\ feedback}$ is assumed to be equal to g_{ie}, the power that must be delivered into the transistor input terminals is twice as large as it would be in case there was no feedback. This means that the "physically correct" feedback losses amount to $- 3$ dB.

The considerations above can be summarized as shown in the following table. Obviously, the sum of the three losses will be independent of the definitions used.

	losses according to adopted definitions	"physically correct" losses
damping ratio of secondary	− 3.0 dB	− 1.77 dB
coupling losses	− 0.51 dB	− 0.09 dB
feedback losses	− 1.34 dB	− 3.0 dB
Total	− 4.85 dB	− 4.85 dB

For determining the gain from the input terminals of the transistor in the amplifier stage under consideration to the input terminals of the next stage use should be made of the "physically correct" feedback losses. Also "physically correct" values of the primary and secondary damping ratio and the coupling losses of the double-tuned bandpass filter in the output circuit of the transistor in this stage should be used.

It follows that the losses referred to as "physically correct" are very well suitable for determining stage gains. They are, however, *not* suitable for a straight-forward amplifier design procedure, as will become apparent from the discussion below:

The "physically correct" losses depend on the magnitude of extra damping which occurs at the input of the transistor due to its feedback.
This extra feedback damping can be expressed as:

$$Re \left(\frac{y_{12} \; y_{21}}{Y_L} \right) \tag{8.6.4}$$

in which Y_L is the load admittance at the output of the transistor. The "physically correct" losses which are applicable to the circuitry at the input side of the transistor are therefore affected by the properties of the circuitry at the output side. This implies that only a iterative process of design of such an amplifier would be possible.

The losses defined according to the adopted definitions are only dependent on feedback as far as the factor referred to as feedback losses is concerned. As already mentioned, this factor expressed the losses in gain due to the feedback compared with a case without feedback.

The factors damping ratio (Φ_p and Φ_s) and coupling losses (Φ_q) are pure bandpass filter properties and are independent of feedback. This makes them conveniently manageable in the practical design of an amplifier. It follows that the more or less algebraic partitioning of the gain components of an amplifier as adopted in this book is very well suitable for a systematic amplifier design procedure. The drawback of being not suitable for calculating actual stage gains is recognized, but accepted because a different method of calculating these gains is possible. This method, based on the transimpedance concept, very well fits into the systematic design procedure and avoids the intricacy of determining the "physically correct" losses.

Although the above discussion is mainly based on an amplifier with double-tuned bandpass filters, it will be obvious that an analogous reasoning holds for amplifiers with single-tuned bandpass filters.

8.6.3 THE VOLTAGE GAIN PER STAGE OF THE AMPLIFIER

The voltage gain per stage of the amplifier can be calculated by using the transimpedance concept introduced below.

Consider the equivalent circuit diagram of part of an amplifier shown in Fig. 8.7. It consists of a transistor four-pole network followed by a double-tuned bandpass filter, also represented as a four-pole network. The latter network is loaded by an admittance Y_L.

The circuit given in Fig. 8.7 can be simplified to that shown in Fig. 8.8. Now the network is divided in two parts between the forward transfer admittance current generator of the transistor and the admittance Y_2. This division is indicated by the dashed line in Fig. 8.8.

The transimpedance of the amplifier part behind the current generator mentioned follows from:

$$Z_{t1} = \frac{-Y_{21}}{Y_1\, Y_2 - Y_{12}\, Y_{21}} \qquad (8.6.5)$$

Considering that this circuit forms a double-tuned bandpass filter this expression reduces to:

$$Z_{t1} = \frac{q}{1 + q^2} \cdot \frac{1}{\sqrt{G_p\, G_s}} \qquad (8.6.6)$$

for the case of resonance and synchronous tuning of primary and secondary. The voltage gain between the input terminals of the transistor and the output terminals of the amplifier now follows from:

$$(V.G.)_1 = |_1 y_{21}| \cdot Z_{t1} \qquad (8.6.7)$$

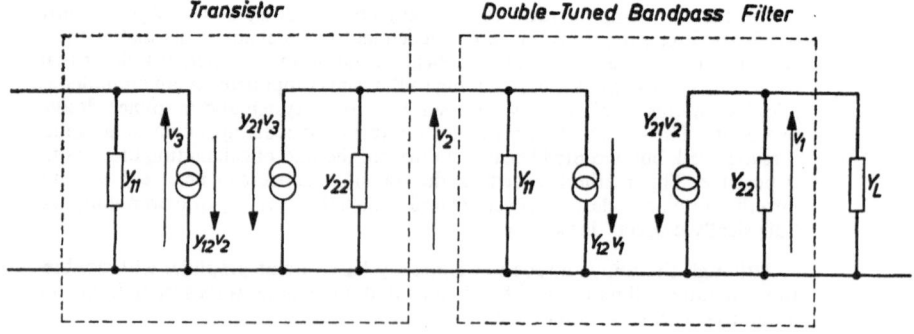

Fig. 8.7 Final part of an I.F. amplifier.

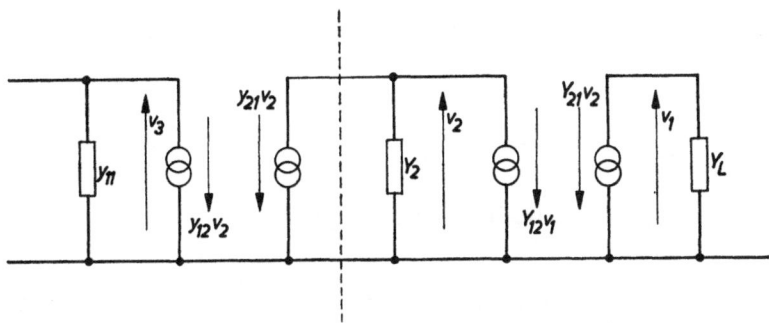

Fig. 8.8 Simplified equivalent circuit diagram of the final part of an I.F. amplifier. The dashed line indicates the place at which the amplifier should be divided in view of the transimpedance concept discussed in the text.

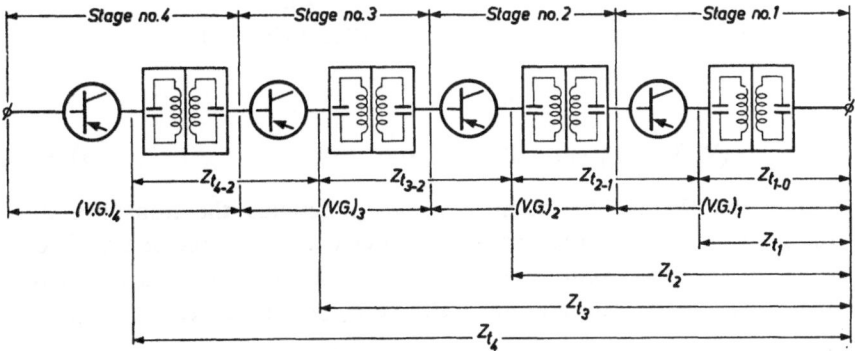

Fig. 8.9 Schematic arrangement of an amplifier for defining the various transimpedances.

The above consideration illustrate the way in which the voltage gain between the output terminals of the amplifier and the input terminals of a given transistor in the amplifier can be calculated when the transimpedance of the part of the amplifier behind the forward transfer admittance current source of this transistor is known.

Fig. 8.9 gives a schematic diagram of an multi-stage amplifier in which the various transimpedances which are of interest are indicated.

The transimpedances Z_{t1}, Z_{t2}, etc., can be determined from the general amplifier determinant according to the method dealt with in Book I. Obviously, only the part of the amplifier determinant that corresponds with the circuit elements contained in the part of the amplifier, for which the transimpedance has to be determined, need be taken into account.

In Table 8.5 the transimpedances Z_t are given for amplifiers in the configurations I, II and III containing up to four stages.

The voltage gain from the input terminals of transistor r to the output terminals of the amplifier is given by:

$$\frac{v_0}{v_r} = |_r y_{21}| \cdot Z_{tr} \cdot \tag{8.6.8}$$

For transistor $(r - 1)_2$, we have accordingly:

$$\frac{v_0}{v_{r-1}} = |_{r-1} y_{21}| \cdot Z_{t\,(r-1)} \cdot \tag{8.6.9}$$

The voltage gain of the stage r then follows from:

$$(\text{V.G.})_r = \frac{v_{r-1}}{v_r} = |_r y_{21}| \cdot \frac{Z_{tr}}{|_{r-1} y_{21}| \cdot Z_{t\,(r-1)}}$$

or:

$$(\text{V.G.})_r = |_r y_{21}| \cdot Z_{t\,r - (r-1)} \tag{8.6.10}$$

In expression (8.6.10) the term $Z_{t\,r-(r-1)}$ denotes the transimpedance of the network between the forward transfer current source of transistor r and the reverse transfer current source of transistor $(r - 1)$. This means that it includes the y_{22} of transistor r as well as the y_{11} of transistor $(r - 1)$.

In Table 8.6 expressions for $(V.G.)_r$ are given for amplifiers with up to four stages. Tuning methods B (or C) are assumed to be applied.

TABLE 8.6 Voltage Gains per Stage of an Amplifier with up to Four Stages (Tuning Method B).

Amplifier	Configuration I	Configuration II	Configuration III
Stage 1 (output stage)	$\lvert _1y_{21}\rvert \cdot \dfrac{1}{G_1} \cdot \dfrac{1}{P_{1M}}$	$\lvert _1y_{21}\rvert \cdot \dfrac{q_1}{\sqrt{G_2 G_1}} \cdot \dfrac{1}{P_{2M}}$	$\lvert _1y_{21}\rvert \cdot \dfrac{1}{G_1} \cdot \dfrac{1}{P_{1M}}$
Stage 2	$\lvert _2y_{21}\rvert \cdot \dfrac{1}{G_2} \cdot \dfrac{P_{1M}}{P_{2M}}$	$\lvert _2{}_{21}\rvert \cdot \dfrac{q_2}{\sqrt{G_4 G_3}} \cdot \dfrac{P_{2M}}{P_{4M}}$	$\lvert _2y_{21}\rvert \cdot \dfrac{q_1}{\sqrt{G_3 G_2}} \cdot \dfrac{P_{1M}}{P_{3M}}$
Stage 3	$\lvert _3y_{21}\rvert \cdot \dfrac{1}{G_3} \cdot \dfrac{P_{2M}}{P_{3M}}$	$\lvert _3y_{21}\rvert \cdot \dfrac{q_3}{\sqrt{G_6 G_5}} \cdot \dfrac{P_{4M}}{P_{6M}}$	$\lvert _3y_{21}\rvert \cdot \dfrac{q_2}{\sqrt{G_5 G_4}} \cdot \dfrac{P_{3M}}{P_{5M}}$
Stage 4	$\lvert _4y_{21}\rvert \cdot \dfrac{1}{G_4} \cdot \dfrac{P_{2M}}{P_{4M}}$	$\lvert _4y_{21}\rvert \cdot \dfrac{q_4}{\sqrt{G_8 G_7}} \cdot \dfrac{P_{6M}}{P_{8M}}$	$\lvert _4y_{21}\rvert \cdot \dfrac{q_3}{\sqrt{G_7 G_6}} \cdot \dfrac{P_{5M}}{P_{7M}}$

CHAPTER 9

EXAMPLES OF AMPLIFIER DESIGN

In this chapter the theoretical and practical aspects of I.F. amplifier design discussed in the preceding chapters will be illustrated by means of a number of examples. These comprise the design of:

— an I.F. amplifier for an A.M. receiver (450 kc/s) equipped with double-tuned bandpass filters,

— an I.F. amplifier for an F.M. receiver (10.7 Mc/s) equipped with double-tuned bandpass filters,

— a combined I.F. amplifier for an A.M./F.M. receiver equipped with double-tuned bandpass filters, and

— an amplifier having a midband frequency of 35 Mc/s, intended for use in a measuring apparatus.

In Chapter 10 a description will be given of the design of a complete I.F. amplifier for a television receiver complying with C.C.I.R. standards. In this example of amplifier design, consideration will also be given to spreads and tolerances in transistors and other components as well as to operational variations.

9.1 An I.F. Amplifier for an A.M. Radio Receiver

I.F. amplifiers employed in domestic A.M. radio receivers usually have a midband frequency of approximately either 270 kc/s or 450 kc/s. In the example discussed below, this frequency is assumed to be 450 kc/s.

The I.F. amplifier for this purpose should provide as high a gain as possible with the fewest stages, at the same time meeting certain requirements with respect to the 3 dB bandwidth and adjacent channel selectivity. In this example a single-stage amplifier with two double-tuned bandpass filters, as usually employed in pocket-size receivers, is considered. It is assumed that the amplifying element in this amplifier is a transistor AF 127 in common emitter connection, biased at $-V_{CE} = 6$ V and $I_E = 3$ mA. The admittance parameters are tabulated below.

The quoted values of Φ_{uM}, t and Θ were calculated from the admittance parameters.

It is moreover assumed that no neutralizing elements are used and the

TABLE 9.1

Admittance parameters of the AF 127	
$-V_{CE} = 6$ V, $I_E = 3$ mA and $f = 450$ kc/s	
g_{ie}	0.5 m℧
C_{ie}	105 pF
$|y_{re}|$	4.0 μ℧
φ_{re}	270°
$|y_{fe}|$	110 m℧
φ_{fe}	0°
g_{oe}	8.0 μ℧
C_{oe}	8.0 pF
ϕ_{u}M	59 dB
t	110 (m℧)²
Θ	270°

design will be calculated for a nominal transistor only. The tuned circuits of which the double-tuned bandpass filters are composed are taken to be identical ($r = 1$) and both bandpass filters are assumed to be critically coupled ($q^2 = 1$).

9.1.1 REGENERATION COEFFICIENT DETERMINED BY STABILITY

Fig. 5.8, Chapter 5, shows that for this type of amplifier with $q^2 = 1.0$, $r = 1.0$ and $\Theta = 270°$, the boundary of stability is $T_g = 3.2$. It should be recognized that there is a considerable discrepancy between the value of T_g obtained from the approximate Eq.(8.3.2) and the more accurate value obtained from the graph mentioned above. Preference is therefore given to the graphical determination.

Since the design is based on the use of a nominal transistor, the criterion for s will be taken to be 4 (cf. sub-section 8.3.1) in order to make sure that the amplifier will be stable under all conditions. The regeneration coefficient is therefore:

$$T = T_g/s = 3.2/4 = 0.8.$$

9.1.2 REGENERATION COEFFICIENT DETERMINED BY RESPONSE CURVE REQUIREMENTS

As indicated in Table 15.1 on p.230 a set of response curves for this amplifier ($n = 1$, type II, $\Theta = 270°$, $q^2 = 1.0$ and $r = 1$) is given on page 239.

The amplitude response curves show that in this respect a value of $T = 1.0$ would be acceptable for this amplifier.

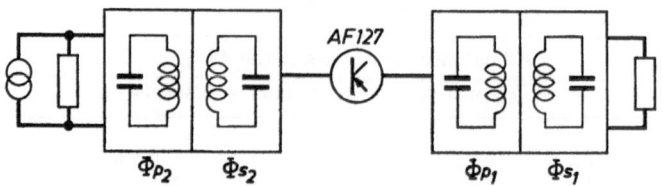

Fig. 9.1 Schematic diagram of the single-stage I.F. Amplifier.

9.1.3 DEFINITE CHOICE OF THE REGENERATION COEFFICIENT

As spreads in transistor parameters may be disregarded in this example, stability and response curve requirements are the only factors which influence the choice of the regeneration coefficient T. Hence $T = 0.8$. Using the notation indicated in Fig. 9.1:

$$\Phi_{p1} \cdot \Phi_{s2} = T/t = 0.8/110 = 7.55 \cdot 10^{-2} \qquad (9.1.1)$$

9.1.4 DESIGN OF THE DOUBLE-TUNED BANDPASS FILTERS

Since the tuned circuits of which the double-tuned bandpass filters are composed are assumed to be identical, we may write:

$$\Phi_{s1} = \Phi_{p1} = \Phi_{s2} = \Phi_{p2} = 7.55 \cdot 10^{-2}.$$

Furthermore:

$$\Phi_p = \Phi_s = 1 - w = 1 - Q/Q_0 = 1 - G^*/G, \qquad (9.1.2)$$

whence:

$$Q/Q_0 = 0.925.$$

There is no difficulty in making the unloaded quality factor of a circuit at a frequency of 450 kc/s equal to $Q_0 = 130$, which gives:

$$Q = 0.925 \cdot 130 = 120.$$

The collector of the transistor will be connected to the top of the primary of bandpass filter 1. At $g_{oe} = 8.0\mu\text{℧}$, the total damping of this primary then becomes:

$$G_2 = g_{oe}/\Phi_{p1} = 105\mu\text{℧}. \qquad (9.1.3)$$

All circuits are assumed to be identical, $G_1 = G_2 = G_3 = G_4 = 105\mu\text{℧}$. This implies that the base terminal of the transistor must be connected to a tap on the secondary of the input bandpass filter, the tapping ratio being:

$$n = \sqrt{g_{oe}/g_{ie}} = 0.127 \tag{9.1.4}$$

(because $n^2 g_{ie}$ should be equal to g_{oe}, so that the condition $G_2 = G_3$ is satisfied).

The tuning capacitance of the circuits is:

$$C = GQ/\omega = 4400 \text{ pF},$$

which gives for the tuning inductance:

$$L = 27 \,\mu\text{H}.$$

9.1.5 THE 3 dB BANDWIDTH

The values of x for positive and negative detuning will be denoted by x_+ and x_- respectively. The Design Chart for the amplifier under consideration (on page 239) then shows that for this amplifier the -3 dB points on the amplitude response curve are located at $x_+ = 0.98$ and $x_- = -1.22$ (according to the curve for $T = 1.0$, which is closest to the actual value of $T = 0.8$).

The 3 dB bandwidth is now:

$$B_{3dB} = (f_0/2Q) \cdot (x_+ - x_-), \tag{9.1.5}$$

so that, for $f_0 = 450$ kc/s and $Q = 120$:

$$B_{3dB} = 4.1 \text{ kc/s}.$$

9.1.6 ADJACENT CHANNEL SELECTIVITY

With a 9 kc/s channel spacing the adjacent channels are at 441 kc/s and 459 kc/s. For $Q = 120$ the normalized detuning at these frequencies is:

$$x_- = -4.8 \text{ and } x_+ = 4.8$$

respectively.

According to the design chart mentioned, this corresponds to a selectivity factor of 125 (22 dB) for both adjacent channels.

9.1.7 TRANSIMPEDANCE

The transimpedance of this amplifier is given by:

$$Z_{to} = \frac{|y_{fe}| \cdot q_1 q_2}{\sqrt{(G_1 G_2 G_3 G_4)} \cdot |\delta_0|} \tag{9.1.6}$$

The corresponding graph of Design Chart 239 shows that $|1/\delta_0|$ is 0.25.

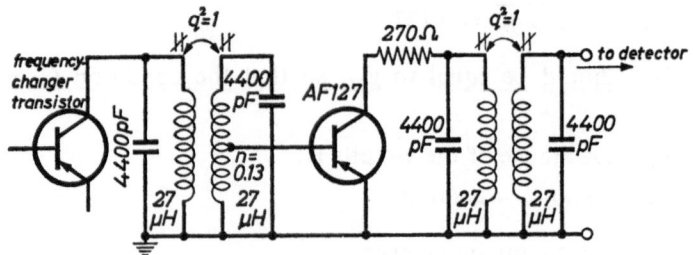

Fig. 9.2. Circuit diagram of the designed 450 kc/s I.F. amplifier. The d.c. biasing network has been omitted for simplicity.

Substituting, $|y_{fe}| = 110$ m℧ and since, as already stated, $q^2 = q_2{}^2 = 1.0$ and $G_1 = G_2 = G_3 = G_4 = 105\mu℧$:

$$Z_{t0} = 2.5 \text{ M}\Omega.$$

If the conversion transadmittance of the mixer transistor is known, this transimpedance figure can be used for calculating the voltage gain between the base of the mixer transistor and the input terminals of the detector.

9.1.8 COMPLETE CIRCUIT DIAGRAM

Fig. 9.2 shows the complete circuit diagram of the I.F. amplifier, the design of which is outlined in the preceding sub-sections. The resistor of 270 Ω in the collector lead of the transistor AF 127 prevents parasitic oscillations[1]:

9.2 An I.F. Amplifier for F.M. Receivers with Double-Tuned Bandpass Filters

The next I.F. amplifier to be discussed is assumed to form part of a full-performance domestic F.M. receiver. The midband frequency of the I.F. amplifier employed in this type of receiver is usually 10.7 Mc/s. The complete receiver should have a high signal-to-noise ratio and hence a large overall gain figure, a (small-signal) 3 dB bandwidth of 170 kc/s, and its adjacent channel selectivity should be better than 60 dB. This implies that the gain of the I.F. amplifier should be as high as possible while fulfilling the bandwidth and adjacent channel selectivity requirements. Double-tuned bandpass filters are therefore employed as interstage coupling networks. It is, moreover, assumed that the I.F. amplifier should not contain more than three transistors.

---

[1] Cf. A. Cense and A. H. J. Nieveen van Dijkum. Parasitic Oscillations in I.F. Stages and Frequency Changers of A.M. Receivers, Electronic Appl. 20.p.41. 1959/60 (No. 2).

This amplifier will be equipped with transistors AF 126 in common emitter connection, biased at $-V_{CE} = 6$ V and $I_E = 1$ mA.

TABLE 9.2

Admittance parameters of the AF 126 at
$-V_{CE} = 6$ V, $I_E = 1$ mA and $f = 10.7$ Mc/s

g_{ie}	1.7 m℧
C_{ie}	60 pF
$|y_{re}|$	100 μ℧
φ_{re}	260°
$|y_{fe}|$	32 m℧
φ_{fe}	335°
g_{oe}	40 μ℧
C_{oe}	3.5 pF
Φ_{uM}	35.8 dB
t	47 (m℧)²
Θ	235°
N	320

The admittance parameters and derived quantities of this transistor at this biasing point are tabulated above. It will be assumed that no neutralization is employed in this amplifier and that the double-tuned bandpass filters are identical as regards q^2, Q_p and Q_s.

9.2.1 GENERAL DESIGN CONSIDERATIONS

An I.F. amplifier for an F.M. receiver may be thought of as consisting of two main parts, namely the stages performing the required I.F. amplification only and the stage which drives the ratio detector circuit. In a three-stage I.F. amplifier, the former part consists of three double-tuned bandpass filters and two transistors, whilst the latter part comprises the driver transistor for the ratio detector and the ratio detector circuit itself, according to the block diagram of Fig. 9.3.

It can be shown that this second part contributes but little to the overall adjacent channel selectivity of the amplifier and that it only slightly affects the top of the response curve. For a 300 kc/s channel spacing, the adjacent channel selectivity of the ratio detector filter is of the order of 3 dB ($\simeq 1.5 \times$).

Furthermore, the input admittance of the ratio detector circuit changes only slightly over the passband of the amplifier. This implies that the loading of the driver transistor on the secondary of the last double-tuned bandpass filter remains almost constant over this passband (y_{ie} is assumed to be constant over the passband, and the admittance due to feedback of the transistor

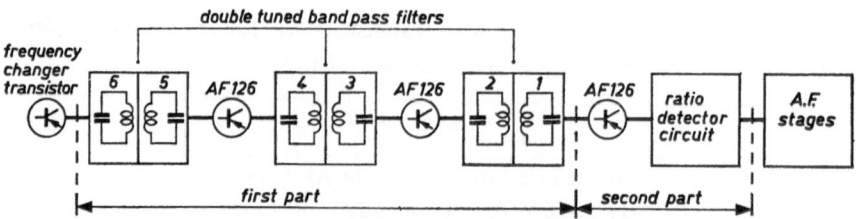

Fig. 9.3 The I.F. amplifier of an F.M. receiver can be thought of as consisting of two main parts, the first part comprising the stages which provide I.F. amplification only and the second part comprising the transistor which drives the ratio detector circuit and the ratio detector itself.

remains substantially constant because this also holds for the input admittance of the ratio detector circuit).

It is thus possible to calculate the response curve of the first part up to this point. To obtain the overall response curve of the complete amplifier, the responses of the first and second parts must be multiplied. The following comments are confined to the design of the first part of the amplifier, an investigation of the second part with the ratio detector circuit being beyond the scope of this book.

9.2.2 REGENERATION COEFFICIENT DETERMINED BY STABILITY

According to Eq.(8.3.2) the approximate boundary of stability for $\Theta = 235°$ equals $T_g = 4.7$. Putting $s = 4$ (thus making allowance for spreads in the transistors) gives for the regeneration coefficient $T = 4.7/4 \simeq 1.2$.

9.2.3 REGENERATION COEFFICIENT DETERMINED BY RESPONSE CURVE REQUIREMENTS

The first part of this amplifier may be considered as a two stage amplifier ($n = 2$) of type II. Table 15.1 on p.230 shows that performance graphs for this type of amplifier are given in Figs 15.12 and 15.13 for $\Theta = 240°$ and $q^2 = 1.0$ and 1.5 respectively. The value of $\Theta = 240°$ is close enough to the actual value of $\Theta = 235°$ as not to introduce an intolerable error.

The amplitude response graphs for $q^2 = 1.0$ and $q^2 = 1.5$ reveal that, as far as response curve distortion is concerned, a value of $T = 1.2$ would be acceptable. Moreover, the graphs show that for $T = 1.0$ and $q^2 = 1.0$ the amplitude response curve is under-critical, whereas for $T = 1.0$ and $q^2 = 1.5$ this characteristic is slightly over-critical. However, since the amplitude response curve of the second part is single-humped, its combination with the first part designed for $T = 1.2$ and $q^2 = 1.5$ is likely to result in a substantially flat-topped response curve. The further design of the first part of the amplifier will therefore be based on this combination of T and q^2.

9.2.4 DESIGN OF THE DOUBLE-TUNED BANDPASS FILTERS

Since all tuned circuits which constitute the double-tuned bandpass filters are assumed to be identical, the losses in each circuit are given by:

$$\Phi_p = \Phi_s = T/t = 0.16.$$

From Eq.(9.1.2):

$$Q/Q_0 = 0.84.$$

At a frequency of 10.7 Mc/s it is quite possible to realize an unloaded quality factor of $Q_0 = 110$, which gives $Q = 92$.

The collector of the AF 126 is connected to the top of the primary of the double-tuned bandpass filter; the total damping of this circuit is then, from Eq.(9.1.2):

$$G_p = g_{oe}/\Phi_p = 40/0.16 = 250\mu\text{U}.$$

The tuning capacitance then becomes (Q being equal to 92):

$$C = GQ/\omega = 340 \text{ pF},$$

and the tuning inductance:

$$L = 1/\omega^2 C = 0.65 \text{ μH}.$$

When dimensioning the primary of the bandpass filter, it should be kept in mind that the collector capacitance of the transistor is voltage dependent, and since this collector capacitance is included in the tuning of the primary it will be clear that the level of the collector a.c. voltage affects the tuning of the primary. To keep this effect as small as possible, the tuning capacitor must have a relatively large value. A capacitance of 340 pF is amply sufficient.

The secondary of the double-tuned bandpass filter has to have a tap to which the base terminal of the transistor can be connected.

In this example an inductive tap is chosen. To ease construction of the tapped secondary coil of the bandpass filter, for which an unloaded quality factor of $Q_0 = 110$ is required, its tuning capacitance should preferably not be larger than 220 pF. With $C = 220$ pF and $Q_0 = 110$, the tuned circuit damping becomes:

$$G_s = \frac{\omega C}{Q_0} = 134 \text{ μU}.$$

The tap on the secondary coil must be such that the damping of the tuned circuit itself, measured at the tap, amounts to:

$$\frac{g_{ie}}{\Phi_s} - g_{ie}.$$

Substituting values this gives:

$$\frac{G_s}{n^2} = 8.9 \text{ m}\mho$$

Here, n denotes the tapping ratio. It follows that:

$$n = 0.12.$$

In this example effects of the spread inductance inherent with a tap on the tuning inductance of the circuit will be neglected. These effects will be considered in the following section in which capacitive and inductive tapping will be compared.

Fig. 9.3 shows that the primary of the input bandpass filter of the I.F. amplifier is loaded by the output conductance g_{oe} of the frequency-changer transistor. It will be assumed that a transistor type AF 125 is used for this purpose: the value of g_{oe} of this transistor at 10.7 Mc/s is 25 $\mu\mho$ and increases to 40 $\mu\mho$ during oscillation. This output conductance must be included in the total damping of the primary of the input bandpass filter.

Moreover, it should be remembered that the inductance of this primary shunts that of the oscillator coil in the tuner. Tests show that the inductance of the primary should not be less than 1.5 µH, to keep the effect of this coil on the oscillator performance sufficiently small. The tuning capacitance then becomes 150 pF.

For $Q = 92$ the total damping is given by:

$$G_p = \omega C/Q = 110 \, \mu\mho$$

At $g_{oe} = 40 \, \mu\mho$ this implies an unloaded quality factor of:

$$Q_0 = (110/70) \times 92 = 145.$$

Provided high-quality materials are used, it should be possible to realize this quality factor. In fact, high-quality materials should be used in any case, because the losses in this primary should be small at 100 Mc/s, to ensure that the oscillator coil of the tuner is not unduly damped.

According to Chapter 8, to determine the value of the loaded quality factor of the double-tuned bandpass filters, the 3 dB bandwidth and the adjacent channel selectivity should be taken as starting points. However, in this particular case it is necessary to follow the reverse procedure, because the loaded quality factor Q should be as large as possible to obtain the very highest adjacent channel selectivity. The maximum value of Q is limited by the maximum realizable value of Q_0 and by the ratio Q/Q_0 imposed by stability conditions. Hence, Q must be determined before the 3 dB band-width and the adjacent channel selectivity are considered.

9.2.5 THE 3 dB BANDWIDTH

According to the curve for $T = 1.0$ (which is closest to the actual value of $T = 1.2$) of Design Chart 15.13, the -3 dB points on the amplitude response curve are located at $x_+ = 1.2$ and $x_- = -1.6$. The 3 dB bandwidth is therefore:

$$B_{3dB} = (f_0/2Q) (x_+ - x_-) = 163 \text{ kc/s.}$$

The bandwidth of the amplifier is thus slightly smaller than required according to the initial specification. This defect can be remedied by making the loaded quality factor of the tuned circuits slightly lower.

9.2.6 THE ADJACENT CHANNEL SELECTIVITY

For a channel spacing of 300 kc/s, the normalized detuning for the upper and lower adjacent channel carriers becomes $x_+ = 5.1$ and $x_- = 5.3$ respectively. According to Fig. 15.13 the selectivity factor for the upper adjacent channel is 830 (58 dB), and that for the lower adjacent channel 1000 (60 dB).

In combination with the adjacent channel selectivity of the second part of the I.F. amplifier, adjacent channel selectivities of 1200 (62 dB) and 1500 (63 dB) respectively are attained.

This is somewhat better than initially specified. It is thus quite permissible to reduce the loaded quality factor of the tuned circuits slightly, as was required to meet the 3 dB bandwidth requirements.

9.2.7 THE ENVELOPE DELAY

The change in envelope delay over the passband of an I.F. amplifier for use in an F.M. receiver should preferably be small. The τ_e graph of Fig. 15.13 for $T = 1.0$ (see sub-section 9.2.5) shows that $\tau_e = 2.5$ rad at $x = 0$. If we consider only the changes between the -3 dB points of the amplitude response curve (that is, between $x_+ = 1.2$ and $x_- = -1.6$), the extreme values of τ_e for positive and negative detuning are 3.0 rad and 3.4 rad respectively.

According to sub-section 2.5.3 of Book I the actual envelope delay is given by:

$$t_e = \tau_e \cdot 2Q/\omega. \tag{9.2.1}$$

Taking the envelope delay at $x = 0$ as reference, the envelope delay reaches an extreme difference of

$$(3.0 - 2.5)2Q/\omega_0 = 1.3\mu s$$

at the lower frequency side of the passband and an extreme difference of

$$(3.4 - 2.5)2Q/\omega_0 = 1.85\mu s$$

at the higher frequency side of the passband.

9.2.8 GAIN

9.2.8.1 *Transducer Gain*

For evaluating the transducer gain of the first part of the I.F. amplifier, the frequency-changer transistor will be regarded as a current-source with parallel damping of 40 $\mu\mho$ delivering power into the I.F. amplifier.

The input conductance g_{in} of the transistor which drives the ratio detector loads the secondary of the last double-tuned bandpass filter. This input conductance may therefore be regarded as the load of the first part of the I.F. amplifier to which this amplifier delivers power. The value of g_{in} depends on the input admittance of the ratio detector circuit. It can be shown that for a commonly used ratio detector circuit, g_{in} may be put equal to 1.6 g_{ie}.

According to Table 8.2. (p. 86) the transducer gain of the first part of the I.F. amplifier is:

$$\Phi_t = \frac{4G_s}{G_6} \cdot \frac{G_L}{G_1} \cdot T^2 \cdot N^2 \cdot q^6 \cdot \frac{1}{|{}_2\delta_0|^2} \tag{9.2.2}$$

In the example under discussion $G_s = 40\ \mu\mho$, $G_L = 2.7\ m\mho$, $G_6 = 110\ \mu\mho$, $G_1 = 8.9 + 2.7 = 11.7\ m\mho$, $T = 1.2$, $N = 320$ and $q^2 = 1.5$, whilst Design Chart 15.13 gives $1/|{}_2\delta_0| = 5.3 \cdot 10^{-2}$.

This gives a transducer gain of: $\Phi_t = 26.6$ dB

Provided the power gain of the second part of the I.F. amplifier is known, the transducer gain of the complete I.F. amplifier can be determined. From this transducer gain it is possible to calculate the transimpedance of the I.F. amplifier. The transducer gain of the complete high-frequency part of the receiver can then be calculated from this transimpedance figure and the transadmittance figure of the tuner.

9.2.8.2 *Gain per Stage*

To calculate the gain per stage of the amplifier, the damping ratios of the primary and secondary circuits of the bandpass filters, the coupling losses of these bandpass filters, and the losses due to real feedback of the transistors must be evaluated.

Damping ratios:

Circuit no. 6 (see Fig. 9.3):

g_{oe} mixer transistor $= 40\ \mu\mho$,
total tuned circuit damping $G_6 = 110\ \mu\mho$,
the primary damping ratio Φ_{p6} then becomes:

$$\Phi_{p6} = \frac{g_{oe}}{G_6} = 0.36 \text{ or } \Phi_{p6} = -4.4 \text{ dB.}$$

Circuits no. 2, 3, 4 and 5:

according to sub-section 9.2.4:

$$\Phi_{s2} = \Phi_{p3} = \Phi_{s4} = \Phi_{p5} = 0.16 \text{ or}$$
$$\Phi_{s2} = \Phi_{p3} = \Phi_{s4} = \Phi_{p5} = -8 \text{ dB}$$

Circuit no. 7:

input damping of 3rd I.F. transistor, $g_{in} = 2.7\ m\mho$,
total damping at the base of this transistor,

$$G_7 = 8,9 + 2.7 = 11.6\ m\mho,$$

damping ratio $\Phi_{s7} = 0.23$ or $\Phi_{s7} = -6.3$ dB.

Coupling losses of the double-tuned bandpass filters:

coupling coefficient $q^2 = 1.5$,

coupling losses $\Phi_q = \left\{\dfrac{2q}{1 + q^2}\right\}^2 = 0.96$ or $\Phi_q = 0.1$ dB·

Losses due to real feedback of the transistors:

According to the curve for Φ_f on Design Chart 15.13, the feedback losses of the two stages together amount to -1.8 dB. Because both stages are identical, we obtain for each stage:

$$\Phi_{f1} = \Phi_{f2} = -0.9 \text{ dB.}$$

The powergain of each of the first two stages can now be calculated.

Power gain of stage 1:

$$\Phi_{uM1} = 35.8 \text{ dB(see p. 103),}$$

$$\Phi_{p2} = -8.0 \text{ dB},$$
$$\Phi_{s7} = -6.3 \text{ dB},$$
$$\Phi_{q} = -0.1 \text{ dB}.$$

Hence it follows:

$$\Phi_1 \quad = 20.5 \text{ dB}.$$

Power gain of stage 2:

$$\Phi_{uM2} = 35.8 \text{ dB},$$
$$\Phi_{p3} \quad = \Phi_{s4} = -8.0 \text{ dB},$$
$$\Phi_{q} \quad = -0.1 \text{ dB}.$$

Hence $\Phi_2 \quad = 18.8$ dB.

Checking, the transducer gain can now be calculated from:

$$\Phi_t = \Phi_{tb3} + \Phi_1 + \Phi_2.$$

Here the quantity Φ_{tb3} denotes the transducer losses of the input double-tuned bandpass filter.

With $\Phi_{t3} = \Phi_{p6} + \Phi_{s5} + \Phi_{q3} = -12.5$ dB, the transducer gain becomes:

$$\Phi_t = -12.5 + 18.8 + 20.5 = 26.5 \text{ dB}.$$

This transducer gain figure corresponds very well with that calculated in the preceding section.

9.2.9 COMPLETE CIRCUIT DIAGRAM

Fig. 9.4 shows the complete circuit diagram of the I.F. amplifier including a ratio detector circuit. For simplicity the d.c. biasing networks of the transistors have again been omitted. As mentioned above, the design of the ratio detector circuit is beyond the scope of this article so that no further details of this circuit are included in Fig. 9.4.

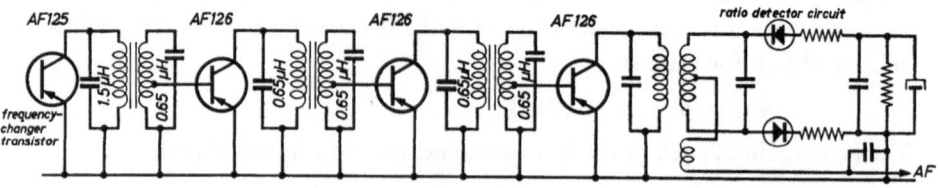

Fig. 9.4 I.F. amplifier for 10.7 Mc/s as designed in Section 9.2. The d.c. biasing networks of the transistors have been omitted for simplicity. The transistors AF126 are biased at $-V_{CE} = 6$ V and $I_E = 1$mA. The tapping ratio $n = 0.153$.

9.3 A Combined I.F. Amplifier for an A.M./F.M. receiver

To illustrate how the I.F. amplifiers for the A.M. and F.M. signals in a radio receiver may be combined, an example will be given of the design of the relevant part of a complete A.M./F.M. receiver[1]). In this example, some consideration will also be given to the parasitic effect of bottoming oscillations[2]) and to measures to prevent this.

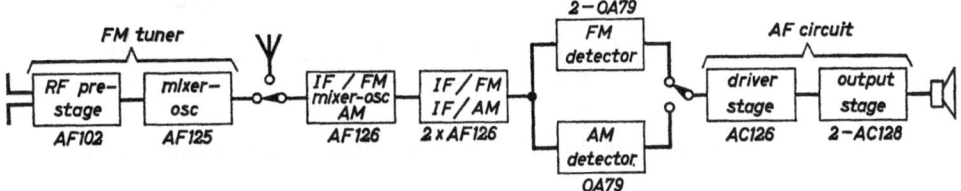

Fig. 9.5. Block diagram of the A.M./F.M. receiver.

Fig. 9.5 shows the block diagram of the receiver. This diagram comprises an F.M. front-end unit consisting of a pre-amplifier with a transistor AF 102, and a self-oscillating mixer with a transistor AF 125. This part is followed by the first I.F. stage for F.M. which is equipped with a transistor AF 126. This transistor is used as a self-oscillating mixer for the A.M. bands. The next two stages are common I.F. amplifiers for A.M. and F.M. with transistors AF 126. The I.F. signal is applied either to the A.M. or to the F.M. detector; these are equipped with diodes OA79.

After detection the A.F. signal is amplified in a two-stage A.F. amplifier, comprising a driver with a transistor AC 126 and a class B output stage with a matched pair of transistors 2-AC 128.

9.3.1 GENERAL DESIGN CONSIDERATIONS OF THE COMBINED I.F. AMPLIFIERS

The intermediate frequencies of the A.M. and F.M. circuits are 452 kc/s and 10.7 Mc/s respectively. The I.F. amplifier for A.M. consists of two stages, that for F.M. of three stages.

To limit the use of switching contacts, a method was devised for combining the A.M. and F.M. circuits. At the output of the I.F. transistors top connection can be applied to both the A.M. and F.M. filters, so that series connec-

---

[1]) A description of this receiver has previously been published in Advance Information bulletins A.I. 111 and A.I. 116 issued by Philips Electronic Market Development department. This A.M./F.M. receiver was designed by Mr. Pölzl, who also prepared the publications mentioned, cooperating with Mr. G. Slingerland. The relevant parts of the original publications have been modified to suit the purpose of illustrating the design method presented in this book.

[2]) Cf. footnote on page 102.

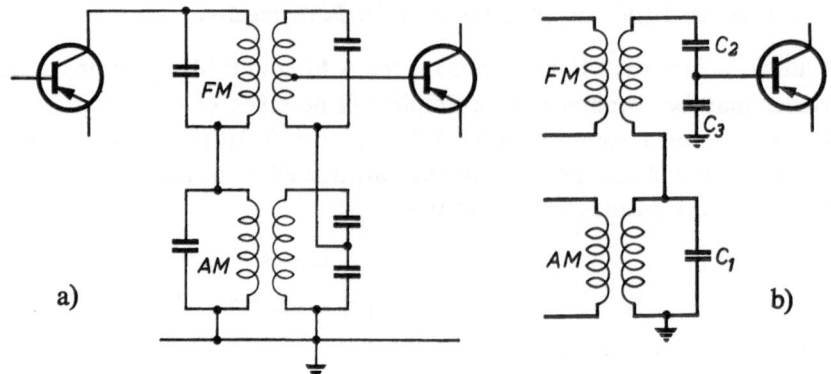

Fig. 9.6. Possible connections of the A.M. and F.M. filters at the base side of the I.F. transistors.

tion of these filters involves no difficulties. The input impedance of transistors is, however, much lower than the output impedance, so that tapped filters must be used at the input, which therefore makes matters less simple. Only two of the conceivable methods of combination of the two tapped filters are practicable. These are (a) series connection of an inductively tapped F.M. filter and a capacitively tapped A.M. filter, see Fig. 9.6.a and (b) parallel connection of both capacitively tapped filters, see Fig. 9.6.b. With an inductively tapped F.M. filter as used with method (a), the coupling factor (k) between the inductors L_1 and L_2 (see Fig. 9.7) is always smaller than unity. This amounts to a spread inductance being connected in series with the transistor lead. This inductance L_{sp} is given by:

$$L_{sp} = L\,(1 - k^2),$$

where

$$k = M/\sqrt{L_1 L_2} \quad (M = \text{mutual inductance}).$$

The spread inductance must be taken into account for calculating the stability condition of the transistor. In practice the mutual inductances of a batch of tapped inductors often vary appreciably depending on the position of the tuning core for correct alignment.

Even if it were possible to circumvent this difficulty it is simpler to use method (b). By using a common tapping for both the A.M. and F.M. filters, this method can be simplified as indicated in Fig. 9.6-b. The number of components required in this case is the same as with method (a).

The transistors AF 126 used in this I.F. amplifier are connected in the common emitter configuration. They are biased at an emitter current of 1 mA and a collector-emitter voltage of the order of 6 V.

Because of the small feedback capacitance of these transistors, no neutrali-

Fig. 9.7. With an inductively tapped F.M. filter, the coupling factor (k) between L_1 and L_2 is always smaller than unity. This amounts to a spread inductance (L_{sp}) being connected in series with the tapping lead.

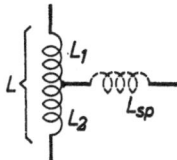

zation is required, since without neutralization a gain of approximately 20 dB per stage and an overall selectivity of more than 60 dB can be obtained with F.M.

Another factor to be considered is the risk of parasitic oscillation occurring in the A.M./I.F. stages due to the bottoming effect[1]). The first A.M./I.F. stage is controlled and, generally speaking, the choice of the collector impedance of a controlled A.M./I.F. stage is not very critical in this respect. With F.M. reception, however, the stage is not controlled and care must be taken to avoid parasitic oscillation which may then be set up in the A.M. circuit if the input signal is large. A limit is therefore set to the maximum value of the collector impedance. Since, for an effective AGC at A.M. a high-ohmic base voltage divider should be used ($100 \text{ k}\Omega + 15 \text{ k}\Omega$ in the receiver under consideration), together with an emitter resistor of $680 \ \Omega$, the maximum permissible collector impedance for preventing parasitic oscillations is about $20 \text{ k}\Omega$ in this case[1]).

In the last A.M./I.F. stage this phenomenon is avoided by keeping the collector impedance of the AF 126 below $10\text{k}\Omega$[1]). In this stage the collector load is formed by the parallel connection of the dynamic impedance of a single-tuned circuit and the reflected impedance of the A.M. detector. The former is $40 \text{ k}\Omega$, so that the latter should not exceed $13 \text{ k}\Omega$.

If no measures were taken, parasitic oscillation might however be set up in this stage with F.M. reception, since the A.M. detector is then disconnected. For this reason the A.M. detector is also connected with F.M. reception, via resistor R_{21} (see Fig. 9.15).

It can be shown that the source admittance Y_s of the I.F. transistor with A.M. reception is not real at resonance but also comprises a capacitive component (see Book I, Chapter 12). This capacitive component affects the transistor four-pole parameters, as follows from Appendix VII of Book I. This gives the four-pole parameters of a transistor with impedances included in base, emitter and collector leads. Since only an impedance in the base lead is of interest here, Z_e and Z_c can be disregarded. It can be calculated that the capacitive component of Y_s considerably increases the output conduct-

---

[1]) Cf. footnote on p. 102.

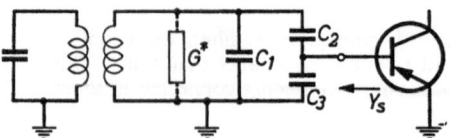

Fig. 9.8. Equivalent circuit diagram of the circuit shown in Fig. 9.6-b at A.M. reception.

ance of the transistor, which results in an improvement of the stability at the expense of gain and selectivity.

This effect becomes particularly noticeable at small values of C_2 (Fig. 9.8) resulting in low value of (Y_s) with an increased argument (in the direction of $90°$) and in a high value of the output conductance. It can be compensated by decreasing the value of the by-pass capacitor of the emitter resistor, so that this resistor is not completely decoupled. It may be calculated from Book I, Appendix VII, that by suitable choice of the impedance Z_e thus created, the effect of Z_b can be decreased. The value of Z_b, moreover, depends on the value of C_2, C_3 and on the filter damping G^*.

With F.M. reception the tuning capacitance of the resonant circuit is mainly determined by C_2. In view of the obtainable quality factor of the filter (Q_0 is about 110), the total tuning capacitance should not exceed 350 pF. Hence C_2 should not become too large. On the other hand, since low values of C_2 increase the output damping of the transistor at A.M. reception, C_2 is made 300 pF. Also, in connection with the design of the double-tuned bandpass filters considered in the next sub-section the capacitances C_1, C_3 and C_e are given values of 1 nF, 1.8 nF and 27 nF respectively.

9.3.2 I.F. AMPLIFIER FOR F.M.

9.3.2.1 *Determination of Regeneration Coefficient*

The transistors used in the I.F. amplifier considered in this section are of the type AF 126 and are biased at $V_{CE} = -6$ V and $I_E = 1$ mV. The relevant admittance parameters are given in the table on page 103.

Considerations in determining the regeneration coefficient of the transistors and the response curve given in sub-sections 9.2.2 and 9.2.3 are also applicable to this amplifier. So we obtain $T = 1.2$ and $q^2 = 1.5$.

9.3.2.2 *Design of the Double-Tuned Bandpass Filters*

The double-tuned bandpass filters used in this combined I.F. amplifier at F.M. are identical with those of the preceding example, except for the tap on the respective secondaries. The capacitive tap used in this combined I.F.

amplifier must be so designed that it delivers an effective tuning capacitance of 350 pF and a tapping ratio of 0.12. The effective tuning capacitance can be calculated from (see Fig. 9.7):

$$C_t = \frac{C_1 C_2 C_3}{C_1 C_2 + C_3(C_1 + C_2)}$$

Taking for $C_1 = 1$ nF, $C_2 = 300$ pF and $C_3 = 1.8$ nF

(see preceding section) we find:

$$C_t = 220 \text{ pF}.$$

In this calculation the input capacitance of the transistor has been neglected $(C_i < C_3)$.

The tapping ratio n is:

$$n = \frac{C_t}{C_3} = \frac{220}{1800} = 0.12.$$

9.3.2.3 *Gain and Frequency Response*

With respect to overall gain, gain per stage, amplitude response and envelope delay, this I.F. amplifier is identical with the one described in Section 9.2.

9.3.3 I.F. AMPLIFIER FOR A.M.

The I.F. amplifier for A.M. signals consists of two stages. Between the self-oscillating mixer stage and the first I.F. stage and also between the first and second I.F. stages, double-tuned bandpass filters have been used. The second I.F. stage is coupled to the detector via a single-tuned bandpass filter. It will be assumed that the unloaded quality factors of the tuned circuits of which the bandpass filters are composed are equal to 90. Furthermore, the bandpass filters are assumed to be critically coupled $(q^2 = 1)$.

On account of the dimensioning of the capacitive tap on the secondaries of the double-tuned bandpass filters for the F.M./I.F. amplifier, it has been necessary to choose a rather low value for the emitter-decoupling capacitor of various transistors. When the receiver is used for A.M. reception this capacitor provides a rather large reactance for the I.F. signal. For a frequency of 452 kc/s and a capacitance of 27 nF this reactance amounts to 13Ω.

In the design of the I.F. amplifier for A.M., therefore, consideration must be given to both the not-completely decoupled emitter resistor of the transistors and the presence of the capacitive tap in the base circuit.

TABLE 9.4

Admittance parameters of the AF 126 at 450 kc/s,

$V_{CE} = -6V, I_E = 1$ mA

Z_e	=	0		Z_e	=	$-j13\ \Omega$
g_{ie}	=	0.25 m℧		g_{ie}'	=	0.12 m℧
C_{te}	=	70 pF		C_{te}'	=	92 pF
$|y_{re}|$	=	4 μ℧		$|y_{re}|'$	=	4 μ℧
φ_{re}	=	270°		φ_{re}'	=	270°
$|y_{fe}|$	=	37 m℧		$|y_{fe}|'$	=	33.2 m℧
φ_{fe}	=	0°		φ_{fe}'	=	30°
g_{oe}	=	1 μ℧		g_{oe}'	=	-2.1 μ℧
C_{oe}	=	4 pF		C_{oe}'	=	52 pF

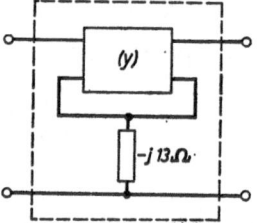

Fig. 9.9. Four-pole network formed by a transistor with an impedance in the emitter lead. The parameters of this four-pole are denoted by y'.

9.3.3.1 *Effects of the Emitter-Decoupling Capacitor*

The transistors of the I.F. amplifier are biased at $I_E = 1$ mA and $V_{CE} = -6$ V. For a frequency of 452 kc/s the admittance parameters of the transistor AF 126 are as given in Table 9.4 ($Z_e = 0$).

The reactance of the emitter-decoupling capacitor together with the transistor fourpole can be regarded as forming a new fourpole as shown in Fig. 9.9.

With the equations derived in Book I, Appendix VII, the admittance parameters of the new fourpole can be calculated. These parameters,[1] which must be used in the further design of the amplifier, are also entered in the Table 9.4.

9.3.3.2 *Design for Stability*

In designing for stability in this amplifier, the effects of the capacitive tap in the base-circuits of both transistors must be taken into account. In Fig. 9.8 the equivalent circuit diagram of the input of the A.M./I.F. transistors is shown. Referring to Chapter 12 of Book I, this circuit can be simplified to

---

[1] To distinguish between the parameters of a transistor with no impedance in base and emitter lead and those of a transistor with an impedance in the emitter lead the former are denoted by the symbol y and the latter by y'.

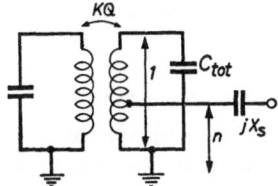

Fig. 9.10. Equivalent diagram of the circuit shown in Fig. 9.8.

that given in Fig. 9.10. The impedance seen when looking into the double-tuned bandpass filter from the base terminal of the I.F. transistor (Z_s in Fig. 9.8) can be calculated as shown in Book I, Section 12.1.4:

$$Z_s = \frac{n^2}{G \cdot y_0} + X_s.$$

In this expression:

$$G = G^* + n^2 g_{ie}',$$

$$y_0 = 1 + j\,x_s + \frac{q^2}{1 + j\,x_p},$$

$$X_s = \frac{1}{j\omega\,(C_2 + C_3 + C_{ie}')},$$

and

$$n = \frac{C_2}{C_2 + C_3 + C_{ie}}.$$

With $C_1 = 1$ nF, $C_2 = 300$ pF, $C_3 = 1.8$ nF and $C_{ie}[1] = 92$ pF the effective tuning capacitance becomes $C_{tot} = 1260$ pF. With an unloaded quality factor of $Q_0 = 90$, we find for the tuned circuit damping $G^* = 39$ μ℧. For the tap ratio n we find $n = 0.137$. The series reactance X_s becomes $X_s = 170\ \Omega$ and the impedance n^2/G can be calculated as $445\ \Omega$.

Furthermore, $q^2 = 1$.

The boundary of stability of an amplifier stage with a complicated base circuit as in this case can best be determined by means of the graphical method presented in Book I, Chapters 5 and 12. For this method the polar diagram of $Z_i = Z_s + 1/g_{ie}'$ in the complex plane must be known (C_{ie}' is contained in the tuning elements). Since $G/n^2 \gg g_{ie}'$, the latter will be neglected in further calculations.

Considering the simplified base circuit given in Fig. 9.10 it is seen that the input impedance Z_i consists of the series connection of an ideal tapped band-pass filter (with impedance n^2/G) and a reactance X_s.

The regeneration coefficient T_g of both amplifier stages can now be determined according to the graphical method mentioned. As follows from the Table 9.4 the regeneration phase angle Θ' of the transistors, including the influences of the emitter decoupling capacitors, equals $\Theta' = 270 + 30 = 300°$.

Construction for T_g for the first and second stages is shown in Figs. 9.11 and 9.12 respectively. The construction has been carried out according to the method presented in Book I, Sections 12.1.4 and 12.1.7. It follows that for the input stage:

$$T_{g1} = 2.7,$$

and for the second stage:

$$T_{g2} = 2.4.$$

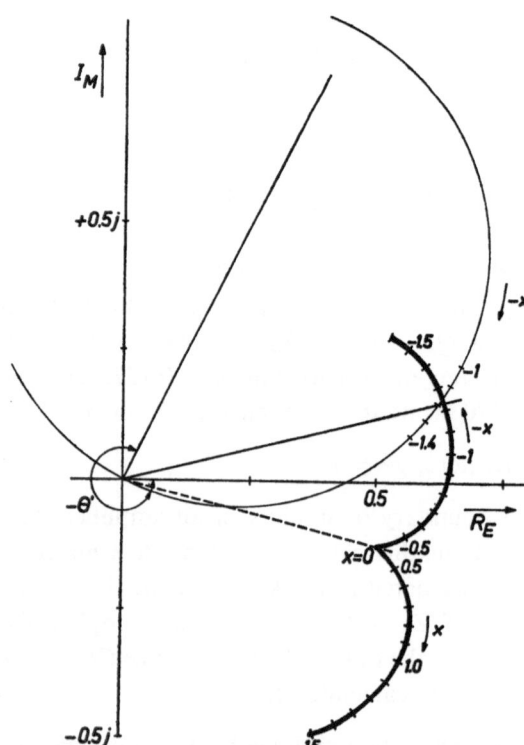

Fig. 9.11. Graphical method of calculating the factor T_g for the first A.M./I.F. stage.

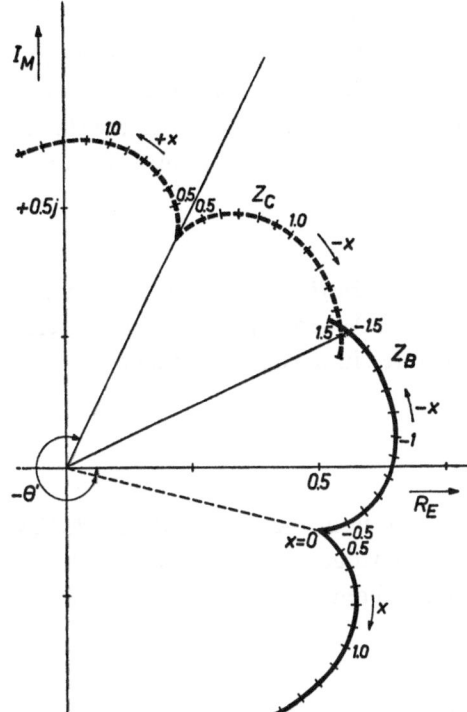

Fig. 9.12. Graphical method of calculating the factor T_g for the second A.M./I.F. stage.

(The value of T_g is the product of the distances measured along a straight line through the origin to the points of intersection of this line with the Y_i and Y_o curves, at the same frequency.)

To determine the stability of the two stages, the value of the regeneration coefficient T must be known.

For the first stage:

$$|y_{fe}|' = 33.4 \text{ m℧}, \quad |y_{re}|' = 4 \text{ μ℧}, \quad G_1 = 2.06 \text{ m℧} \quad \text{and} \quad G_2 = g_{oe}' + G^*{}_p.$$

The damping G_p^* of the primary of the bandpass filter in the collector lead of the first I.F. transistor follows from the tuning capacitance of 4.7 nF and the unloaded quality factor of 130. Hence, $G^* = 147 \text{ μ℧}$ and $G_2 = 145 \text{ μ℧}$. For T we obtain: $T_1 = 0.45$. Then the stability factor follows from

$$s_1 = \frac{T_{g1}}{T_1} = 5.9.$$

For the second stage:

Here the collector circuit consists of a single-tuned bandpass filter loaded by a detector. According to Section 9.3.3 its damping is 100 $\mu\mho$. For T_2 we then obtain $T = 0.65$ and the stability factor becomes $s_2 = 3.7$.

In the determination of the boundary of stability of the two stages the effect of reduction of the stability of each stage due to cascading the two stages has been disregarded. This is allowable because it turns out that the stability factor of each stage when considered individually is sufficiently large.

9.3.3.3 *Gain and Frequency Response*

The gain and frequency response of the A.M./I.F. amplifier under consideration in this section will not be calculated. When required, these properties can be ascertained by the method illustrated in the design example in Section 9.1.

9.3.4 COMPLETE CIRCUIT DIAGRAM

The complete circuit diagram of the high-frequency part of the A.M./F.M. receiver is given in Fig. 9.13. Component values are only given for the I.F. part of the receiver. For further details reference is made to the publication mentioned in footnote 1 on page 111.

9.4 A Selective Amplifier for 35 Mc/s

As the fourth example, the design of a selective amplifier for 35 Mc/s, intended for a measuring apparatus, will be discussed. The design requirements of this amplifier are as follows:

midband frequency   : 35 Mc/s

input impedance   : 50 Ω

output circuit   : diode detector with a 100 μA meter in series with the load resistance

sensitivity   : 70 μV e.m.f. in a 50 Ω source for full scale deflection of the output meter.

linearity   : linear for input signals causing the output meter to deflect between 20% and 100%.

amplitude response curve: flat topped

—   3 dB bandwidth: 0.7 Mc/s

—   20 dB bandwidth: $\leqslant$ 2.5 Mc/s.

Also, spreads in transistor parameters must not appreciably affect amplifier performance. V.H.F. transistors Type AFZ12 are used throughout, biased at $-V_{CE} = 12$V and $I_E = 1$mA, with the exception of the output transistor, which is biased at $I_E = 2$mA (see sub-section 9.4.2.1). With these biases this type of transistor will show little spread in the admittance parameters at a frequency of 35 Mc/s. The stability factor s will moreover be chosen fairly large because the transistor characteristics then have little effect on the response curve of the amplifier.

After the stability requirements of the amplifier have been satisfied, design for maximum gain is the most important point. In this particular example the design for gain will be carried out by the stage-by-stage method. The design for response curve requirements consists merely of determining the loaded quality factors of the various tuned circuits, because the only stringent requirement imposed on the response curve is that of the 3 dB bandwidth.

In designing this amplifier it will be assumed that no neutralization networks are employed.

Table 9.5 gives the admittance parameters of the transistor AFZ12 at $-V_{CE} = 12$ V and $I_E = 1$ mA and 2 mA for a frequency of 35 Mc/s.

TABLE 9.5

Admittance parameters of the AFZ12 at 35 Mc/s $-V_{CE} = 12$ V and $I_E = 1$mA and 2mA.				
I_E	1	mA	2	mA
g_{ie}	1.6	m℧	2.6	m℧
C_{ie}	33	pF	37	pF
$\|y_{re}\|$	85	μ℧	85	μ℧
φ_{re}	270°		270°	
$\|y_{fe}\|$	37	m℧	70	m℧
φ_{fe}	346°		340°	
g_{oe}	10	μ℧	20	μ℧
C_{oe}	2.2	pF	2.3	pF
Φ_{uM}	43.0	dB	43.7	dB
t	3.15 (m℧)²		5.95 (m℧)²	
N	435		825	
Θ	256°		250°	

9.4.1 REGENERATION COEFFICIENT

According to Eq. (8.2.4) the boundary of stability for this amplifier with $\Theta = 256°$ ($I_E = 1$ mA) is approximately:

$$T_g = 2/(1 + \cos 256°) = 2.6.$$

Putting $s = 5$ for reasons of interchangeability, the regeneration coefficient becomes $T = 2.6/5 = 0.5$.

The boundary of stability for the output transistor biased at $I_E = 2$ mA is $T_g = 3.0$. To obtain the same value of T for all stages of the amplifier, the output stage will also be designed for $T = 0.5$, which gives $s = 6$ for this stage.

9.4.2 THE OUTPUT STAGE OF THE AMPLIFIER

9.4.2.1 *Collector Circuit*

The output stage of the amplifier (stage 1) must be so designed that it is capable of delivering a signal sufficiently large to produce a current of 100 μA through the load resistance of the detector. The voltage excursion at the detector should be such that at full scale deflection of the meter no limiting occurs due to the knee voltage of the transistor being reached. Furthermore, the operation of the detector should be linear with signals producing more than 20 μA current through the load resistance.

A single-tuned bandpass filter will be used as coupling device between the transistor and the detector. This type of circuit is preferred to a double-tuned bandpass filter because of its higher transimpedance, so that a smaller signal need be applied to the base of the transistor to obtain full scale deflection of the output meter.

If a germanium diode OA79 is used for detection, linear operation will be obtained for signals exceeding 1.5 V at the detector input. This implies that the maximum signal should be at least 7.5 V. Choosing a detector load resistance of $R_L = 82$ kΩ and assuming the detector efficiency to be $\eta = 90\%$, this corresponds to a peak input voltage of at least 9.1 V at the detector, i.e. an r.m.s. value of 6.4 V. Since $-V_{CE} = 12$ V, this peak voltage will not cause the transistor to be driven into the knee region.

The input damping of the detector is:

$$2\eta/R_L = 22 \text{ μ℧}.$$

The unloaded quality factor Q_0 of the single-tuned circuit may be given a value of 100. With a tuning capacitance of 10 pF this corresponds to a damping of:

$$G^* = \omega C/Q_0 = 23 \text{ μ℧}.$$

Together with the output damping of the transistor of $g_o = 20$ μ℧, the total damping at the collector of the output transistor thus becomes:

$$G_1 = 22 + 23 + 20 = 65 \text{ μ℧}.$$

With $|y_{fe}| = 70$ m℧ the voltage gain of the output transistor is therefore:

$$|y_{fe}|/G_1 = (70 \times 10^{-3})/(65 \times 10^{-6}) = 1080.$$

so that the signal voltage at the base of the output transistor should be:

$$v_{b1} = 6400/1080 \simeq 6 \text{ mV}.$$

This value is low enough to avoid non-linearity due to curvature of the transistor input characteristic.

9.4.2.2 *Base Circuit*

The collector circuit of the output transistor has been designed for maximum voltage gain without taking the stability requirements into account. These requirements must therefore be incorporated in the design of the base circuit. Now

$$T = M/G_1G_2.$$

in which G_2 denotes the total damping of the base circuit. Hence, with $T = 0.5$, $M = 5.95$ (m℧)^2 and $G_1 = 65$ μ℧,

$$G_2 = 184 \text{ m℧}.$$

This gives for the losses Φ_{s1} in the secondary of the double-tuned bandpass filter 1:

$$\Phi_{s1} = g_{ie}/G_2 = 2.6/184 = 1.41 \cdot 10^{-2},$$

which corresponds to -18.5 dB.

Power Gain

To evaluate the power gain of the output stage, that is to say the power gain between the input terminals of the detector and the transistor input terminals, it is necessary to know the maximum unilateral gain Φ_{uM} of the transistor, the mismatch losses Φ_{mm} between the transistor output damping and the input damping of the detector, the insertion losses Φ_i of the single-tuned circuit and, finally, the losses Φ_{f1} due to the real feedback of the output transistor.

The mismatch losses are, according to Table 8.2:

$$\Phi_{mm} = \frac{4g_{oe} \cdot g_{\text{in det}}}{(g_{oe} + g_{\text{in det}})^2}.$$

$$= \frac{4 \times 20 \times 22 \times 10^{-12}}{(20 + 22)^2 \times 10^{-12}} = 0.995 \simeq 1.0,$$

whence $\Phi_{mm} = 0$ dB.

The insertion losses Φ_i are given by Eq.(5.6.8):

$$\Phi_i = (1-G^*/G_1)^2$$
$$= (1-23/65)^2 = 0.417,$$

whence $\Phi_i = -3.8$ dB.

According to Tables 8.3 and 8.4 the losses due to real feedback are:

$$\Phi_{f1} = (1 + q_1{}^2) \cdot \frac{1}{(1 + q_1{}^2 - T_1 \cos \Theta)^2} \cdot$$

Putting $q^2 = 1.5$ (see following sub-section), this gives:

$$\Phi_{f1} = 0.87,$$

whence $\Phi_{f1} = -0.6$ dB.

The power gain of the output stage is, therefore:

$$\Phi_1 = \Phi_{uM1} \cdot \Phi_i \cdot \Phi_{mm} \cdot \Phi_{f1}$$
$$= 43.7 - 3.8 - 0 - 0.6 = 39.3 \text{ dB}.$$

9.4.3 TRANSDUCER GAIN OF THE COMPLETE AMPLIFIER

According to sub-section 9.4.2.1 the power to be delivered into the input damping of the detector is:

$$P_L = v_0{}^2 G_L = 6.4^2 \times 22 \times 10^{-6} = 9.02 \times 10^{-4} \text{ W}.$$

The amplifier is required to match a 50 Ω source and to provide sufficient gain for full scale deflection of the output meter to be obtained with a source having an e.m.f. of 70 µV.

The power available from this source is then:

$$P_{S \text{ av}} = v_S{}^2/4R_S$$
$$= 70^2 \cdot 10^{-12}/(4 \cdot 50) = 2.45 \cdot 10^{-11} \text{W},$$

which gives for the required transducer gain:

$$\Phi_t = P_L/P_{S \text{ av}} = 3.7 \cdot 10^7,$$

corresponding to $\Phi_t = 75.7$ dB.

The amplifier will probably require four stages with double-tuned band-pass filters as interstage coupling devices. No bandpass filter will be used at the input of the amplifier. Fig. 9.14 shows the schematic diagram of the proposed amplifier.

Because of the single-tuned bandpass filter at the output of the amplifier, the coupling of the double-tuned bandpass filters must be made slightly

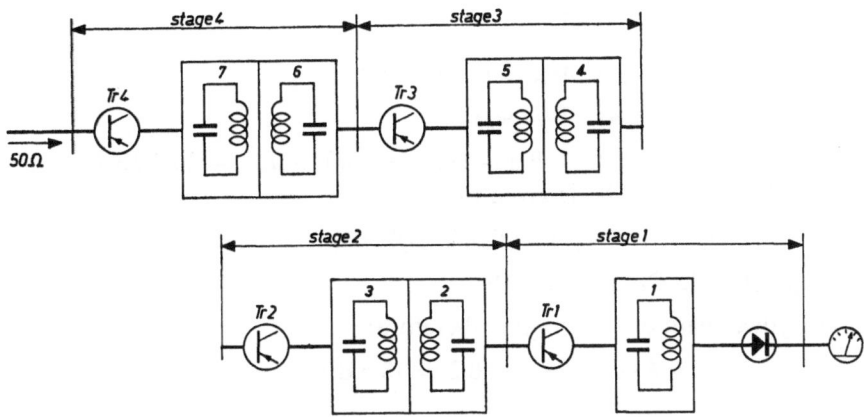

Fig. 9.14. Block diagram of the selective amplifier for a measuring apparatus.

overcritical to obtain an overall response curve with a flat top. For this reason it was decided to choose $q^2 = 1.5$ and $r = 1$.

9.4.4 POWER GAIN OF STAGE 2

For stage 2 of the amplifier the losses Φ_{p1} in the primary of double-tuned bandpass filter 1 and those in the secondary of double-tuned bandpass filter 2, Φ_{s2} will be made equal. Then:

$$\Phi_{p1} \cdot \Phi_{s2} = T/t,$$

or $\qquad \Phi_{p1} = \Phi_{s2} = 0.5/196 = 0.0505,$

$\qquad$ or: $\Phi_{p1} = \Phi_{s2} = -13.0$ dB.

The power gain of stage 2 includes the transducer losses Φ_{tb}, of double-tuned bandpass filter 1. According to Section 8.6.1 and Table 8.2:

$$\Phi_{tb1} = \Phi_{p1} \cdot \Phi_{s1} \cdot \Phi_{q1},$$

Now $\Phi_{p1} = -13.0$ dB, $\Phi_{s1} = -18.5$ dB (see sub-section 9.4.2.2) and $\Phi_q = -0.2$ dB (because $q^2 = 1.5$), whence $\Phi_{tb} = -31.5$ dB.

The feedback losses Φ_{f2} can be calculated from:

$$\Phi_{f2} = (1 + q^2)^2 \cdot (P_{3M}/P_{5M})^2$$

$$= (1 + 1.5)^2 \cdot (2.67/6.82)^2 = 0.96.$$

corresponding to $\Phi_{f2} = -0.2$ dB.

The power gain of stage 2 is therefore:

$$\Phi_2 = 43.0 - 31.5 - 0.2 = 11.3 \text{ dB}.$$

9.4.5 POWER GAIN OF STAGE 3

The power gain Φ_3 of stage 3 is given by:

$$\Phi_3 = \Phi_{uM3} \cdot \Phi_{i3} \cdot \Phi_{f3}.$$

Now $\Phi_{uM3} = 43.0$ dB and $\Phi_{f3} = (11.0 + 13.0 + 0.2) = 24.2$ dB (see subsection 9.4.4), whilst Φ_{f3} is given by:

$$\Phi_{f3} = (1 + q^2)^2 \cdot (P_{5M}/P_{7M})^2 = 0.96,$$

corresponding to $\Phi_{f3} = -0.2$ dB.

The power gain of this stage is therefore:

$$\Phi_3 = 43.0 - 24.2 - 0.2 = 18.6 \text{ dB.}$$

9.4.6 POWER GAIN OF THE INPUT STAGE

The power gain of stages 1, 2, and 3 together totals:

$$\Phi = 39.3 + 11.3 + 18.6 = 69.2 \text{ dB.}$$

The insertion losses of the double-tuned bandpass filter between stages 4 and 3 (which is made identical with the other double-tuned bandpass filters) are similarly -26.2 dB. If the losses due to real feedback are also assumed to be -0.2 dB, the power gain from the base terminals of the input transistor to the input damping of the detector becomes:

$$\Phi = 43.0 + 69.2 - 26.2 - 0.2 = 85.8 \text{ dB.}$$

Now the input impedance of the amplifier must be matched to 50 Ω, which is achieved by shunting a resistor R_1 across the transistor input terminals, as shown in Fig. 9.15. The value of R_1 follows from the condition:

$$1/R_1 + g_{11} = 1/50,$$

which, for $g_{11} = 1.6$ m℧, gives $R_1 = 54,5$ Ω.

Part of the power delivered to the input of the amplifier is lost in this resistor. This loss amounts to:

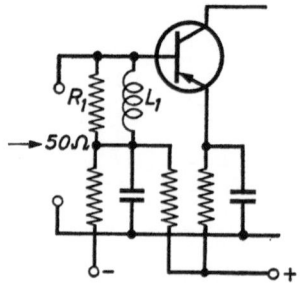

Fig. 9.15. Circuit arrangement for matching the input transistor to 50 Ω. L_1 has been provided to compensate the input capacitance of the transistor.

$$g_{11}/(g_{11} + 1/R_1) = 0.08.$$

which corresponds to -11.0 dB.

The transducer gain of the complete amplifier therefore amounts to:

$$\Phi_t = 85.2 - 11.0 = 75.8 \text{ dB}.$$

which is in accordance with the required transducer gain of 75.7 dB. By making the emitter current of the input transistor variable by means of a pre-set potentiometer, the transducer gain can easily be adjusted to the required value. This preset control may also serve to compensate spreads in transistors and other components of the amplifier.

9.4.7 AMPLITUDE RESPONSE CURVE

To ascertain the amplitude response curve of the amplifier, the relevant set of graphs given in Chapter 15 may be consulted. The amplifier is of type III with $n = 3$, $q^2 = 1.5$ and $\Theta \cong 255°$. According to Table 15.1 the performance graphs are given in Fig. 15.72. The curve for $T = 0.75$ is closest to the chosen value of $T = 0.5$.

This graph shows that for $q^2 = 1.5$ and $T = 0.75$ the amplitude response curve is in fact almost flat topped; the -3 dB points are located at a detuning of $x_+ = 0.9$ and $x_- = -1.3$. The 3 dB bandwidth of the amplifier is given by:

$$B_{3dB} = (f_0/2Q)/(x_+ - x_-),$$

whence for $B_{3dB} = 0.7$ Mc/s:

$$Q = 55 \text{ and } Q_1 = 0.75 \cdot 55 = 41,$$

Fig. 15.72 is based on $m = Q_1/Q = 0.75$. This figure further shows that, for the -20 dB bandwidth, $x_+ = 3.0$ and $x_- = -2.9$, which gives:

$$B_{20dB} = 1.9 \text{ Mc/s}.$$

Since the requirement was that the 20 dB bandwidth should not exceed 2.5 Mc/s, all amplitude response curve requirements of the amplifier are fulfilled with $Q = 55$.

9.4.8 DIMENSIONING OF THE TUNED CIRCUITS

9.4.8.1 *Circuit 1*

According to sub-section 9.4.2.1, the data of tuned circuit 1 at the output are $G_1 = 65 \,\mu\text{U}$, $Q_0 = 100$ and $C_1 = 10$ pF. This gives $L_1 = 2 \,\mu\text{H}$ and $Q_1 = 35$, which is sufficiently close to the value of $\Phi_1 = 41$ calculated in sub-section 9.4.7.

9.4.8.2 *Circuit 2*

The tuned circuit connected to the base of the output transistor should have a damping ratio of -18.3 dB (see sub-section 9.4.2.2). This implies that the total damping at the base tapping should be 184 m℧.

The maximum value of the tuning capacitance that can be used in practice at this frequency is 100 pF. With $Q = 55$ this gives a total damping of 400 μ℧. The tap for the base connection of the output transistor should therefore be so chosen that:

$$n = 0.4/184 = 1/21.5.$$

If a capacitive tap is used according to the circuit of Fig. 9.16, C_2 may be given a value of 2000 pF at $C_1 = 100$ pF. In that case $Q_0 = Q$ is very nearly 55.

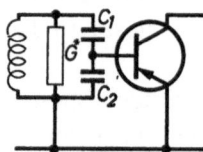

Fig. 9.16. Capacitive tapping of the tuned circuit connected to the base of the transistor.

9.4.8.3 *Circuits 3, 5 and 7*

The total damping of the primaries of the double-tuned bandpass filters is:

$$G_p = g_{oe}/\Phi_p = 20/0.05 = 400 \text{ μ℧}.$$

With $Q = 55$ and $C = 100$ pF this corresponds to

$$L = 0.2 \text{ μH} \quad \text{and} \quad Q_0 = 58.$$

9.4.8.4 *Circuits 4 and 6*

Since the losses of circuits 4 and 6 are equal to those of circuits 5 and 7, the tap on the former should be so chosen that:

$$n = \sqrt{g_{oe}/g_{ie}} = \sqrt{20/1600} = 1/8.95.$$

A capacitive tap with a tuning capacitance of 100 pF can be realized by connecting $C_1 = 120$ pF and $C_2 = 1000$ pF in series as shown in the circuit of Fig. 11.8.

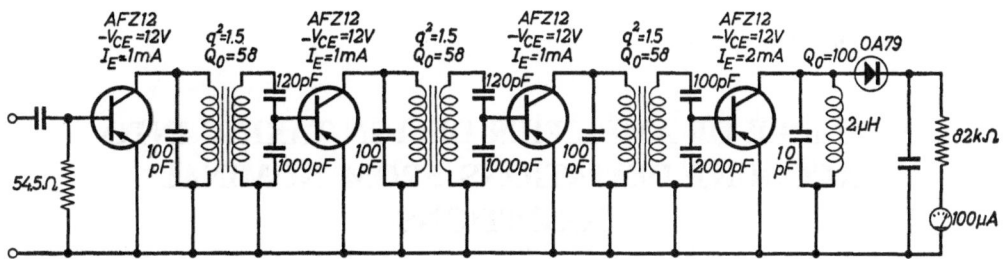

Fig. 9.17. Circuit diagram of the 35 Mc/s amplifier for a measuring apparatus. The d.c. biasing networks have been omitted for simplicity.

9.4.8.5 *Parasitic effects due to capacitive taps*

Parasitic effects due to the use of capacitive taps on the secondary circuits of the double-tuned bandpass filters are negligibly small in this case. This is due to the large values of C_2 employed.

9.4.9 COMPLETE CIRCUIT DIAGRAM

Fig. 9.17 shows the complete circuit diagram of the amplifier. The d.c. biasing networks have again been omitted.

CHAPTER 10

EXAMPLE OF AMPLIFIER DESIGN TAKING INTO ACCOUNT DEVIATIONS FROM NOMINAL CONDITIONS

10.1 General

In this chapter the design of two different I.F. amplifiers intended for the vision channel of a television receiver according to the C.C.I.R. standards will be described. The difference between the two amplifiers lies in the methods by which automatic gain control is achieved. In one amplifier the system of reverse gain control is used, whereas in the other, forward gain control is applied.

The I.F. amplifiers each consist of three stages. The amplifier with reverse gain control, referred to as the "*reverse AGC amplifier*", is equipped with three transistors AF 179. The "*forward AGC amplifier*" has a transistor AF 181 in the control stage and transistors AF 179 in the second and third stages.

Double-tuned bandpass filters are used as interstage coupling networks throughout the amplifiers except between the final stage and the video detector[1]. At this place a single-tuned bandpass filter is used in both amplifiers.

The supply voltage for the amplifiers is 12V.

In Fig. 10.1 the arrangements of the two amplifiers, which are of type III (see Chapter 9), are shown.

Fixed-component neutralization will be applied to the various transistors in the amplifiers, merely to ease alignment. In the following sections the design of both amplifiers will be carried out for nominal conditions. By

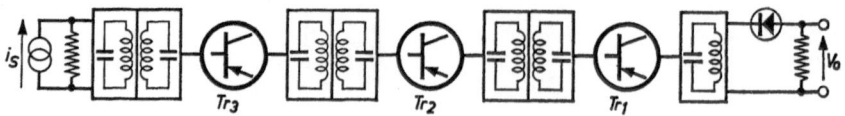

Fig. 10.1. Schematic arrangements of the amplifiers with reverse gain control and with forward gain control.

---

[1] The "video detector" of a television receiver is sometimes referred to as the "second detector".

means of measurements on practical amplifiers constructed according to this design, variations in performance which occur when operating conditions deviate from the nominal case will be investigated.

10.2 Choice of Transistor Biasing Points

The choice of the biasing point of a transistor in an amplifier is always a compromise between attainable stage gain and transistor dissipation. The latter determines to a large extent the maximum ambient temperature at which the amplifier is allowed to operate without exceeding the permissible junction temperature of the transistors used.

The recommended biasing point of the transistors AF 179 and AF 181 (in non-controlled condition) is $V_{CE} = -10$ V and $I_E = 3$ mA. At higher

TABLE 10.1   Transistor Admittance Parameters

Quantity	AF 181	AF 179		Unit
	$V_{CE}=-10V, I_E=3mA$	$V_{CE}=-10V, I_E=3mA$	$V_{CE}=-11V, I_E=4mA$	
g_{ie}	10	6.5	7.8	m℧
C_{ie}	49	35	36	pF
$\lvert y_{re} \rvert$	70	100	97	μ℧
φ_{re}	260	260	260	°
$\lvert y_{fe} \rvert$	73	80	93	m℧
φ_{fe}	322	325	321	°
g_{oe}	50	100	120	μ℧
C_{oe}	3.0	1.6	1.6	pF
Φ_{uM}	34.2	33.9	33.6	dB
N	1040	800	960	
Θ	232	225	221	°
t	10.2	12.3	9.7	
T_g	5.3	6.9	8.3	

biasing points the attainable stage-gain of the AF 181 rapidly decreases (see Fig. 10.26) because of its suitability for forward-gain control. The attainable stage gain of the AF 179 only slightly increases at emitter currents above 3 mA. This will become apparent from the graphs in Chapter 4 showing the dependence of the admittance parameters of this transistor on biasing point at a frequency of 35 Mc/s.

The output stage of each amplifier, which is an AF 179, is biased at $V_{CE} = -11$ V and $I_E = 4$ mA. This higher biasing point has been chosen because of the linearity of this stage with large signal excursions.

The average values of the admittance parameters of the transistors AF 179 and AF 181 at the chosen biasing points are set out in Table 10.1. The quantities Φ_{uM}, N, Θ and t as well as the approximate value of T_g, which are required for the design of the amplifiers, are also included in this table.

10.3 Design for Nominal Conditions

The nominal design of the two amplifiers will be based on the following initial assumptions:

— fixed-component neutralization will be applied in all stages.
— the damping ratio of the tuned circuits (forming the double-tuned band-pass filters) to which transistors with a fixed biasing point are connected should be equal to 3 dB at least.
— the damping ratio of the tuned circuits to which the gain control transistors are connected should be −5 dB at least.
— the tuning capacitances of the various tuned circuits should be equal to or larger than 10 pF.

For the reasoning behind these initial assumptions, reference is made to Chapter 8.

10.3.1 NEUTRALIZATION

When fixed components are used to neutralize the internal feedback of the transistors the values of these components must be carefully chosen. For this choice, spreads in the internal feedback of the transistors as well as tolerances of the components of the neutralizing network must be considered.

From the discussion of transistor parameter spreads and neutralizing network tolerances in Book I, Chapter 11, it became apparent that the nominal values of the neutralizing network components should be related to the minimum value of the transistor feedback admittance. Then equal stability

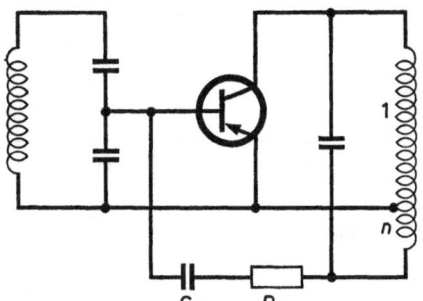

Fig. 10.2. Simplified circuit diagram showing the components of the neutralizing
network.

factors of the various stages of the amplifier are obtained in cases in which
extreme spreads and tolerances occur simultaneously.

In the following sub-sections it will be seen that in the present amplifier
designs it is not necessary to neutralize the internal feedback of the transistors to achieve stability. Without a neutralizing network the various stages of
both amplifier designs are sufficiently stable.

The reason for using neutralizing networks is to increase ease of alignment
of the amplifier, which is of special importance in large scale manufacturing.
It is therefore not necessary to try to incorporate all possible spreads and
tolerances in the dimensioning of the neutralizing network. Dimensioning
for the strictly nominal case will provide sufficient improvement in all practical cases.

Using the Eqs. (6.1.2) and (6.1.3) for the series connected components C_N
and R_N of the neutralizing network (see Fig. 10.2), and the values of $|y_{re}|$
from Table 10.1 the values for C_N, R_N and the tapping ratio n given in
Table 10.2 are obtained. For C_N and R_N, values from the E12 standard
range are chosen which lie closest to the calculated values.

10.3.2 DETERMINATION OF BANDPASS FILTER PARAMETERS

As the arrangement of the amplifier (type III) and the number of stages
($n = 3$) has been fixed and the average value of Θ is known from Table 10.1
($\Theta = 225°$), the amplitude response and envelope delay curves of the amplifier (without wavetraps) can be determined from the Design Charts on pages
308 and 309 (III, $n=3$, $\Theta = 225°$). These Design Charts are valid for $q^2 = 1$,
2 and 3 respectively.

It follows that a good response curve is obtained for the combination
$T=1$ and $q^2 = 2$.

The 3 dB points of this response curve are at $x_- = -1.7$ and $x_+ = 1.0$.
For a 3 dB bandwidth of 4.5 Mc/s, which is required for the amplifier under

TABLE 10.2   Neutralizing Network Components

Stage	Amplifier with reverse AGC	Amplifier with forward AGC
Input Stage	Transistor: AF 179 $C_N = 3.3\,\text{pF}$ $R_N = 220\,\Omega$ $n = 0.13$	Transistor: AF 181 $C_N = 2.2\,\text{pF}$ $R_N = 330\Omega$ $n = 0.13$
Middle Stage	Transistor: AF 179 $C_N = 3.3\,\text{pF}$ $R_N = 220\Omega$ $n = 0.13$	
Output Stage	Transistor: AF 179 $C_N = 2.7\,\text{pF}$ $R_N = 270\,\Omega$ $n = 0.16$	

consideration (see Chapter 2), it then follows for the average (loaded) quality factor of the double tuned bandpass filters:

$$Q = \frac{(x_-) + (x_+)}{2} \cdot \frac{f_o}{B_3\,\text{dB}} \approx 10.$$

The loaded quality factors of primary and secondary of all bandpass filters will be made equal. We therefore obtain for each double-tuned bandpass filter:

$$Q_p = Q_s = 10,$$

and $q^2 = 2$.

10.3.3 THE BIASING NETWORKS OF THE VARIOUS TRANSISTORS

The biasing networks of the transistors in the various stages are designed with transistor dissipation, transistor parameter spreads and resistance tolerances in mind. A maximum ambient temperature at which the amplifiers should operate satisfactorily, taking into account also extreme spreads of biasing network resistors and transistor d.c. properties, is chosen as $t_{amb} = 45\,°\text{C}$.

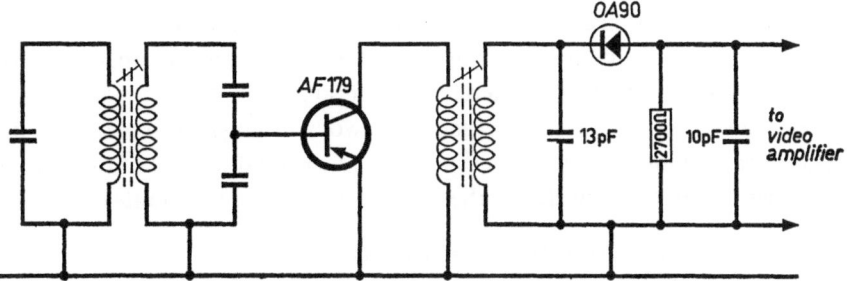

Fig. 10.3. Simplified circuit diagram of the output stage.

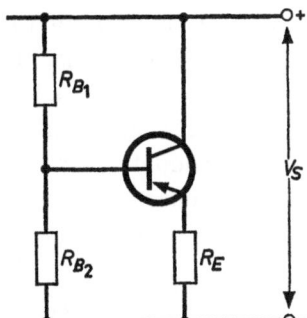

Fig. 10.4. Biasing network of the output stage.

The values of the resistors of the various biasing networks (see Fig. 10.4) are given in the circuit diagram in Fig. 10.7.

10.3.4 *The Output Stage*

The output stage of the I.F. amplifier, an AF 179 biased at $V_{CE} = -11$ V and $I_E = 4$ mA, is coupled to the video detector by means of a single-tuned bandpass filter.

To obtain d.c. separation between the detector circuit and the collector circuit of the final I.F. transistor, a bifilar double-winding coil is used as tuning inductance. This d.c. separation is required in view of the arrangement of the circuits following the video detector.

10.3.4.1 *The Output Circuit*

The output circuit of the final I.F. stage consists of a single-tuned band-pass filter loaded by a detector circuit. Investigations have shown that the input damping of such a detector circuit with the component values indicated in Fig. 10.3 and with a diode OA 90 amounts to 350 μ℧ (measured at a frequency of 38 Mc/s and with an input voltage of 2V r.m.s.).

Taking into account the effects of the bifilar winding and the screening can, a tuning inductance for this single-tuned bandpass filter can be realized having an unloaded quality factor of $Q_0 = 60$. With a tuning capacitance of 13 pF (see below) this results in a damping of 50 µ℧.

The coefficient of coupling between the two windings of the tuning inductance can be made as good as $k = 0.95$. The total damping of the output circuit, seen from the collector of the transistor, therefore amounts to:

$$(0.95)^2 \, (350 + 50) \, \text{µ℧} = 360 \, \text{µ℧}.$$

The biasing point of the output transistor has been chosen for optimum loading by the total output dircuit impedance. In determining this biasing point is was assumed that the signal voltage at the collector could be driven to so large a value that the instantaneous collector-emitter voltage of the transistor would reach a lowest value of 1.5 V, without affecting the linearity of the output stage to an appreciable extent. For the same reason the current excursion should not exceed 85 % of the d.c. current.

With a collector-emitter voltage of $V_{CE} = -11$ V and an emitter current $I_E = 4$ mA, the optimum collector impedance becomes:

$$\frac{11 - 1.5}{0.8 \cdot 5.4} \, \text{k}\Omega = 2.8 \, \text{k}\Omega,$$

which equals the actual output circuit damping of 360 µ℧.

The output damping of the transistor is $g_{oe} = 120$ µ℧. The total tuned circuit damping therefore equals $360 + 120 = 480$ µ℧. The damping ratio of the output circuit, that is the ratio of the total tuned circuit damping and the contribution of the transistor to this damping, then becomes $\frac{120}{480} = 0.25$ or -6 dB. The damping ratio of -6 dB is well in excess of the minimum requirement of -3 dB.

The amplitude response curve of the collector bandpass filter should not have too large an effect on the overall response curve of the amplifier. The 3 dB bandwidth of this circuit is therefore chosen at 6 Mc/s. The circuit is tuned to the midband frequency of the amplifier, i.e. 36 Mc/s. Its quality factor should therefore be equal to $Q = 6$. The tuning capacitance follows from:

$$C = \frac{Q \cdot G_{total}}{\omega}$$

$$= \frac{6 \cdot 480 \cdot 10^{-6}}{2\pi \cdot 36 \cdot 10^6} = 13 \text{ pF.}$$

This gives a tuning inductance of: $L = 1.6$ µH.

As appears from Fig. 10.3 the tuning capacitance of the single-tuned band-pass filter has been connected across the second winding of the bifilar wound coil. In this way, a better efficiency of the video detector is obtained because now the effective "reservoir" capacitance is larger (the current peaks through the detector diode are delivered by this capacitor). If the tuning capacitance were placed across the first winding of the coil a stray inductance would be present between this capacitance and the diode, which would reduce the detector efficiency.

10.3.4.2 The input circuit

The secondary of the last double-tuned bandpass filter of the amplifier forms the input tuned circuit of the output transistor. This circuit is required to have a damping ratio of -3 dB in the nominal case.

According to Table 10.1 the input damping of the output transistor is $g_{ie} = 7.8$ m℧. In order to fulfil the 3 dB condition, the damping of the tuned circuit G^* seen from the transistor terminals should be made equal to g_{ie}. The total damping then amounts to $2\,g_{ie} = 15.6$ m℧.

The design of the capacitive tap of the tuned circuit at the base of the output transistor will not be dealt with here. It can be carried out along the same lines as discussed in the preceding chapter.

Proper values of the components of this capacitive tap are given in the complete circuit diagrams of the two amplifiers in Fig. 10.7.

10.3.4.3 Stability

The approximate boundary of stability of the output stage follows from:

$$T_g = \frac{2}{1 + \cos \Theta}$$

For $\Theta = 221°$, $T_g = 8.3$.

The total damping at the input side of the transistor is $G_1 = 15.6$ m℧ and that at the output side of the transistor $G_2 = 480$ µ℧. Together with the values for $|y_{re}|$ and $|y_{fe}|$ from Table 10.1 the regeneration coefficient becomes:

$$T = \frac{|y_{re}| \cdot |y_{fe}|}{G_1 G_2} = 1.2.$$

Without neutralization the stability factor in the nominal case therefore amounts to $s = \dfrac{T_g}{T} = 6.9$. With the neutralizing network, which is employed to improve alignment of the amplifier (see sub-section 10.3.3.), the stability factor becomes still larger.

10.3.5 THE SECOND STAGE

The transistor in the second stage of both amplifiers is of the type AF 179. Its nominal biasing point is $V_{CE} = -10$ V, $I_E = 3$ mA.

10.3.5.1 *The Damping Ratios*

The -3 dB condition applies also to the damping ratios of the tuned circuits connected to input and output terminals of this transistor. This means that the total damping at the input terminals becomes $G_1 = 2\, g_{ie}$ and at the output terminals $G_2 = 2\, g_{oe}$. Substituting values from Table 10.1 we obtain $G_1 = 13$ m℧ and $G_2 = 200$ μ℧.

There is no difficulty in realizing the input circuit; again a capacitive tap is used.

When constructing the output tuned circuit, we have to take into account that its tuning capacitance should be 10 pF or larger.

With the loaded quality factor of this circuit obtained in sub-section 10.3.2 of $Q = 10$, the minimum damping becomes:

$$G = \frac{\omega C}{Q} = 230 \text{ μ℧}.$$

This damping is larger than that required for the -3 dB condition. The design should therefore be carried out for this larger value. The damping ratio of the circuit then becomes:

$$\Phi_p = \frac{100}{230} = 0.435 \text{ or } \Phi_p = -3.5 \text{ dB}.$$

10.3.5.2 *Stability*

The value of the regeneration coefficient T_g at the boundary of stability of the second stage becomes $T_g = 6.9$. The actual regeneration coefficient, disregarding the neutralizing network, can be calculated as $T = 2.7$. This gives a stability factor of $s = 2.5$.

With the neutralizing network, the actual stability factor of the stage will be considerably larger.

10.3.6 THE INPUT STAGE OF THE AMPLIFIER WITH REVERSE A.G.C.

In the input stage of the amplifier with reverse A.G.C. a transistor AF 179 is employed. In the non-controlled condition its biasing point is $V_{CE} = -10$ V, $I_E = 3$ mA.

10.3.6.1 *The Damping Ratios*

The damping ratios of the tuned circuits around the transistor in the input stage should at least be equal to −5 dB. This larger value has been chosen to avoid the amplitude response curve of the amplifier being too much affected by variations of the transistor parameters when the gain of this transistor is controlled.

With reference to Table 10.1 we then find for the total input damping of the transistor: $G_1 = 3.16\, g_{ie} = 20.6$ m℧. The total output damping becomes $G_2 = 316$ µ℧.

10.3.6.2 *Stability*

Disregarding neutralization, the regeneration coefficient of this stage becomes: $T = 1.06$. With $T_g = 6.9$ the stability factor is $s = 5.7$.

10.3.7 THE INPUT STAGE OF THE AMPLIFIER WITH FORWARD A.G.C.

In the input stage of the amplifier with forward gain control a transistor AF 181, which is specially designed for this type of gain control, is used. In the non-controlled condition its biasing point is $V_{CE} = -10$V and $I_E = 3$ mA.

10.3.7.1 *The Damping Ratios*

To this stage also the −5 dB condition applies. The total damping at the base of this transistor therefore becomes $G_1 = 3.16\, g_{ie} = 31.6$ m℧ and $G_2 = 3.2\, g_{oe} = 160$ µ℧. The minimum damping of the collector tuned circuit, however, amounts to 230 µ℧ (see sub-section 10.3.5). This gives a damping ratio of

$$\Phi_p = \frac{50}{230} = 0.22, \text{ or } \Phi_p = -6.6\text{dB}.$$

10.3.6.2 *Stability*

It can be calculated that for this stage $T = 0.68$, disregarding neutralization. With $T_g = 5.3$ the stability factor becomes $s = 7.8$.

10.3.8 THE INPUT DOUBLE-TUNED BANDPASS FILTER

As already mentioned in the preceding sub-sections, the secondary of the input double-tuned bandpass filter has a damping ratio of −5dB.

For the purpose of defining a transducer gain figure for the complete amplifier we assume that the amplifier is driven by a current source with a damping of 250 μ℧. The damping of the primary circuit itself is 30 μ℧. The damping ratio therefore becomes:

$$\Phi_p = \frac{250}{280} = 0.89, \text{ or } \Phi_p = -0.5 \text{ dB.}$$

For $Q = 10$ (see Section 10.3.2), the tuning capacitance becomes 12 pF. The coefficient of coupling of this bandpass filter also equals $q^2 = 2$. Due to this non-critical coupling, losses of 0.5 dB occur. Moreover, in the final design of the amplifier wave traps are connected to the input bandpass filter. These cause additional losses of 1.5 dB. So $\Phi_q = -2$ dB.

10.3.9 THE GAIN OF THE AMPLIFIER

In this section the gain of the two amplifier designs under consideration will be determined in terms of *transimpedance, transducer gain* and *gain per stage* of the amplifiers. In this calculation it will be assumed that the neutralizing networks are designed in such a way that for transistors with nominal parameters the feedback is exactly cancelled.

10.3.9.1 *The Output Stage*

According to Table 10.1, $\Phi_{uM1} = 33.6$ dB. The insertion losses in the single-tuned bandpass filter can be calculated from:

$$\Phi_i = \left\{ \frac{g_{oe} + k \, g_{\text{in det}}}{g_{oe} + G^* + k \, g_{\text{in det}}} \right\}^2.$$

Substituting values from sub-section 10.2.4.1 we obtain:

$$\Phi_i = -0.9 \text{ dB.}$$

Mismatch losses occur between the input damping of the detector and the output damping of the transistor. These losses can be calculated from:

$$\Phi_{mm} = \frac{4 \, g_{oe} \cdot k^2 \, g_{\text{in det}}}{(g_{oe} + k^2 \, g_{\text{in det}})^2}.$$

Substituting values:

$$\Phi_{mm} = -1.0 \text{ dB.}$$

The power gain of the output stage then becomes:

$$\Phi_1 = \Phi_{uM} + \Phi_i + \Phi_{mm}$$

or $\qquad \Phi_1 = 31.7$ dB.

The calculated voltage gain between the input terminals of the detector and the input terminals of the transistor is 200 times.

10.3.9.2 The Second Stage

Again according to Table 10.1, $\Phi_{uM} = 33.9$ dB. The double-tuned band-pass filter at the output of the second transistor in the amplifier has a primary damping ratio of $\Phi_p = -3.6$ dB. That of the secondary is $\Phi_s = -3$ dB (see sub-section 10.3.5.1).

The coupling losses of this bandpass filter follow from:

$$\Phi_q = (\frac{2q}{1 + q^2})^2.$$

For $q^2 = 2$, $\Phi_q = -0.5$ dB.

The power gain of the second stage then becomes:

$$\Phi = \Phi_{uM} + \Phi_p + \Phi_s + \Phi_q,$$

or: $\Phi = 26.8$ dB.

The voltage gain of this stage is 21 times.

10.3.9.3 The Input Stage of the Amplifier with Reverse A.G.C.

For this stage $\Phi_{uM} = 33.9$ dB, $\Phi_p = -5$ dB, $\Phi_s = -3$ dB and $\Phi_q = -0.5$ dB. This gives for the power gain of this stage $\Phi = 25.4$ dB; and for the voltage gain 18.6 times.

10.3.9.4 The Input Stage of the Amplifier with Forward A.G.C.

Now $\Phi_{uM} = 34.2$ dB, $\Phi_p = -6.6$ dB, $\Phi_s = -3$ dB and $\Phi_q = -0.5$ dB. This gives for the power gain $\Phi = 24.1$ dB; and for the voltage gain 19.7 times.

10.3.9.5 The Input Double-Tuned Bandpass Filter

According to sub-section 10.3.7, $\Phi_p = -0.5$ dB, $\Phi_s = -5$ dB and $\Phi_q = -2$ dB. This gives $\Phi_{tb3} = -7.5$ dB.

10.3.9.6 The Gain of the Complete Reverse A.G.C. Amplifier

The transducer gain of the complete amplifier follows from:

$$\Phi_t = \Phi_1 + \Phi_2 + \Phi_3 + \Phi_{tb3},$$

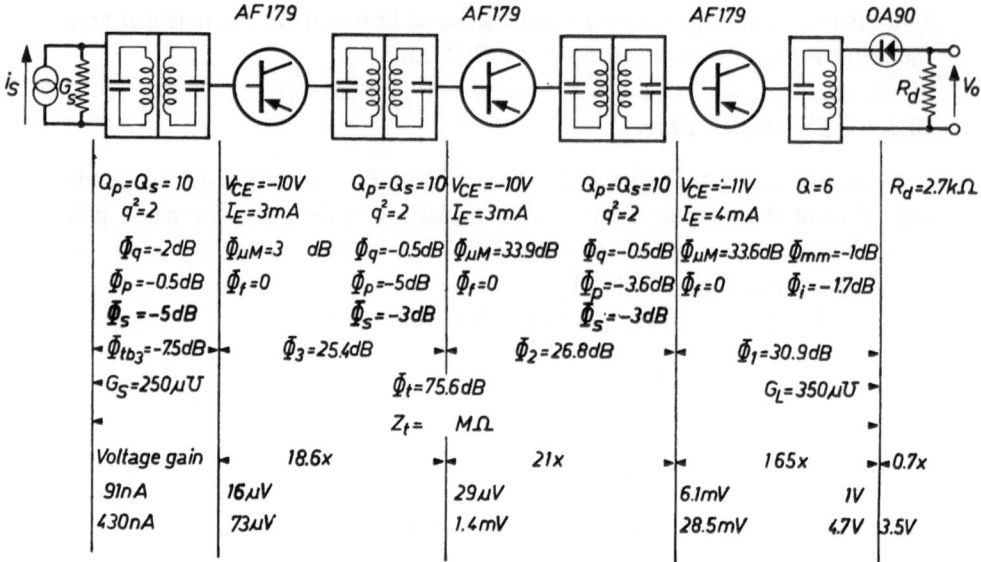

Fig. 10.5. Diagram indicating overall gain and the stage gains of the reverse AGC amplifier.

or: $\qquad \Phi_t = 31.7 + 26.8 + 25.4 - 7.5 = 76.4$ dB.

The transimpedance of the I.F. amplifier can be calculated from:

$$Z_t = \sqrt{\frac{\Phi_t}{4 G_S G_L}} \cdot$$

In this expression G_L is the load damping — in this case the input damping of the detector ($G_L = 350$ μ℧). The quantity G_S is the damping of the current source at the input of the amplifier ($G_S = 250$ μ℧). Substituting figures we obtain:

$$Z_t = 13.2 \text{ M}\Omega.$$

In Fig. 10.5 the gain of the complete amplifier and the various parts of which it is constituted are indicated schematically. Also, voltage levels are given at various points of the amplifier for an output from the video detector, of 3.5 V.

10.3.9.7 The Gain of the Complete Forward A.G.C. Amplifier

The transducer gain of the complete forward A.G.C. amplifier now becomes:

$$\Phi_t = 31.7 + 26.8 + 24.1 - 7.5 = 75.2 \text{ dB}.$$

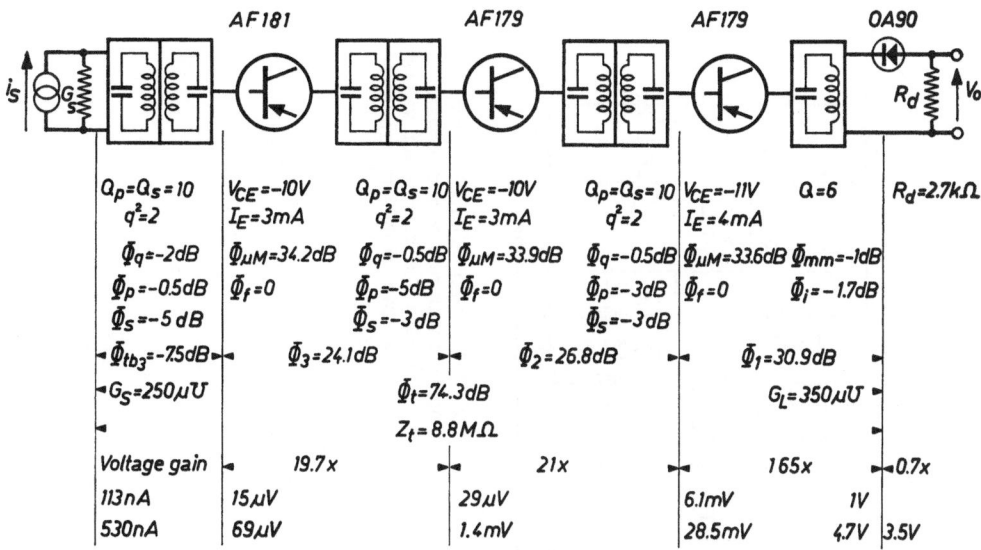

Fig. 10.6. Diagram indicating the overall gain and the stage gains of the forward AGC amplifier.

This corresponds to transimpedance of

$$Z_t = 10.6 \text{ M}\Omega.$$

In Fig. 10.6 the overall gain of this amplifier and the gains per stage are indicated schematically. The voltage levels at the various points are those which produce an output of 3.5 V from the video detector.

10.4 Complete circuit diagram

In Fig. 10.7 a complete circuit diagram of the vision I.F. amplifier is given. This circuit diagram applies to the amplifier with reverse gain control as well as to the amplifier with forward gain control. The component values which are the same in both amplifiers are given in the diagram. For component values which are different, reference is made to Table 10.3.

It can be seen from the circuit diagram that the resistor across the secondaries of the various double-tuned bandpass filters is connected across the lower capacitance of the capacitive tap.

This resistor is required because of the sharing of the total damping of the circuit between the transistor and the external circuit, as discussed in the preceding sub-sections. The damping of the tuned circuit itself is not sufficiently large ($Q_0 \approx 80$) to meet the various damping ratio conditions. Connection of this resistor across the tap has, in comparison with the con-

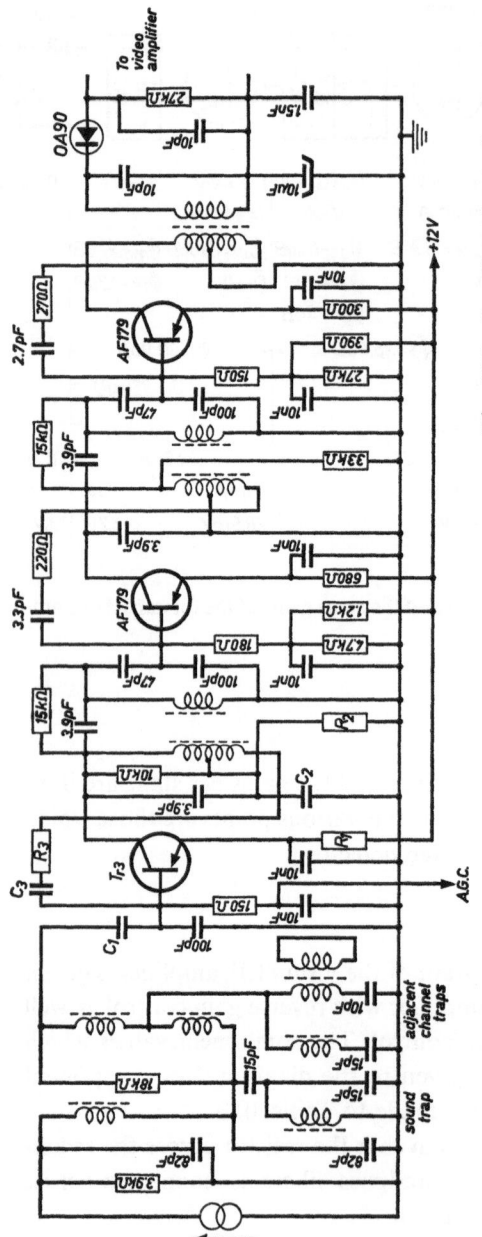

Fig. 10.7. Complete circuit diagram of the amplifiers with reverse and forward AGC; the differences in component values between the two amplifiers are given in Table 10.3.

TABLE 10.3 Component values

Component	Reverse AGC amplifier	Forward AGC amplifier
C_1	56 pF	47 pF
C_2	0 [1]	10 nF
C_3	3.3 pF	2.2 pF
R_1	2 kΩ	330 Ω
R_2	0 [1]	180 Ω
R_3	220 Ω	330 Ω
Transistor Tr_3	AF 179	AF 181

[1] Not used in Amplifier with reverse AGC

nection across the whole secondary, the advantage of a smaller effect on the "spread capacitance" of the amplifier stage. This has been considered in detail in Book I, Chapter 12.

A disadvantage of this method of obtaining sufficient damping is that the capacitive tap is heavily loaded. This affects, obviously, the tapping ratio, and also causes some asymmetry of the transimpedance curve (as a function of frequency) of the bandpass filter.

The asymetry can, however, easily be compensated by means of a properly dimensioned complex coupling between primary and secondary of the bandpass filter. This has been achieved by parallel connection of the resistance and the capacitance as the coupling elements of the bandpass filters in Fig. 10.7

The complex coupling also compensates for the skewness of the response curve due to the decrease of $|y_{fe}|$ of the transistor over the passband of the amplifier. Wave-traps are included in the amplifier to meet the requirements given in Chapter 2 for the suppression of the adjacent channel carriers and the own sound carrier. The design of these wave-traps is regarded as outside the scope of this book.

10.5 Measurements

In this section measurements will be described which are carried out on practical amplifiers with reverse AGC and forward AGC, constructed according to the designs developed in the preceding sections. The purpose of presenting the results of measurement is twofold:

— to check whether the measured performance of the amplifiers complies with that which was calculated.

— to obtain detailed information for studying the consequences of spreads, tolerances and variations occurring in the amplifiers.

In the following sub-sections the various measurements will be described separately for the reverse AGC and the forward AGC amplifiers. To obtain information on the effects. of spreads in transistor properties, the measurements are carried out with several different sets of transistors inserted in the amplifiers.

10.5.1 THE REVERSE AGC AMPLIFIER

10.5.1.1 *The Amplitude Response Curve*

The amplitude response curve of the reverse AGC amplifier equipped with a set of transistors with approximately average values of parameters is

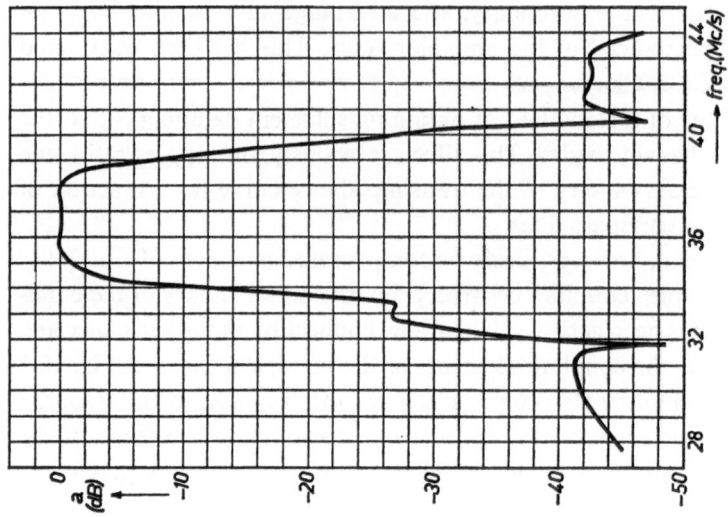

Fig. 10.9. Amplitude response curve of the reverse AGC amplifier including the wave traps.

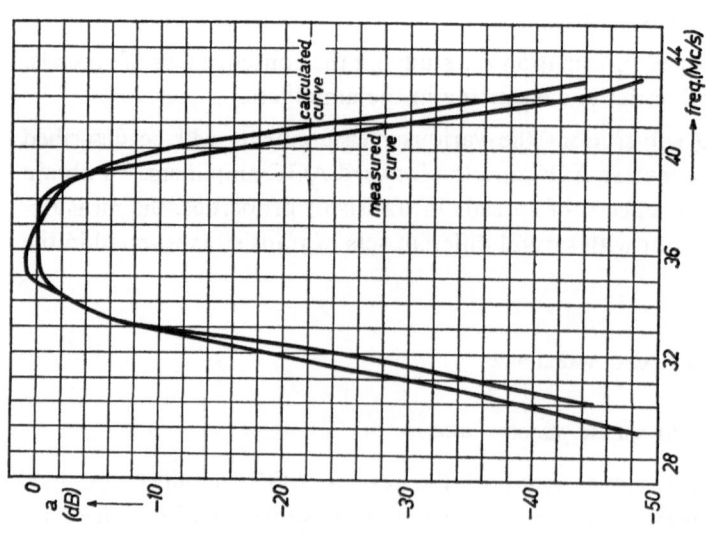

Fig. 10.8. Amplitude response curve of the reverse AGC amplifier with the wave traps disconnected. The calculated curve is derived from the proper Design Chart.

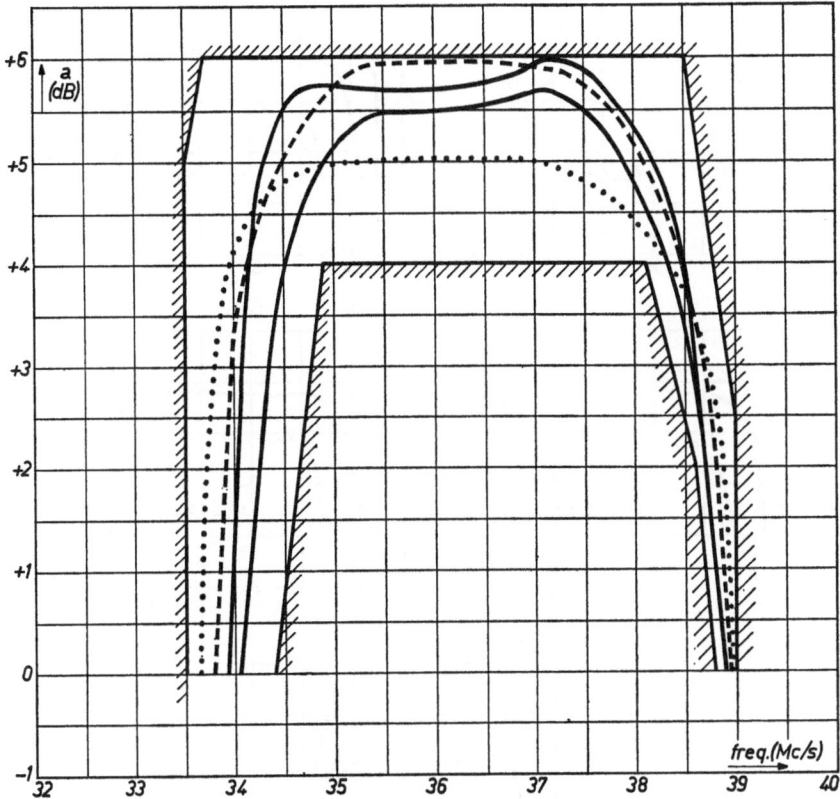

Fig. 10.10. Amplitude response curves of the reverse AGC amplifier measured with different sets of transistors. The amplifier has been realigned after insertion of each set of transistors. The tolerances indicated by shading are those that can be accepted according to the C.C.I.R. norm curve.

shown in Fig. 10.8. This curve applies to the non-controlled condition of the amplifier with the wave-traps disconnected. The calculated curve in Fig. 10.8 is derived from the Design Chart on which the design of the amplifier was based (see sub-section 10.3.2). It appears that the measured amplitude response curve agrees very well with the calculated one.

Fig. 10.9 shows the amplitude response curve of the amplifier with the wave-traps normally connected. It follows that the response curve adequately meets the requirements demanded of I.F. amplifiers for television receivers, as presented in Chapter 2.

In an amplifier with wave-traps, spreads in transistor parameters only affect the top of the amplitude response curve. Response curves are measured for a number of different sets of transistors. For four sets of transistors the

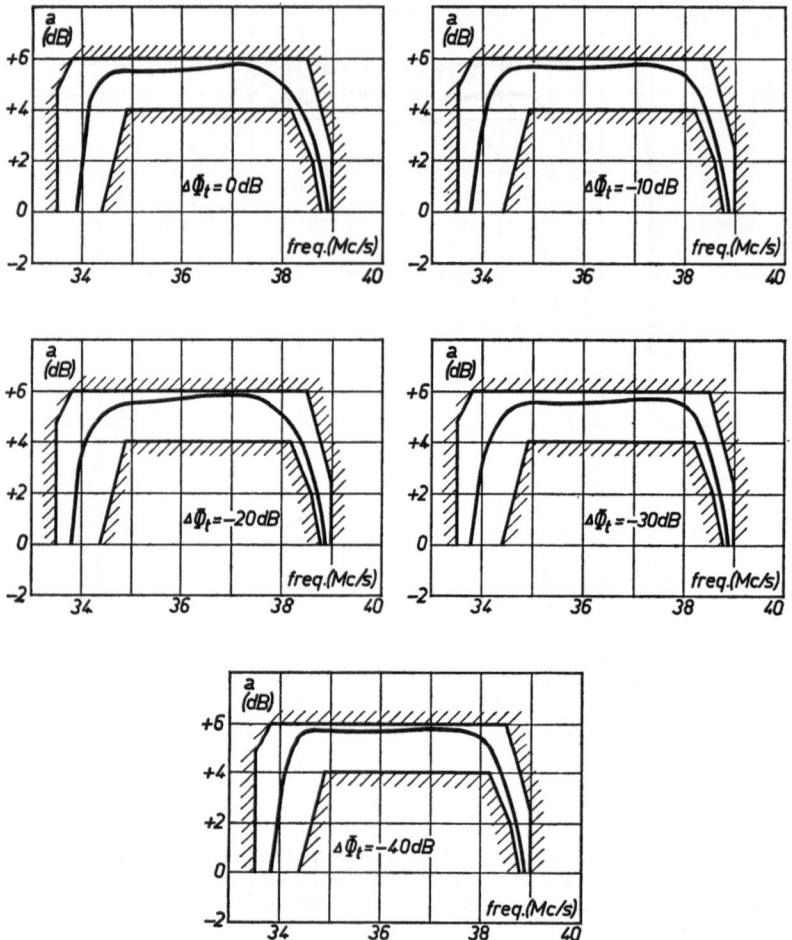

Fig. 10.11. Variation of the top of the amplitude response curve of the reverse AGC amplifier for various levels of gain control.

tops of these curves are shown in Fig. 10.10. Also the tolerances that can be accepted according to the C.C.I.R. norm curve[1]) are indicated. It will be seen that the measured curves remain well within the tolerance field.

The variation of the tops of amplitude response curves during (reverse) gain control of the amplifier are shown in Fig. 10.11.

A small variation during the first 20 dB of gain control can be observed. This is due to the decrease of the input and output dampings of the first transistor when its emitter current is controlled downwards.

---

[1]) See reference in Chapter 2.

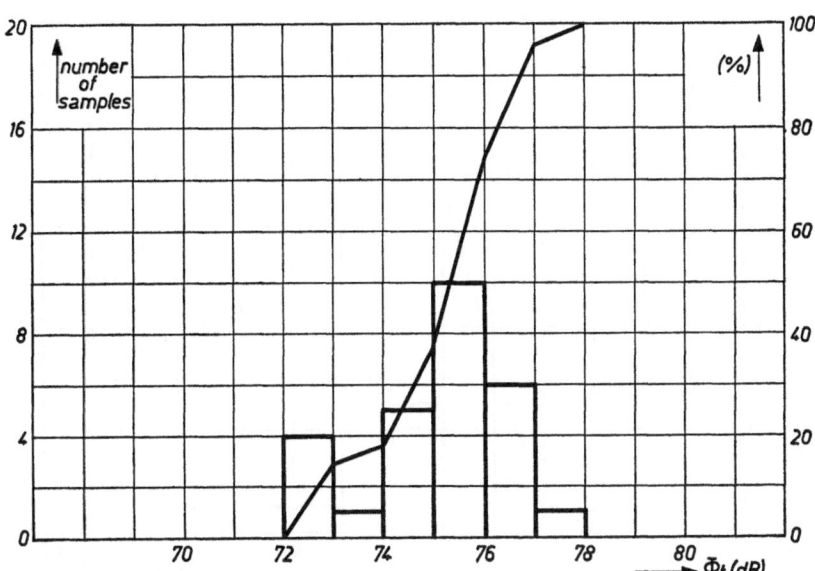

Fig. 10.12. Gain distribution of the reverse AGC amplifier for different sets of transistors.

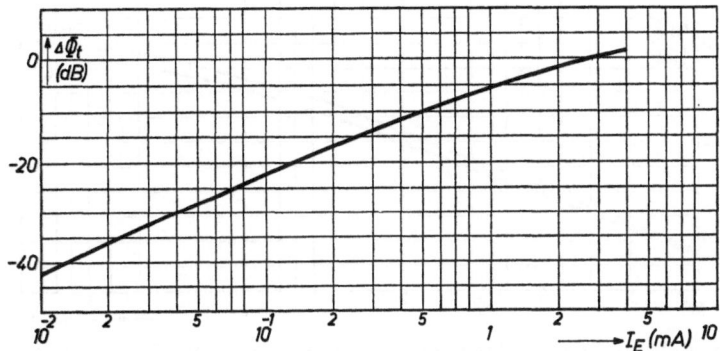

Fig. 10.13. Relative variation of gain of the reverse AGC amplifier as a function of the emitter current of the first transistor.

10.5.1.2 *The Gain*

For a number of different sets of transistors the transducer gain of the amplifier has been measured.

The measured values are entered in the histogram shown in Fig. 10.12. A cumulative curve has also been drawn. The 50 % value of 75.4 dB corresponds fairly well with the calculated gain figure of 76.4 dB.

In Fig. 10.13 the variation of the transducer gain of the amplifier as a function of the emitter current of the first transistor is shown.

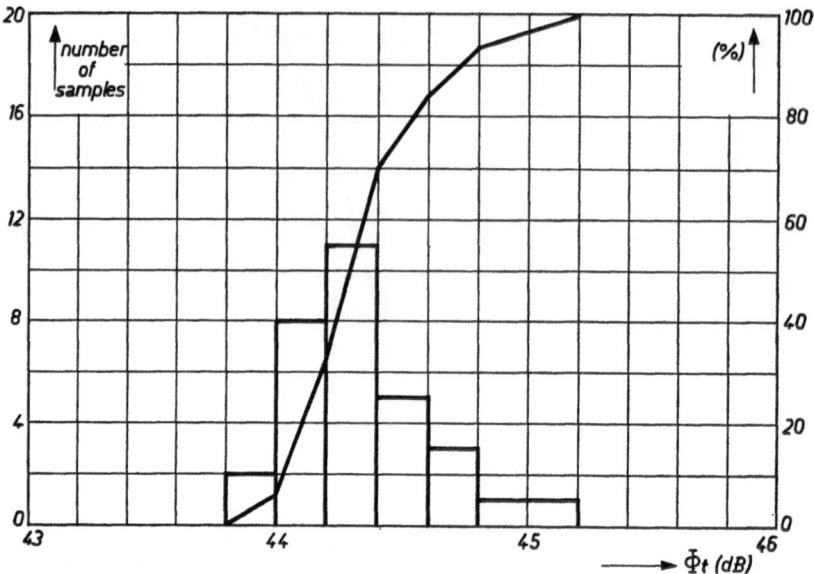

Fig. 10.14. Gain distribution of the reverse AGC amplifier at an emitter current of 40 μA of the input transistor for different transistors in the input stage.

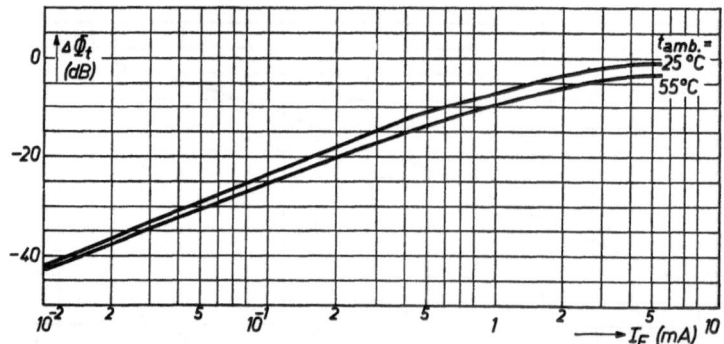

Fig. 10.15. Relative variation of the gain of the reverse AGC amplifier as a function of the emitter current of the first transistor with the ambient temperature as parameter.

Fig. 10.14 shows the spreads in transducer gain of the amplifier with various transistors in the input stage at an emitter current of 40 µA of this first transistor. The second and third stages are equipped with average transistors. The spread in Φ_t in Fig. 10.14 is therefore attributable only to the spreads in gain of the AGC transistor when its emitter current is controlled down to the stated value.

10.5.1.3 *Dependence of Gain on Ambient Temperature*

In Fig. 10.15 the dependence of the transducer gain of the amplifier on the

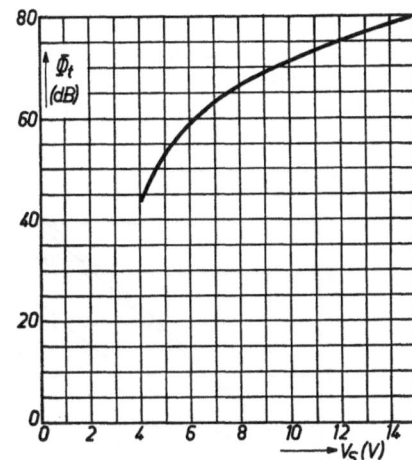

Fig. 10.16. Dependence of the gain of the reverse AGC amplifier on the supply voltage.

ambient temperature is given for an average set of transistors. It is seen that the gain over the whole control range of the amplifier decreases by approximately 2 dB at increasing ambient temperatures.

10.5.1.4 *Dependence of Gain on the Supply Voltage*

Due to variations of the supply voltage of the amplifier, variation of the operating point of the transistors occurs which brings about variation of gain. The dependence of the overall gain of the reverse AGC amplifier on the supply voltage is shown in Fig. 10.16.

10.5.1.5 *Linearity of the Amplifier*

The output voltage from the video detector of the amplifier is shown in Fig. 10.17 as a function of the relative input voltage for different values of gain control. Fig. 10.18 shows the corresponding gamma distortion curve. The different levels of composite video signal according to the C.C.I.R. system with negative modulation are indicated. It can be seen that when the gain of the amplifier is controlled down from -40 dB to -50 dB, a decrease of linearity occurs, which is to be attributed to overloading of the input stage.

10.5.1.6 *Further Measurements*

The signal-to-noise ratio of the amplifier for a detector output voltage of 3.5 V as a function of gain control is shown in Fig. 10.19.

Fig. 10.20 shows the input source current as a function of the gain control that produces cross modulation of 1 %. This cross modulation figure is de-

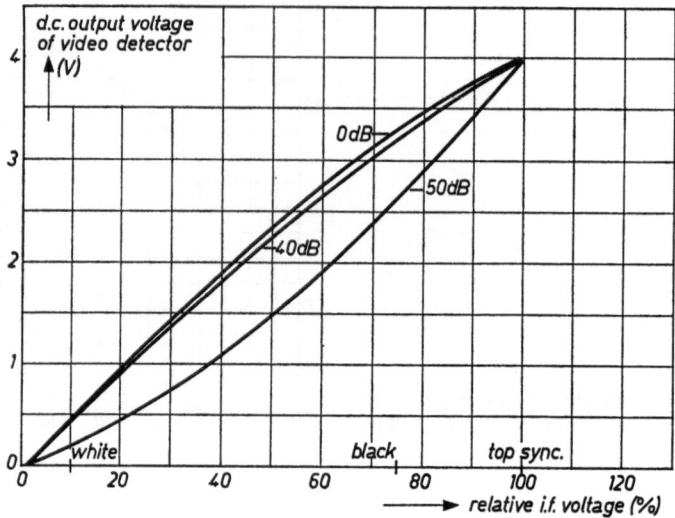

Fig. 10.17. Linearity of the reverse AGC amplifier for different levels of gain control.

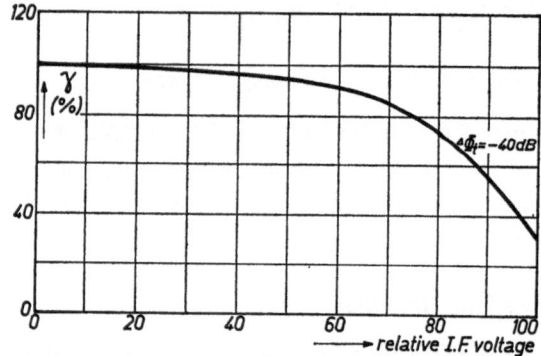

Fig. 10.18. Gamma distortion curve of the reverse AGC amplifier for 40 dB gain control.

fined as the percentage of amplitude modulation occuring on the unmodula-
ted desired signal carrier due to an interfering carrier, which is amplitude
modulated to a depth of 100% by a 1 kc/s signal.

10.5.2 THE FORWARD AGC AMPLIFIER

10.5.2.1 *The Amplitude Response Curve*

In Fig. 10.21 the amplitude response curve of the forward gain control amplifier is shown for various levels of gain control. The tops of the response
curves for gain control levels down to 50 dB are shown in Fig. 10.22.

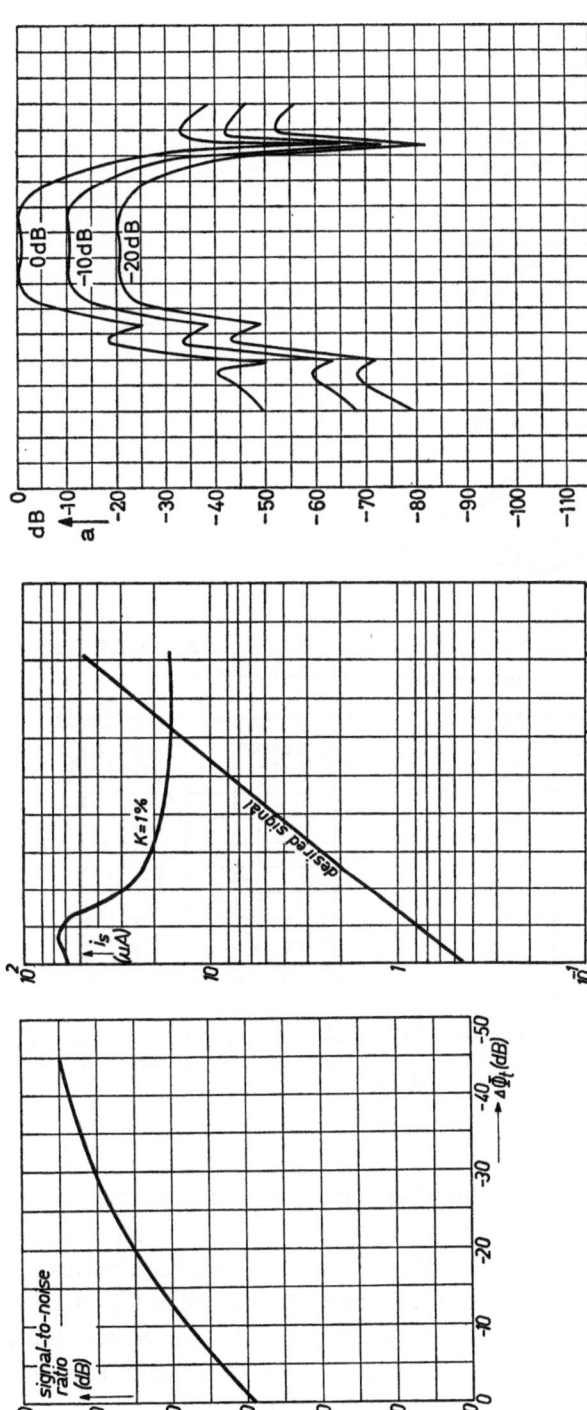

Fig. 10.21. Amplitude response curve of the forward AGC amplifier at different levels of gain control.

Fig. 10.20. Source current of the reverse AGC amplifier producing 1% cross modulation.

Fig. 10.19. Signal-to-noise ratio of the reverse AGC amplifier as function of the gain control for a video detector output voltage of 3.5 V.

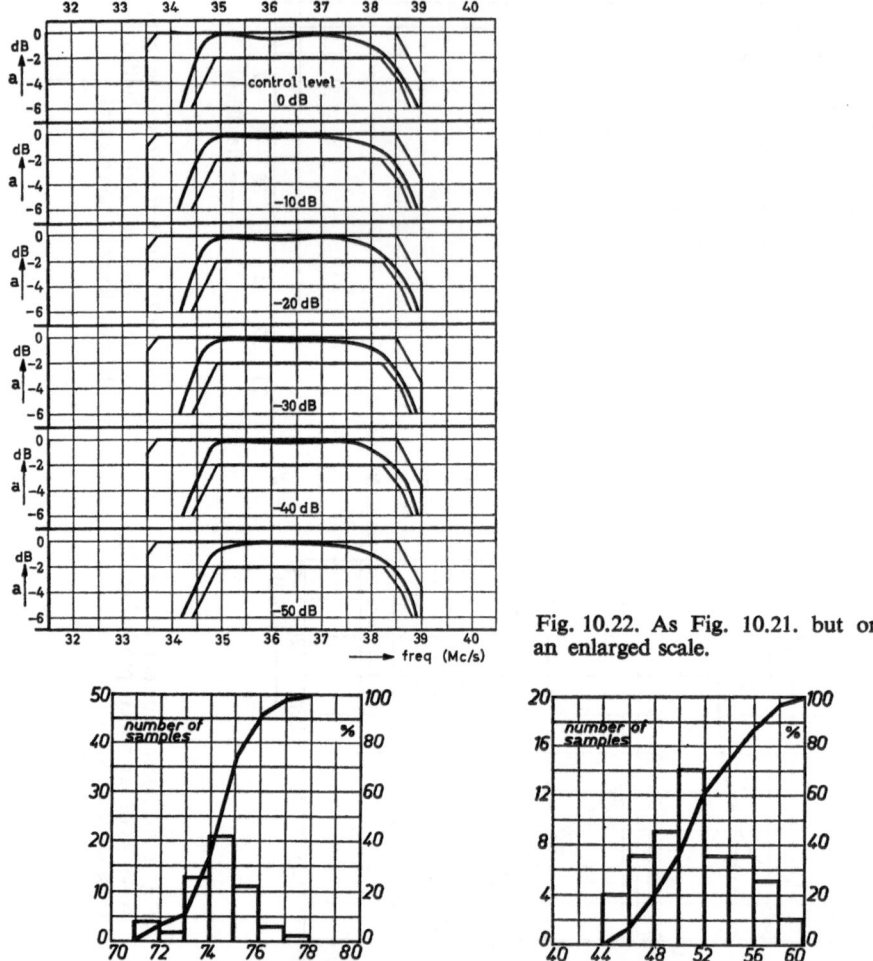

Fig. 10.22. As Fig. 10.21. but on an enlarged scale.

Fig. 10.23. Gain distribution of the forward AGC amplifier in non-controlled condition.

Fig. 10.24. Gain distribution of the forward AGC amplifier in controlled condition.

10.5.2.2 *The Gain*

The histogram in Fig. 10.23 shows the spread in gain measured in the amplifier with a number of different transistors AF181 in the input stage. During these measurements, the second and third stages are equipped with average transistors AF179. This means that the spread shown in Fig. 10.23 is attributable to spread in gain of the input transistor only. The commulative curve indicates that the 50 % value of the gain is 74.4 dB. Again this gain figure agrees very well with that calculated (75.2 dB).

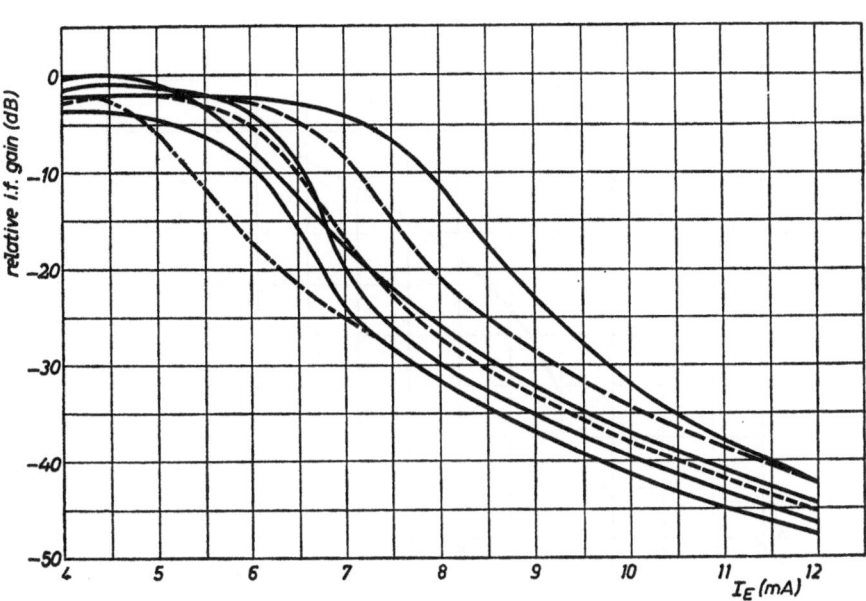

Fig. 10.25. Gain control curves for various transistors AF181.

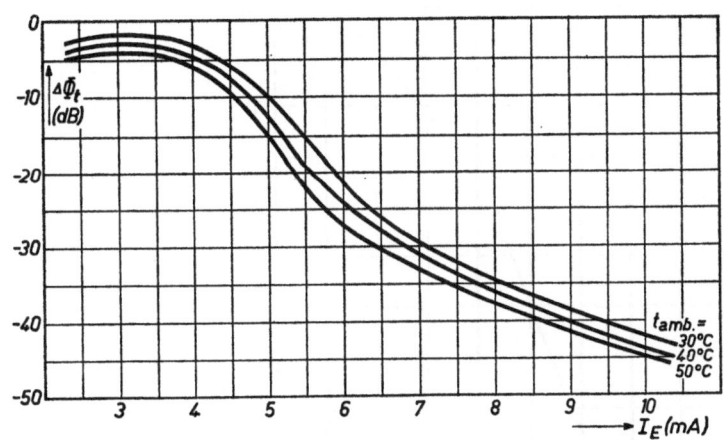

Fig. 10.26. Relative variation of the gain of the forward AGC amplifier as a function of the emitter current of the first transistor for different ambient temperatures.

In Fig. 10.24 the gain measured in the amplifier is shown for a current setting of 8 mA of the input transistor ($R_E + R_C = 510 \, \Omega$, $V_S = 12$ V). This setting corresponds to an average gain reduction of approximately 25 dB. During these measurements various transistors AF181 are used in the input

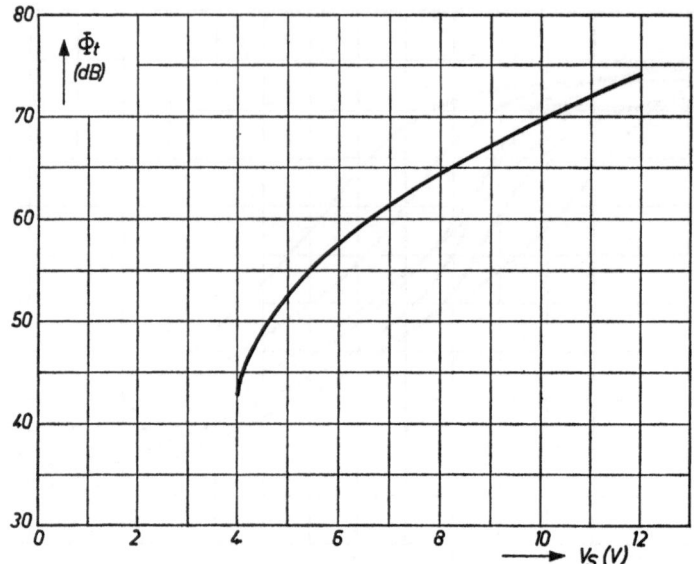

Fig. 10.27. Dependence of gain of the forward AGC amplifier in non-controlled condition on the supply voltage.

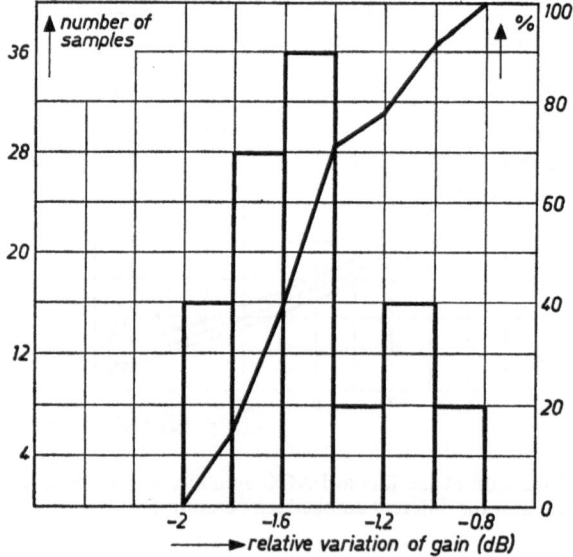

Fig. 10.28. Distribution of gain of the forward AGC amplifier in controlled condition for a supply voltage variation of + 10%.

stage; in the second and third stages average AF179's are used. Gain control curves as a function of the emitter current of various transistors AF181 are shown in Fig. 10.25.

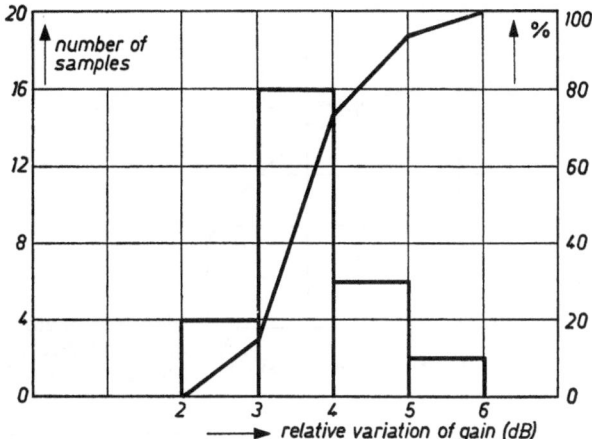

Fig. 10.29. As Fig. 10.28, but for a supply voltage variation of −20%.

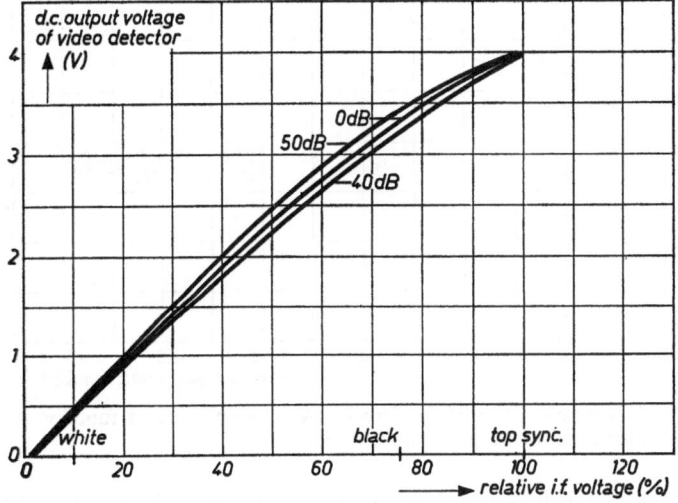

Fig. 10.30. Linearity of the forward AGC amplifier for different levels of gain control.

10.5.2.3 *Dependence of Gain on Ambient Temperature*

In Fig. 10.26 the gain of the amplifier is shown as a function of the emitter current of the first transistor with the ambient temperature as parameter. The measurements are carried out with an average set of transistors.

10.5.2.4 *Dependence of Gain on Supply Voltage*

The dependence of the gain of the amplifier on the supply voltage, measu-

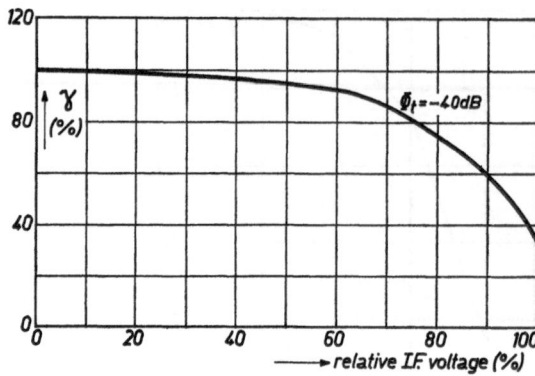

Fig. 10.31. Gamma distortion of the forward AGC amplifier for a gain control level of 40 dB.

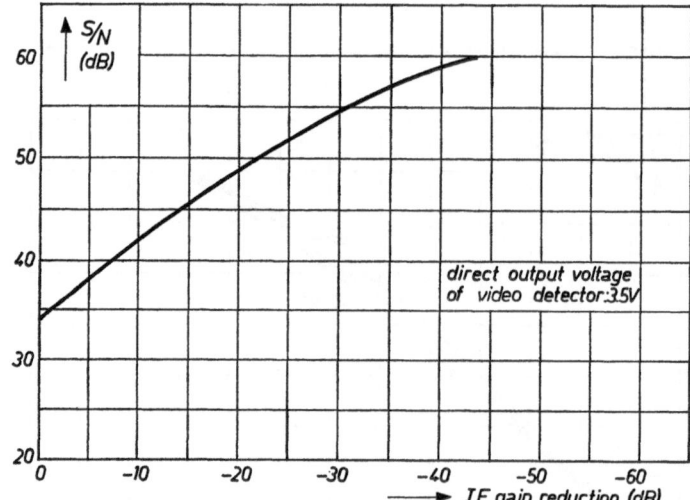

Fig. 10.32. Signal-to-noise ratio of the forward AGC amplifier as a function of the gain control for a video detector output voltage of 3.5 V.

red in the non-controlled condition of the amplifier, is given in Fig. 10.27.

The dependence of the gain on the supply voltage in the controlled condition will be affected by the spreads in the control characteristics of the AF181. The histograms in Figs. 10.28 and 10.29 present these spreads for a supply voltage variation of $+10\%$ and -20% respectively.

The measurements are carried out with several transistors AF181 in the input stage and average transistors AF179 in the second and third stages.

10.5.2.5 *Linearity of the amplifier*

The linearity of the forward AGC amplifier for the same conditions as in

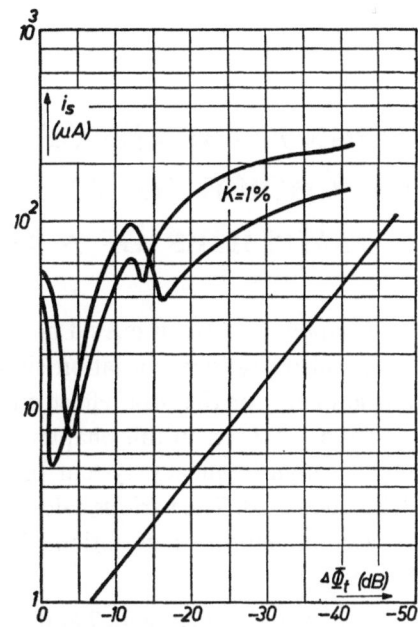

Fig. 10.33. Source current of the forward
AGC amplifier for 1% cross modulation as
a function of the gain control.

sub-section 10.5.1.5 is shown in Fig. 10.30. Fig. 10.31 shows the corresponding gamma distortion.

10.5.2.6 *Further Measurements*

The signal-to-noise ratio of the amplifier and the permissible source current for 1 % cross modulation are given in Figs. 10.32 and 10.33 respectively
as a function of the gain control. The same conditions as mentioned in sub-
section 10.5.1.6 are applicable.

CHAPTER 11

DESIGN EXAMPLES OF INCORPORATING SPREADS THE CHOICE OF AN AGC-SYSTEM FOR AN I.F. AMPLIFIER OF A TELEVISION RECEIVER

In Chapter 3 it was discussed how, in a practical amplifier, spreads, tolerances and operational variation should be combined in order to obtain a realistic figure for the total effect on performance of these deviations from nominal conditions. In this chapter overall spreads in the performance of a complete high-frequency system of a television receiver, including AGC, will be calculated. The various calculations will elucidate the different aspects of the combination of deviations from nominal conditions referred to in Chapter 3.

As considered in Chapters 7 and 10, gain control in a vision I.F. amplifier of a television reciever may be achieved by either increasing or decreasing the emitter current of the first transistor with respect to the normal value. These systems are known as forward and reverse AGC respectively. For both systems of gain control the relevant circuits may be represented by the simplified block diagram shown in Fig. 11.1. Fig. 11.2 shows the gain control curves of tuner and I.F. amplifier. The AGC reference circuit together with the AGC amplifier causes the AGC in the I.F. amplifier to start at a certain input voltage. The AGC amplifier is so designed that after a certain amount of gain variation the AGC action is transferred from the I.F. amplifier to the tuner as indicated in Fig. 11.2. In a practical system this cross-over, as it is called, takes place more gradually than is indicated in Fig. 11.2.

Which of the two methods of gain control is employed in a particular I.F

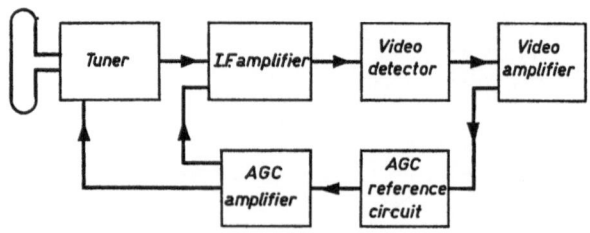

Fig. 11.1. Schematic representation of an AGC circuit of a television receiver.

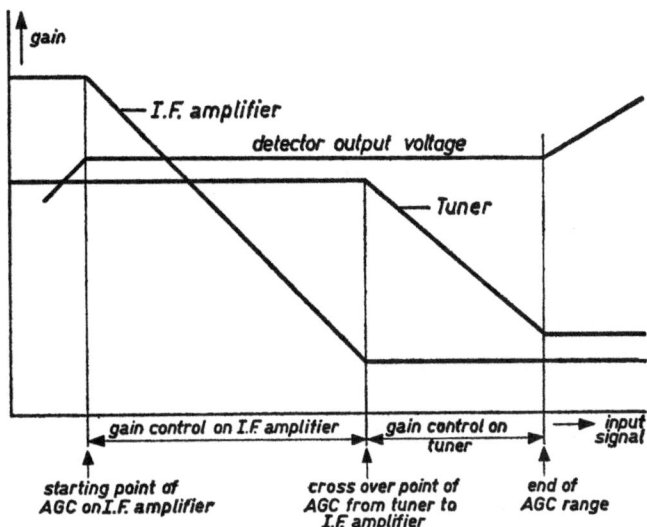

Fig. 11.2. Idealized gain control curves of a television receiver.

amplifier design will be dependent on the relevant properties of either of the systems. Generally that system will be chosen which gives, in combination with a tuner, the largest signal-to-noise ratio at the cross-over point of gain control from I.F. amplifier to tuner, without overloading of any of the stages of the whole system. This may be clarified by Fig. 11.3, in which various voltage levels in tuner and I.F. amplifier are depicted as a function of gain control. The different voltage levels in this figure are *not* drawn to scale; the relative values as shown may therefore not be compared.

In Fig. 11.3 lines are drawn indicating the nominal level of the signal at the aerial terminals of the tuner and the noise level of the tuner. It follows that the signal-to-noise ratio of the tuner becomes larger when the starting point of its AGC is further delayed.

Also a line is drawn indicating the permissible signal at the aerial terminals. It is assumed that above this level unacceptable overloading of the tuner occurs[1]. The permissible signal level is constant with no AGC control and is assumed to increase when AGC is applied.

In practical tuner and I.F. amplifier designs spreads will occur in the nominal signal level as well as in the permissible signal level. To ensure that no overloading occurs in practice an allowance for these spreads must be made between the lines for the nominal signal level and the permissible signal level. In Fig. 11.3 this allowance is indicated by the arched area.

---

[1]) This overloading may occur either in the R.F. stage or in the mixer stage.

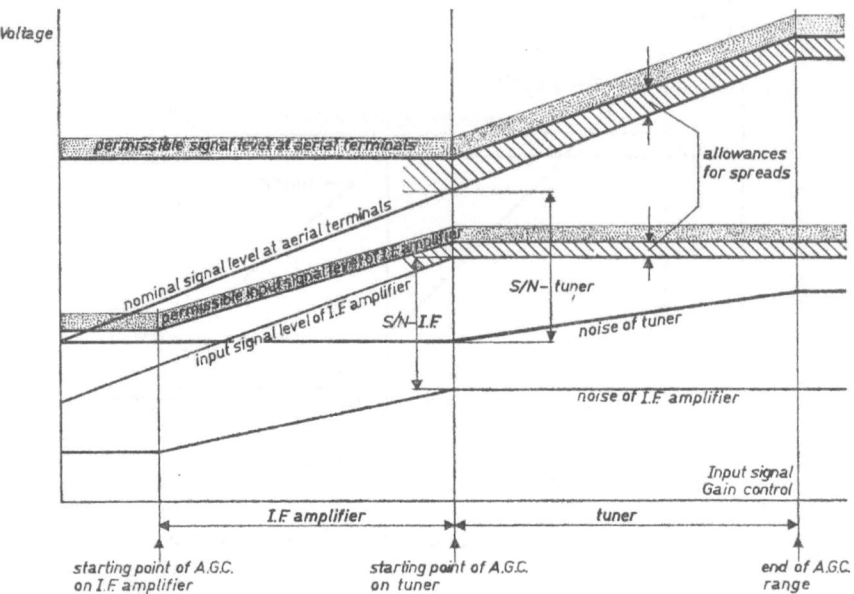

Fig. 11.3. Diagram indicating actual and permissible voltage levels during gain control in tuner and I.F. amplifier of a television receiver.

It follows from the above that the starting point of the AGC on the tuner may be delayed until the distance between the nominal signal level and that which is permissible reaches the margin that must be allowed in view of spreads. Then the ultimate signal-to-noise ratio of the complete receiver at the end of the AGC range will reach the largest value, provided the signal-to-noise ratio of the I.F. amplifier remains larger than that of the tuner over the whole AGC range. If the latter condition is not fulfilled the signal-to-noise ratio of the I.F. amplifier can be improved by applying gain control to the first two stages instead of to the input stage only.

The above considerations are equally applicable to the I.F. amplifier. Lines are shown in Fig. 11.3 which indicate the various voltage levels of interest in this amplifier. It will often happen that the margin between permissible signal level and nominal signal level is reached earlier in the I.F. amplifier than in the tuner. In that case the cross over point of the AGC from I.F. amplifier to tuner is determined by the properties of the I.F. amplifier.

The starting point of AGC as indicated in Fig. 11.3 is reached as soon as the voltage level at the output of the (video) detector reaches a predetermined value. When the input signal of the system increases, its gain is decreased in such a way that the output voltage mentioned remains nearly constant, until the end of the AGC range is reached (see Fig. 7.1).

The end of the AGC range is reached either when the gain of the relevant transistors cannot be reduced any further or when the drive power of the AGC amplifier (see Fig. 11.1) reaches the maximum value.

In this chapter we will investigate how spreads, tolerances and variations in amplifier properties, components and environmental conditions respectively affect the relative mertis of the forward and reverse AGC systems. For this investigation, use will be made of the results of the measurements made on forward AGC and reverse AGC amplifiers described in the predecing chapter.

The tuner which will be used in combination with these I.F. amplifiers is a V.H.F. tuner equipped with the transistor AF180 in the R.F. stage and transistors AF178 in the mixer and oscillator stages. A short description of this tuner will be given in Section 11.3.

The investigation to be carried out will furthermore provide information on what spreads may be expected in practice in the overall sensitivity of a combination of tuner and I.F. amplifier.

11.1 The Various Signal Levels in I.F. Amplifier and Tuner

In the analysis the signal levels at various points in tuner and I.F. amplifier are to be considered. These points are listed in the first column of Table 11.1. In the second column the factors on which the signal levels are dependent are set out.

11.1.1 THE PERMISSIBLE SIGNAL LEVELS

For determining the allowable range of gain control of the I.F. amplifier, permissible signal levels at the input terminals of the 1st I.F. transistor, at the input terminals of the mixer transistor and at the aerial terminals must be known. These signal levels are limited by crossmodulation and/or modulation distortion. In our analysis, a television system with negative modulation of the picture signal and frequency modulation of the sound signal is considered (CCIR system E). For this system the following signal levels are assumed to be permissible:

a) At the input terminals of the 1st I.F. transistor at approximately 30 dB gain reduction and at the top of the response curve: 10 mV with a reverse gain controlled AF179, 30 mV with a forward gain controlled AF181.

b) At the input terminals of the mixer transistor at the picture carrier: 20 mV in Bands I and III (in a properly designed circuit).

TABLE 11.1   Factors affecting signal levels of interest in I.F. amplifier and tuner

Signal level at:	Dependent on:
A.1 output terminals of video detector	a· whether contrast control is applied via AGC circuit or not and reference level AGC b. sensitivity of video amplifier
A.2 input terminals of video detector	a. type of detector b. values of R_d and C_d providing required bandwith and value of source capacitance seen by the detector c. requirements for detection of inter-carrier sound signal d. presence of a separate sound detector
A.3 input terminals of 2nd and 3rd I.F. transistors	a. amount of power to be delivered by last I.F. stage for adequate drive of video detector 1. with single-tuned bandpass filter 2. with double-tuned bandpass filter b. gain of 2nd and 3rd I.F. stages
A.4 input terminals of 1st I.F. transistor	gain of 1st I.F. transistor: a. in the non-controlled condition b. with forward gain control c. with reverse gain control
A.5 input terminals of mixer transistor	a. gain of mixer transistor
A.6 aerial terminals (e.m.f. of a 300 Ω source)	a. gain of R.F. stage of tuner

c) At the aerial terminals, at the picture carrier, assuming that distortion is caused by the R.F. transistor only, and expressed in the e.m.f. of a 300 Ω source for the condition of no gain control on the RF stage:

20 mV in Band I and

40 mV in Band III.

The signal levels quoted refer to the peak value of the synchronizing pulses of the signal.

The permissible signal levels stated are believed to be realistic values for the relevant types of transistors under conditions as mentioned. In this chapter, however, the different signal levels are used merely for the purpose of illustrating the method of combining deviations from nominal conditions. It is not therefore of importance whether these signal levels are in fact permissible.

11.2 Spreads, Tolerances and Operational Variations

The signal levels as well as the factors on which they are dependent are subject to the effects of various spreads, tolerances and operational vari-

TABLE 11.2   Spreads

Spreads:	Conditions:
A. *Transistor high frequency properties*	
a. $\Delta\Phi$ due to spreads in transistor y-parameters and a.c. circuit components of 2nd and 3rd stage of I.F. amplifier	under nominal biasing conditions and $t_{amb} = 25$ °C.
b. ditto for 1st stage of I.F. amplifier	1. in non-controlled condition under nominal biasing and $t_{amb} = 25$ °C 2. in controlled condition at nominal cross-over point of I.F. and tuner gain control and $t_{amb} = 25$ °C.
c. $\Delta\Phi_t$ of I.F. amplifier as a function of I_E	1. in the vicinity of nominal value in the non-controlled condition 2. in the vicinity of the nominal value at the nominal cross-over point
d. $\Delta\Phi_t$ of I.F. amplifier as a function of t_{amb}	over entire control range
e. $\Delta\Phi$ due to spreads in transistor properties and a.c. circuit components of tuner	1. mixer stage 2. R.F. stage
f. $\Delta\Phi$ of tuner as a function of I_E	1. mixer stage 2. R.F. stage
B. *D.C. properties of transistors* Spreads due to: a. I_{CBO} spreads b. α' spreads c. V_{BE} spreads	 at higher ambient temperatures ditto

TABLE 11.3   Tolerances

Tolerance:	Conditions:
A. *Resistances* Effects of resistance tolerances ΔR on: a. operating point of 2nd and 3rd I.F. transistor	
b. operating point of 1st I.F. transistor	1. in non-controlled condition 2. in controlled condition
c. reference level AGC circuit	
d. reference level of cross-over point AGC in I.F. amplifier and tuner	
e. gain of mixer transistor	
f. gain of R.F. transistor tuner	

ations. These deviations from nominal conditions and the conditions under which they are to be investigated are set out in Tables 11.2, 11.3 and 11.4 respectively.

In these tables, variations in gain are denoted by $\Delta\Phi$ and resistance tolerances by Δ_R.

TABLE 11.4   Operation Variations

Variation:	Conditions:
A. Contrast control	if contrast control is achieved via the AGC of tuner and I.F. amplifier
B. Voltages Effects of spreads in supply voltage on: a. operating points of I.F. transistors b. reference level AGC circuit c. cross-over point of AGC in I.F. amplifier and tuner d. gain of mixer transistor e. gain of R.F. transistor tuner	
C. Ambient temperature	only in those cases in which variation of ambient temperature is not included under the heading Spreads (Table 11.2)

11.3 Data of the I.F. amplifier and the tuner

To determine the relative merits of the forward and reverse gain control methods in the I.F. amplifiers a comparative analysis will be made of:

a) a combination of a tuner and an I.F. amplifier suitable for reverse gain control, and

b) a combination of the same tuner and an I.F. amplifier suitable for forward gain control.

Both combinations have approximately the same sensitivity in the noncontrolled condition and the same amplitude response curve. The reverse gain controlled I.F. amplifier as well as the forward gain control I.F. amplifier are described in Chapter 10.

11.3.1 THE TUNER [1])

A description of the tuner together with the data required for the investigation is given below. The tuner consists of an R.F. stage, a mixer stage and a separate oscillator.

The circuit diagram is given in Fig. 11.4.

The R.F. stage is equipped with a transistor AF180 in the common base configuration. Forward gain control is applied to this transistor to prevent overloading of the mixer transistor or the first I.F. transistor.

---

[1]) A detailed description of this tuner can be found in Philips Advance Information Bulletin no. 206: J. W. B. A. Francois and A. H. J. Nieveen van Dijkum, *A transistor V.H.F. television tuner with AF180 and AF178*, Philips Electronic Market Division Department, Eindhoven, Netherlands, July 1963.

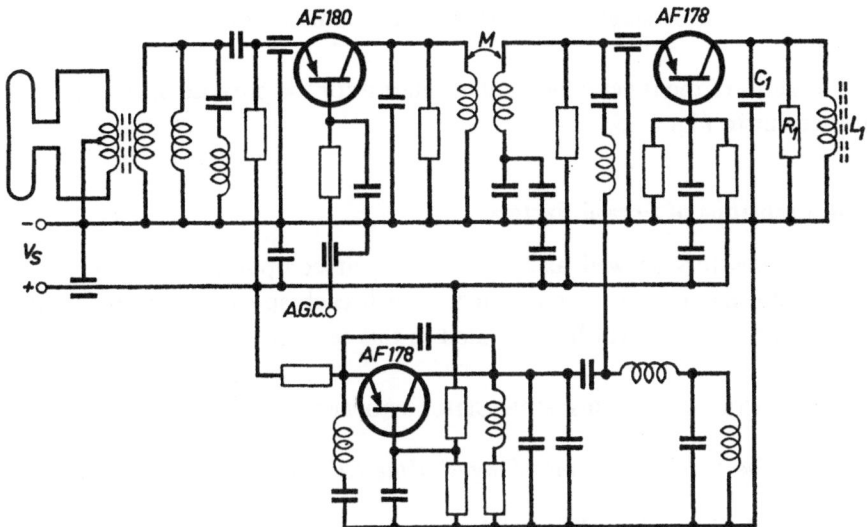

Fig. 11.4. Circuit diagram of a single channel of a V.H.F. tuner.

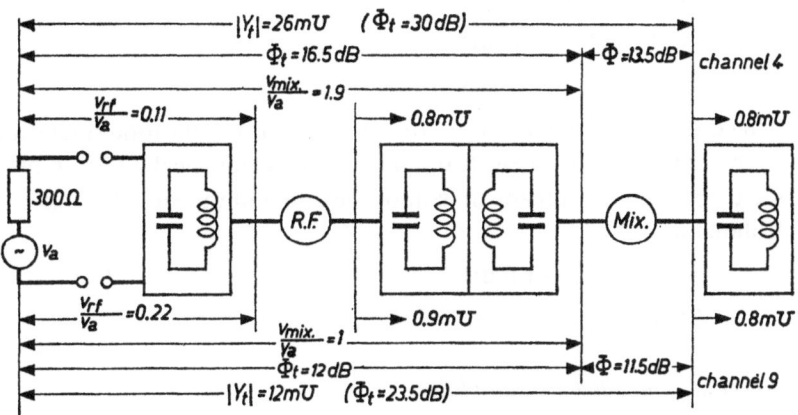

Fig. 11.5. Schematic representation of the various gains in the V.H.F. tuner shown in Fig. 11.4.

The nominal emitter current of the R.F. transistor is made 2.5 mA. Gain control is obtained by reducing the AGC line voltage, which results in an increase of the emitter current. By means of an emitter series resistance the collector emitter voltage is decreased at the same time, which limits the dissipation in this transistor to a safe value.

The mixer stage is equipped with a transistor AF178 biased at an emitter current of 1.5 mA.

In the oscillator stage a transistor AF178 is used; the emitter current is 2 mA.

The gains of the tuner which are of interest for the purpose of this chapter are set out in Fig. 11.5.

11.4 Evaluation of Spread Factors

The transistors as well as the other circuit components are subject to manufacturing spreads. The extent to which each spread factor affects the gain of the amplifier will be evaluated in this section.

11.4.1 THE TRANSISTOR HIGH-FREQUENCY PROPERTIES

11.4.1.1 *The Second and Third Stages of the I.F. Amplifier*

Reverse gain controlled I.F. amplifier

Measured spread in gain of amplifier with $3 \times$ AF179 in non-controlled condition (see Fig. 10.12):

$$5\% \text{ limit} : -2.8 \text{ dB}$$
$$95\% \text{ limit} : +1.9 \text{ dB}$$

Assuming that the individual stages contribute equally to the total spread and, moreover, that the total spread is obtained from a statistical addition of the individual spreads, the spreads in gain of the 2nd and the 3rd I.F. stages can be calculated as:

$$5\% \text{ limit} : -1.6 \text{ dB}$$
$$95\% \text{ limit} : +1.1 \text{ dB}$$

Forward gain controlled I.F. amplifier

As the 2nd and 3rd stages of the forward gain controlled I.F. amplifier are equipped with transistors AF179, the same spread factors as above are obtained.

11.4.1.2 *The Input Stage of the I.F. Amplifier*

Reverse gain controlled I.F. amplifier

1. Non-controlled condition:

The spread in gain is equal to that of the second and third stages. Hence:

5 % limit : -1.6 dB

95 % limit : $+1.1$ dB

2. Controlled condition:

To determine the spread in power gain, the nominal cross-over point has been assumed at approximately 30 dB gain reduction. Then emitter current I_E of the control transistor is in the order of 40 μA, see Fig. 10.13 on page 149. Measured spread in gain for transistors AF179 at $I_E = 40$ μA (see Fig. 10.14):

5 % limit : -0.7 dB

95 % limit : $+0.7$ dB

Forward gain controlled I.F. amplifier

1. Non-controlled condition:

Measurements on an amplifier with a transistor AF181 in the control stage and average transistors AF179 in the second and third stages have resulted in the following power gain spread figures for the AF181 (see Fig. 10.23):

5 % limit : -2.7 dB

95 % limit : $+2.1$ dB

2. Controlled condition

Measurements on an amplifier with a large number of transistors AF181 in the control stage have given a figure for a spread in gain between different transistors AF181 of ± 6 dB at approximately 30 dB gain reduction. Then the emitter current is approximately 9 mA, see Fig. 10.25.

Therefore:

5 % limit : -6 dB

95 % limit : $+6$ dB

11.4.1.3 *The Tuner*

1. *Mixer stage*

Measurements have shown that spreads in power gain are:

	Band I	Band III
5 % limit	-0.7 dB	-1.0 dB
95 % limit	$+0.7$ dB	$+1.0$ dB

2. R.F. Stage

By measurement the spreads in power gain are:

	Band I	Band III
5% limit	— 1.5 dB	— 2.0 dB
95% limit	+ 1.5 dB	+ 2.0 dB

11.4.1.4 The Detector Circuit

The investigations are based on a video detector circuit with a diode OA90 and $R_d = 2700\Omega$, $C_d = 10$ pF. The tuning capacitance of the circuit seen by the detector at its input terminals amounts to 10 pF. Measurements have given the efficiency η of this detector circuit at an input voltage $V_d = 4$ V as:

$$\eta = 0.53 \pm 0.8 \text{ dB}$$

11.4.2 THE DEPENDENCE OF GAIN ON EMITTER CURRENT

To facilitate determinination of the deviations of gain from nominal conditions due to spreads in transistor d.c. properties, resistance tolerances and supply voltage variation, it is required to determine the dependence of the gain on emitter current variation under various conditions.

Reverse gain controlled I.F. amplifier

1. Non-controlled condition:

Measurements have shown that in an amplifier with three AF179's, the transducer gain variation for a 10 % variation in supply voltage V_S amounts to (see Fig. 10.16):

$$\pm 1.5 \text{ dB} / \pm 10\% \, \Delta V_S$$

Usually the voltage V_{RE} across the d.c. emitter resistance R_E is large compared with the base-emitter voltage V_{BE} of the transistor. We may therefore assume that variation of supply voltage results in proportional variation of emitter current. The variation of power gain per stage for a variation of emitter current I_E therefore becomes:

$$\pm 0.5 \text{ dB} / \pm 10\% \, \Delta I_E$$

2. Controlled condition:

At relatively low emitter currents (of the order of 40 µA) it may be assumed that the power gain of the stage is proportional to $|y_{fe}|$. Hence it can be calculated that the variation of power gain with emitter current equals (see also Fig. 10.13):

$$\pm 0.8 \text{ dB} / \pm 10\% \, \Delta I_E$$

Forward gain controlled I.F. amplifier

1. Non-controlled condition:

Considerations and measurements similar to those mentioned above have indicated that the variation of power gain of the control stage (AF181) with emitter current, taking the nominal operating point at $-V_{CE} = 10\text{V}$, $I_E = 3$ mA, equals:

$$-0.5 \text{ dB} / -10\% \, \Delta I_E$$
$$+0.3 \text{ dB} / +10\% \, \Delta I_E$$

2. Controlled condition:

At approximately 30 dB gain reduction, the variation of power gain of the control stage lies, according to measurements (see Fig. 10.25), between -3.2 dB $/ +10\% \, I_E$ and -4.8 dB $/ +10\% \, \Delta I_E$.

In our calculations we will take into account a variation of:

$$-5.0 \text{ dB} / +10\% \, \Delta I_E$$

11.4.3 THE D.C. PROPERTIES OF THE TRANSISTORS

The transistor AF 179

1. Non-controlled condition:

Calculations for $R_B = 2R_E$ using (see Fig. 11.6) limit values for $-I_{CBO}$ and $-I_B$ have shown that the emitter current may spread between

$$+5.7\% \text{ and } -7.3\%$$

with respect to the nominal value of 3 mA.

With reference to the previous section, this may be expressed as:

$$\Delta \Phi = +0.3 \text{ dB to } -0.4 \text{ dB}$$

due to $-I_{CBO}$ and $-I_B$ variations

According to the publication data the spreads in V_{BE} at an emitter current of $I_E = 3$ mA are given as:

$$V_{BE} = > 290 - 330 - < 370 \text{ mV}.$$

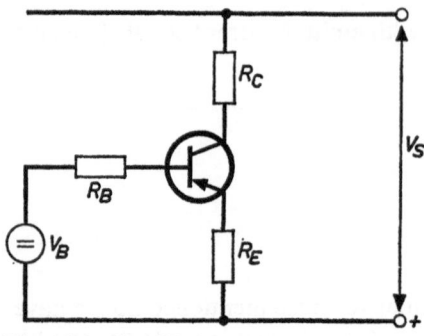

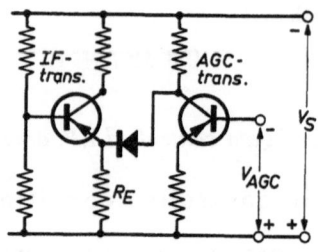

Fig. 11.6. Simplified transistor biasing net-   Fig. 11.7. AGC circuits for a reverse gain-
work.   controlled amplifier stage.

In considering V_{BE} spread we assume that the relative spreads at higher ambient temperatures are equal to those at $t_{amb} = 25\,°C$. For a voltage across the emitter resistance R_E of 2 V we find:

$$\Delta I_E = +2\% \ / \ -\Delta V_{BE},$$

and $\qquad \Delta I_E = -2\% \ / \ +\Delta V_{BE}.$

The effect of V_{BE} spread on power gain can therefore be neglected.

2. Controlled condition:

Spreads are largely dependent on the type of AGC drive circuit. When a circuit as shown in Fig. 11.7 is employed, no spreads due to transistor d.c. properties are involved. This is because at the AGC cross-over point, the condition exists that $R_E \gg R_B$.

The transistor AF181

1. Non-controlled condition:

Calculations for $R_B = 3R_E$ using limit values for $-I_{CBO}$ and $-I_B$ indicate that the emitter current may spread between

$$+ 6\% \text{ and } -10\%$$

According to the previous section:

$$\Delta\Phi = +0.2 \text{ dB to } -0.5 \text{ dB}$$

$$\text{due to } -I_{CBO} \text{ and } -I_B \text{ variations}$$

Effects of V_{BE} spreads on power gain may be neglected because they are of the same order of magnitude as for the AF179.

2. Controlled condition:

In the controlled condition of the AF181 spreads in α' need not be taken into account when determining spreads in power gain. This is because the α' spreads are implicitly included in the power gain test for each transistor at approximately 30 dB gain reduction as may be seen from the publication data. The d.c. circuit in which this power gain test is carried out, is as shown in Fig. 11.6; $V_B = -3.6$ V, $R_E = 330$ Ω, $Ra = 1000$ Ω, $R_C = 180$Ω.

In the controlled condition, only variations in I_{CBO} need therefore be considered. Taking limit values at $T_j = 75$ °C, we find an increase in emitter current of approximately 8 %. This gives a power gain variation of

$$\Delta\Phi = -4 \text{ dB due to } I_{CBO} \text{ variations}$$

The Tuner

It may be calculated that for the R.F. stage and the mixer stage of the tuner (in non-controlled condition) the variations in gain due to $-I_B$ and $-I_{CBo}$ variations amount to

$$\Delta\Phi = +0.3 \text{ dB to } -0.3 \text{ dB due to } I_{CBO}$$

and $-I_B$ variations.

The effects on the gain of spreads in the base-emitter voltage of both transistors can be neglected.

11.5 Evaluation of Resistance Tolerances

To investigate the effects of resistance tolerances, we will assume tolerances Δ_R of 5 % and 10 %.

If V_{BE} is small compared with the voltage V_{RE} across the d.c. emitter resistance R_E, spreads Δ_R cause proportional spreads ΔI_E in the emitter current. The spreads in I_E may with reference to section 11.4.2 be expressed in corresponding spreads in power gain.

In I.F. amplifiers and tuners d.c. biasing of the transistors is usually achieved by means of three resistances each with a tolerance of Δ_R%. To obtain the total spread of the three resistances the individual spreads must be added statistically; hence total spread amounts to $\sqrt{3}\,\Delta_R$%. The same applies to corresponding spreads in emitter current.

11.5.1 THE SECOND AND THIRD STAGES OF THE I.F. AMPLIFIER

For the 2nd as well as the 3rd stage of the I.F. amplifier, we obtain for the variation in power gain, taking into account the results from sub-section 11.4.2:

	$\Delta_R = 5\%$	$\Delta_R = 10\%$
ΔI_E	8.7 %	17.3 %
$\Delta\Phi$	0.4 dB	0.9 dB

11.5.2 THE INPUT STAGE OF THE I.F. AMPLIFIER

Reverse gain controlled I.F. amplifier

1. Non-controlled condition:

In the non-controlled condition the biasing point of the transistor of the control stage is the same as that of the transistors in the 2nd and 3rd stages.

We will assume that biasing of the control-stage transistor is also affected by spreads of three resistances which may be in the circuitry of either this transistor or the AGC amplifier transistor.

The spreads in power gain due to resistance tolerances then equal the figures obtained in sub-section 11.5.1.

2. Controlled condition:

Again assuming that biasing is provided by three resistances we find:

	$\Delta_R = 5\%$	$\Delta_R = 10\%$
$\Delta\Phi$	0.7 dB	1.4 dB

Forward gain controlled I.F. amplifier

1. Non-controlled condition:

According to sub-section 11.4.2 the variation of power gain is different in magnitude for positive and negative variation of I_E.
Since performance in least favourable conditions is under investigation, we assume:

$$0.5 \text{ dB } / \ 10\% \ \Delta I_E$$

The effects of resistance tolerances then equal the figures of the previous item.

2. Controlled condition:

Taking into account the spreads of three resistances we find, with sub-section 11.4.2, for worst conditions $(-\Delta_R)$:

	$\Delta_R = 5\%$	$\Delta_R = 10\%$
$\Delta\Phi$	4.4 dB	8.7 dB

11.5.3 THE TUNER

The results obtained by measurement in Band I and Band III, for both the R.F. (non-controlled) and the mixer stages, are:

$$\Delta_R = 5\% \quad \Delta_R = 10\%$$
$$\Delta\Phi \qquad 0.4 \text{ dB} \qquad 0.8 \text{ dB}$$

11.5.4 THE REFERENCE LEVELS OF THE AGC CIRCUIT

The reference level for the AGC circuit as well as the voltage level determining the cross-over point of I.F. amplifier and tuner gain control is subject to spreads of two resistances, assuming a circuit like that shown in Fig. 11.8 to be employed. Again the individual spreads must be added statistically to obtain the overall spread.

$$\Delta_R = 5\% \quad \Delta_R = 10\%$$

spread of voltage level   0.6 dB   1.1 dB

11.6 Evaluation of Operational Variation

11.6.1 THE CONTRAST CONTROL

For present purposes, only contrast control via the AGC system of the I.F. amplifier and tuner need be considered. It is then required that by means of the contrast control, a variation of (at least) 12 dB in output signal of the video detector is achieved. Measurements of and considerations regarding the operation of a typical detector circuit (see below) at signal levels of interest, indicate that for a variation of 12 dB in output voltage a variation of 10.9 dB in input voltage is required. The discrepancy between the output and input voltage variations is due to the increase of detector efficiency at higher input voltages.

When considering the effect of contrast control on the voltage levels at the input terminals of 1st I.F. transistors, two cases must be distinguished, viz.:

a) contrast control takes place when only the I.F. amplifier gain is controlled;

b) contrast control takes place when only the tuner gain is controlled.

In case a) the signal voltage at the base of the 1st I.F. transistor remains the same. The AGC circuit changes the gain of the I.F. amplifier by 10.9 dB.

In case b) a variation of 10.9 dB in the tuner gain is obtained by contrast control. This implies a 10.9 dB increase of the input voltage at the base of the 1st I.F. transistor.

Hence:

> Due to contrast control:
> 10.9 dB increase of base voltage of 1st I.F. transistor with respect to the setting for minimum contrast.

11.6.2 THE VARIATIONS OF SUPPLY VOLTAGE

In our investigations supply voltage variations of $+10\%$ and -20% will be taken into account.

11.6.2.1 *The I.F. Amplifiers*

Reverse gain controlled I.F. amplifier

1. Non-controlled condition:

According to sub-section 11.4.2.:

$$\Delta\Phi = \pm\,0.5\;\text{dB}\;/\;\pm\,10\%\;\Delta V_S$$

2. Controlled condition:

According to sub-section 11.4.2:

$$\Delta\Phi = \pm\,0.8\;\text{dB}\;/\;\pm\,10\%\;\Delta V_S$$

Forward gain controlled I.F. amplifier

3. Non-controlled condition:

According to sub-section 11.4.2:

$$\Delta\Phi = +0.3\;\text{dB}\;/\;+10\%\;\Delta V_S$$
$$\Delta\Phi = -0.2\;\text{dB}\;/\;-10\%\;\Delta V_S$$

2. Controlled condition:

For ascertaining the variations of power gain of the AF181 in the first stage of a forward gain controlled amplifier at a gain reduction of 30 dB, the transducer gain of the complete amplifier has been measured at supply voltage variations of $+10\%$ and -20% for various transistors AF181. Only the supply voltage of the control stage has been varied, and at the nominal value of supply voltage, the gain reduction has been adjusted at 30 dB for each AF181.

Hence (see Figs. 10.28 and 10.29):

$$\text{at }\Delta V_S = +10\% \qquad \Delta\Phi$$
$$5\%\text{ limit} \qquad\qquad -0.9\;\text{dB}$$

$$
\begin{array}{ll}
95\% \text{ limit} & -1.9 \text{ dB} \\
\text{at } \Delta V_S = -20\% & \Delta\Phi \\
5\% \text{ limit} & +2.5 \text{ dB} \\
95\% \text{ limit} & +5.5 \text{ dB}
\end{array}
$$

Only the spread figures which indicate the lowest gain need be taken into account, because in these cases the voltage levels at the transistor input terminals are maximum.

11.6.2.2 *The Tuner*

The results obtained by measurement in Band I and Band III, for both the R.F. and the mixer stages, give:

$$\Delta\Phi = \pm 0.5 \text{ dB} \; / \; \pm 10\% \, \Delta V_S$$

11.6.2.3 *The Reference Levels of the AGC Circuits*

Because of the linear relation between the reference level ACG, the cross-over point and the supply voltage:

$$\text{voltage level} = \pm 0.8 \text{ dB} \; / \; \pm 10\% \, \Delta V_S$$

11.6.3 THE VARIATIONS IN AMBIENT TEMPERATURE

Experiments show that for a rise in ambient temperature from 25 °C to 55 °C the transducer gain of both the reverse controlled and forward controlled amplifiers decreases 2 dB over the entire control range (see Figs. 10.13 and 10.26); we therefore assume a variation per stage of:

$$-0.7 \text{ dB} \; / \; 30°\text{C} \, \Delta t_{amb}$$

11.7 Combination of the Various Spreads, Tolerances and Operational Variations

After having evaluated the various deviations from nominal conditions, the effects must be combined in a suitable manner to obtain the total spreads to which the overall gain and the voltage levels of interest in tuner and I.F. amplifier are subject. In our analysis the following method of combining the spreads has been adopted (see also Chapter 3):

a) Spread factors due to manufacturing tolerances are added *statistically*. Examples are spreads in high-frequency properties of the transistors and resistance tolerances.

b) Spread factors due to variations of certain properties are added *linearly*. Examples are the changes in amplifier gain due to variations in ambient temperature and to variations of supply voltage.

Spreads or operational variations always have a positive and a negative effect on a performance figure. In the following, however, we will confine ourselves to considering only those spreads, etc. which lead to higher levels of signal on the various stages of I.F. amplifier and tuner. The reason for this restriction is that we are mainly interested in evaluating the nominal and worst-case performance of two different AGC systems.

The deviations from the nominal level of the input voltage of the video detector are set out in Table 11.5. The total deviations are determined for a nominal value of the supply voltage (i.e. when the I.F. amplifier and the tuner are fed from a stabilized power supply) and for deviations of $+10\%$ and -20% from the nominal value. It is worth noticing that the spreads entered in Table 11.5 are applicable to an elementary AGC system. For more sophisticated AGC systems the corresponding spreads may be larger.

Table 11.6 gives the deviations of the gains of the second and third stages of the I.F. amplifier for the same conditions. The deviations of the gain of the first stage of the reverse AGC and forward AGC amplifiers are entered in

TABLE 11.5

Total spread or variation of voltage level at input terminals of video detector

Spreads to be added:		linearly		statistically	
Description of spread:		ΔR $=5\%$	ΔR $=10\%$	ΔR $=5\%$	ΔR $=10\%$
Effects of resistance tolerances on reference level AGC	dB	—	—	+0.6	+1.1
Effects of supply voltage on reference level AGC for					
$\Delta V_S: = +10\%$	dB	+0.8	+0.8	—	—
$\Delta V_S: = -20\%$	dB	−1.6	−1.6	—	—
Effects of detector efficiency	dB	—	—	+0.8	+0.8
Total spread of input voltage v_d video detector for:					
$\Delta V_S = +10\%$	dB	+0.8	+0.8	+1.0	+1.4
$\Delta V_S = 0$	dB	0	0	+1.0	+1.4
$\Delta V_S = -20\%$	dB	−1.6	−1.6	+1.0	+1.4

TABLE 11.6

Total spread of power gain per stage of 2nd and 3rd stages of I.F. amplifier

Spreads to be added:		linearly		statistically	
Description of spread:		ΔR $=5\%$	ΔR $=10\%$	ΔR $=5\%$	ΔR $=10\%$
due to transistor high-frequency properties and a.c. circuit components:	dB	—	—	−1.6	−1.6
due to resistance tolerances:	dB	—	—	−0.4	−0.9
due to supply voltage variations:					
$\Delta V_S = +10\%$	dB	+0.5	+0.5	—	—
$\Delta V_S = -20\%$	dB	−1.0	−1.0	—	—
due to $-I_{CBO}$ and $-I_B$ variations:	dB	—	—	−0.4	−0.4
due to V_{BE} variations:	dB	—	—	0	0
due to temperature variations:	dB	−0.7	−0.7		
Total spread per stage for:					
$\Delta V_S = +10\%$	dB	−0.2	−0.2	−1.7	−1.9
$\Delta V_S = 0$	dB	−0.7	−0.7	−1.7	−1.9
$\Delta V_S = -20\%$	dB	−1.7	−1.7	−1.7	−1.9

Tables 11.7 and 11.8 for the non-controlled and the controlled condition respectively. The deviations of the gain of the mixer stage and the R.F. stage (in non-controlled condition) of the tuner are given in Table 11.9.

The deviation from the nominal gain of the various stages as determined in Tables 11.5 to 11.9 can now be used to evaluate the deviation from nominal gain of the complete I.F. amplifier, the complete tuner and of the whole high frequency system of the television receiver under consideration.

11.8 Spreads in Gain of the I.F. Amplifiers

Using the results of the evaluation of spreads in Tables 11.6 and 11.7 the spreads in overall gain of the complete I.F. amplifier, expressed either in terms of transimpedance or in terms of transducer gain, can be determined. The results obtained for the reverse AGC amplifier and the forward AGC amplifier, in the non-controlled condition, are given in Table 11.10. Only the spread effects which have a negative effect on gain are evaluated. With a good approximation it may be assumed the positive spread effects have the same magnitude.

TABLE 11.7

Total spreads of power gain of 1st stage of I.F. amplifier — Non-controlled condition

Type of amplifier:		Reverse-gain controlled				Forward-gain controlled			
Spreads to be added:		linearly		statistically		linearly		statistically	
Description of spread:		$\Delta R = 5\%$	$\Delta R = 10\%$	$\Delta R = 5\%$	$\Delta R = 10\%$	$\Delta R = 5\%$	$\Delta R = 10\%$	$\Delta R = 5\%$	$\Delta R = 10\%$
due to transistor high-frequency properties and a.c. circuit components:	dB	—	—	-1.6	-1.6	—	—	-2.7	-2.7
due to temperature variations:	dB	-0.7	-0.7	—	—	-0.7	-0.7	—	—
due to resistance tolerances:	dB	—	—	-0.4	-0.9	—	—	-0.5	-0.9
due to supply voltage variations: $\Delta V_S = +10\%$	dB	+0.5	+0.5	—	—	+0.5	+0.5	—	—
$\Delta V_S = -20\%$	dB	-1.0	-1.0	—	—	-1.0	-1.0	—	—
due to $-I_{CBO}$ and $-I_B$ variations:	dB	—	—	-0.4	-0.4	—	—	-0.5	-0.5
due to V_{BE} variations:	dB	—	—	0	0	—	—	0	0
Total spread for: $\Delta V_S = +10\%$	dB	-0.2	-0.2	-1.7	-1.9	-0.2	-0.2	-2.8	-2.9
$\Delta V_S = 0$	dB	-0.7	-0.7	-1.7	-1.9	-0.7	-0.7	-2.8	-2.9
$\Delta V_S = -20\%$	dB	-1.7	-1.7	-1.7	-1.9	-1.7	-1.7	-2.8	-2.9

TABLE 11.8

Total spreads of power gain of the 1st stage of I.F. amplifier — Controlled condition

Type of amplifier		Reverse-gain controlled				Forward-gain controlled			
Spreads to be added:		linearly		statistically		linearly		statistically	
Description of spread:		$\Delta_R=5\%$	$\Delta_R=10\%$	$\Delta_R=5\%$	$\Delta_R=10\%$	$\Delta_R=5\%$	$\Delta_R=10\%$	$\Delta_R=5\%$	$\Delta_R=10\%$
due to transistor high-frequency properties and a.c. circuit components:	dB	—	—	-0.7	-0.7	—	—	-6.0	-6.0
due to temperature variations:	dB	-0.7	-0.7	—	—	-0.7	-0.7	—	—
due to resistance tolerances:	dB	—	—	-0.7	-1.4	—	—	-4.4	-8.7
due to supply voltage variations: $\Delta V_S = +10\%$	dB	+0.8	+0.8	—	—	-1.9	-1.9	—	—
$\Delta V_S = -20\%$	dB	-1.6	-1.6	—	—	+2.5	+2.5	—	—
due to $-I_{CBO}$ and $-I_B$ variations:	dB	—	—	0	0	—	—	-4.0	-4.0
due to V_{BB} variations:	dB	—	—	0	0	—	—	0	0
Total spread for: $\Delta V_S = +10\%$	dB	+0.1	+0.1	-1.0	-1.6	-2.6	-2.6	-8.5	-11.4
$\Delta V_S = 0$	dB	-0.7	-0.7	-1.0	-1.6	-0.7	-0.7	-8.5	-11.4
$\Delta V_S = -20\%$	dB	-2.3	-2.3	-1.0	-1.6	-1.8	+1.8	-8.5	-11.4

TABLE 11.9

Total spread of power gain of mixer stage and R.F. stage of tuner in non-controlled condition

Spreads to be added:		Mixer stage				R.F. stage			
		linearly		statistically		linearly		statistically	
Description of spread:		$\Delta_R=5\%$	$\Delta_R=10\%$	$\Delta_R=5\%$	$\Delta_R=10\%$	$\Delta_R=5\%$	$\Delta_R=10\%$	$\Delta_R=5\%$	$\Delta_R=10\%$
due to transistor high-frequency properties[1]:	dB	—	—	-1.0	-1.0	—	—	-2.0	-2.0
due to temperature variations[2]:	dB	—	—	—	—	—	—	—	—
due to resistance tolerances:	dB	—	—	-0.4	-0.8	—	—	-0.4	-0.8
due to supply voltage variations of:									
$\Delta V_S = +10\%$	dB	+0.5	+0.5	—	—	+0.5	+0.5	—	—
$\Delta V_S = 0$	dB	0	0	—	—	0	0	—	—
$\Delta V_S = -20\%$	dB	-0.5	-0.5	—	—	-0.5	-0.5	—	—
due to $-I_{CBO}$, $-I_B$ and V_{BE} variations[2]:									
Total spread for:									
$\Delta V_S = +10\%$	dB	+0.5	+0.5	-1.1	-1.3	+0.5	+0.5	-2.1	-2.2
$\Delta V_S = 0$	dB	0	0	-1.1	-1.3	0	0	-2.1	-2.2
$\Delta V_S = -20\%$	dB	-1.0	-1.0	-1.1	-1.3	-1.0	-1.0	-2.1	-2.2

[1] In Band III. In Band I these spread figures are slightly lower.

[2] Not investigated.

TABLE 11.10

Spreads in Gain of the I.F. Amplifier at Full Gain

Type of AGC:		Reverse		Forward	
Resistance tolerance:		$\Delta_R = 5\%$	$\Delta_R = 10\%$	$\Delta_R = 5\%$	$\Delta_R = 10\%$
$\Delta V_S = +10\%$	dB	−3.5	−3.9	−4.3	−4.6
$\Delta V_S = 0$	dB	−5.0	−5.4	−5.8	−6.1
$\Delta V_S = -20\%$	dB	−8.0	−8.4	−8.8	−9.1

TABLE 11.11

Spreads in Gain of the I.F. Amplifier at Reduced Gain

Type of AGC:		Reverse		Forward	
Resistance tolerance:		$\Delta_R = 5\%$	$\Delta_R = 10\%$	$\Delta_R = 5\%$	$\Delta_R = 10\%$
$\Delta V_S = +10\%$	dB	−2.9	−3.4	−11.8	−14.5
$\Delta V_S = 0$	dB	−4.7	−5.2	−10.9	−13.6
$\Delta V_S = -20\%$	dB	−8.3	−8.8	−10.4	−13.1

The spreads in gain have also been determined with the gain of the input stages reduced by approximately 30 dB. The results are entered in Table 11.11.

It follows from Table 11.10 that a spread of approximately 6 dB must be expected in the gain of an I.F. amplifier of the type considered, in the non-controlled condition. Table 11.11 indicates that the spread of the forward AGC amplifier in the controlled condition is much larger than that of the reverse AGC amplifier. From Table 11.8 it becomes apparent that this larger spread is mainly to be attributed to the large spread in gain at a certain value of the emitter current and to the large slope of the gain versus emitter current characteristic of the AGC transistor.

It follows, moreover, from Tables 11.10 and 11.11 that no considerable improvements can be obtained by using resistances with a tolerance of 5% instead of 10%.

Also the spreads in voltage level at the base of the first transistor of the reverse AGC and forward AGC amplifiers can now be determined. These

TABLE 11.12

Spreads of Input Voltage of First I.F. Transistor at a Fixed Cross-Over Point

Type of AGC:			Reverse				Forward			
Spreads to be added:			linearly		statistically		linearly		statistically	
Resistance tolerance:	ΔV_S	unit	$\Delta_R=5\%$	$\Delta_R=10\%$	$\Delta_R=5\%$	$\Delta_R=10\%$	$\Delta_R=5\%$	$\Delta_R=10\%$	$\Delta_R=5\%$	$\Delta_R=10\%$
Due to spreads in gain of the I.F. Amplifier (from Tables 11.6, 11.7 and 11.8):	+10%	dB	0.3	0.3	2.6	3.1	3.0	3.0	8.8	11.5
	0	dB	2.1	2.1	2.6	3.1	2.1	2.1	8.8	11.5
	−20%	dB	5.7	5.7	2.6	3.1	1.6	1.6	8.8	11.5
Due to spreads in input voltage of video detector (from Table 11.5):	+10%	dB	0.8	0.8	1.0	1.4	0.8	0.8	1.0	1.4
	0	dB	0	0	1.0	1.4	0	0	1.0	1.4
	−20%	dB	−1.6	−1.6	1.0	1.4	−1.6	−1.6	1.0	1.4
Due to spreads in the fixed cross-over point:	+10%	dB	0.8	0.8	0.6	1.1	0.8	0.8	0.6	1.1
	0	dB	0	0	0.6	1.1	0	0	0.6	1.1
	−20%	dB	−1.6	−1.6	0.6	1.1	−1.6	−1.6	0.6	1.1
Due to contrast expansion:		dB	1.0	1.0	—	—	1.0	1.0	—	—
Total spreads (separated):	+10%	dB	2.9	2.9	2.9	3.6	5.6	5.6	8.9	11.7
	0	dB	3.1	3.1	2.9	3.6	3.1	3.1	8.9	11.7
	−20%	dB	3.5	3.5	2.9	3.6	−0.6	−0.6	8.9	11.7
Total spreads (combined):	+10%	dB	$\Delta_R=5\%$ 5.8		$\Delta_R=10\%$ 6.5		$\Delta_R=5\%$ 14.5		$\Delta_R=10\%$ 17.3	
	0	dB	6.0		6.7		12.0		14.8	
	−20%	dB	6.4		7.1		8.3		11.1	

TABLE 11.13

Spreads of Input Voltage of First I.F. Transistor at an Adjusted Cross-Over Point

Type of AGC		Reverse		Forward	
Resistance tolerance:		$\Delta_R = 5\%$	$\Delta_R = 10\%$	$\Delta_R = 5\%$	$\Delta_R = 10\%$
$\Delta V_S = +10\%$	dB	3.9		6.6	
$\Delta V_S = 0$	dB	4.1		4.1	
$\Delta V_S = -20\%$	dB	4.5		0.4	

spreads comprise the spreads in gain in the controlled condition as given in Table 11.11 together with the spreads in the reference levels of the AGC circuit and in the voltage level at which cross-over of the AGC from I.F. amplifier to tuner takes place. The results of this evaluation are given in Table 11.12.

The point of cross-over of the AGC from tuner to I.F. amplifier is determined by a network usually consisting of two resistors in the AGC amplifier. If this network contains a pre-set potentiometer, the cross-over point can be adjusted at a certain value of the (reduced) gain of the I.F. amplifier. Such an adjustment, obviously, requires a dynamic gain measurement on each amplifier.

By means of a pre-set potentiometer for the cross-over point the spreads in voltage levels at the base of the first I.F. transistor at this cross-over point can be reduced considerably. The only spreads of these voltage levels that remain are due to variations of the ambient temperature, variations of the supply voltage and to maladjustment of the gain at the cross-over point. With an allowance of 1 dB for this possible maladjustment the spreads in input voltages become as given in Table 11.13.

11.9 The Output Voltage of the Video Detector

For determining the possible AGC range it will be assumed that the AGC operation starts at an output of the video detector of 3.5 V. This output refers to the top sync pulse level of the video signal of a television system with a transistorized video amplifier; a voltage gain of 30 may easily be reached in this amplifier. This means that the output of 3.5 V from the video detector is sufficient to provide full drive for a cathode ray tube, which requires a composite video signal of about 100 V peak-to-peak.

With respect to non-linearity of the last stage of the I.F. amplifier it fol-

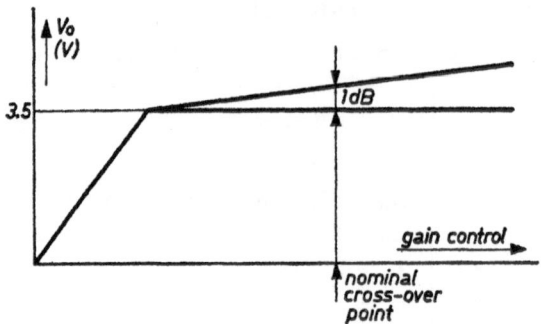

Fig. 11.8. Practical AGC characteristic of an I.F. amplifier.

lows from Figs. 17, 18, 30 and 31 in Chapter 10 that an output voltage of 3.5 V can easily be handled.

If contrast control of the television receiver is achieved via the AGC circuit, it must be checked whether at minimum contrast the input to the video detector is sufficiently large in view of inter-carrier mixing. It can be calculated that for a contrast control of 12 dB, the minimum input to the video detector becomes 1.35 V r.m.s. which is sufficiently large, as experiments have proved.

For determining voltage levels during automatic gain control in I.F. amplifier and tuner, it must furthermore be taken into account that the output of the video amplifier slightly increases over the AGC range (see Fig. 11.8). For this increase until the cross-over point of the AGC from I.F. amplifier to tuner is reached, we will include an allowance of 1 dB in our calculations (see Table 11.12).

11.10 The Possible AGC Range of the I.F. Amplifier

According to Figs. 10.5 and 10.6 the voltages at the bases of the input transistors of the I.F. amplifiers for an output of 3.5 V are, under nominal conditions:

— reverse AGC amplifier: 73μV,

— forward AGC amplifier: 69μV.

Taking into account the voltage levels that are assumed to be permissible as referred to in sub-section 11.1.1, the spreads as given in Tables 11.12 and 11.13, and the 1 dB allowance for contrast expansion, the possible AGC range of the I.F. amplifier can be calculated. The results are given in Table 11.14.

It is evident from the table that the possible control range of the I.F. ampli-

TABLE 11.14

Possible AGC Range of I.F. Amplifier

Type of AGC:		Reverse		Forward	
Resistance tolerance:		$\Delta_R = 5\%$	$\Delta_R = 10\%$	$\Delta_R = 5\%$	$\Delta_R = 10\%$
with a fixed cross-over point:					
non-stab. power supply:	dB	36.9	36.6	38.3	35.4
stabilized power supply:	dB	36.7	36.0	40.7	37.9
with an adjusted cross-over point:					
non-stab. power supply:	dB	37.2		46.1	
stabilized power supply:	dB	37.6		48.6	

fier is larger with forward AGC than with reverse AGC. Furthermore, it follows that with forward gain control a considerable improvement in control range can be obtained by adjusting the gain at the cross-over point by means of a pre-set potentiometer.

Whether the AGC range as calculated here may actually be used depends on the signal handling capabilities of the mixer and the R.F. stages of the tuner. If overloading of the tuner occurs earlier than that of the I.F. amplifier, the AGC range of the latter must be made smaller.

11.11 The Overall Gain of Tuner and I.F. Amplifier

To obtain the overall gain of tuner and I.F. amplifier the transimpedance of the I.F. amplifier must be multiplied by the transadmittance of the tuner.

According to Figs. 10.5 and 10.6 the nominal transimpedance figures are:

for the reverse AGC amplifier : $|Z_t| = 11.0$ MΩ, and

for the forward AGC amplifier: $|Z_t| = 8.8$ MΩ.

The transadmittances of the tuner are given in Fig. 11.4. These are:

for Channel 4 : $|Y_t| = 26$ m℧, and

for Channel 9 : $|Y_t| = 12$ m℧.

TABLE 11.15

Overall Sensitivity of Tuner and I.F. amplifier

Type of AGC in I.F. amplifier	Reverse	Forward
In Band I Channel 4	7 μV	9 μV
In Band III Channel 9	15 μV	19 μV

It can now be calculated what input signal is required to produce a signal of 1 V r.m.s. at the input terminals of the video detector. These input voltages, given in Table 11.15, are expressed in terms of aerial e.m.f. at the picture carrier frequency. At this frequency the overall amplitude response curve is 6 dB down with respect to the mid-band frequency.

11.12 Spreads in Overall Gain of Tuner and I.F. Amplifier

By combining the spreads in gain of the I.F. amplifier in the non-controlled condition as determined in Table 11.10 and the spreads in power gain of the mixer and R.F. stages of the tuner as given in Table 11.9, the spreads in overall gain are obtained. The results are given in Table 11.16. Again only spread effects which are negative with respect to gain are determined. With a good approximation it may be assumed that the positive spread effects are of the same magnitude.

TABLE 11.16

Spreads in Overall Sensitivity of Tuner and I.F. Amplifier

I.F. Amplifier with	reverse AGC		forward AGC	
resistance tolerance	$\Delta_R=5\%$	$\Delta_R=10\%$	$\Delta_R=5\%$	$\Delta_R=10\%$
$\Delta V_S = +10\%$	− 3.4 dB	− 3.8 dB	− 4.0 dB	− 4.4 dB
$\Delta V_S = 0$	− 5.9 dB	− 6.3 dB	− 6.5 dB	− 6.9 dB
$\Delta V_S = -20\%$	−10.9 dB	−11.3 dB	−11.5 dB	−11.9 dB

It is evident that with respect to overall sensitivity no advantage is gained by using resistances with a tolerance of 5 %.

Furthermore, it follows from Tables 11.15 and 11.16 that there is no remarkable difference in performance between the arrangement with the reverse AGC and the forward AGC I.F. amplifier.

11.13 Possible Delay of AGC Cross-Over Point

During gain control of the I.F. amplifier the signal levels at the R.F. transistor and the mixer transistor of the tuner increase. After a certain amount of I.F. gain control, signal levels at which overloading occurs will therefore be reached. This means that, as far as overloading of the mixer stage or R.F. stage of the tuner is concerned, the transfer of the AGC from I.F. amplifier to tuner may be delayed up to this point. In this section the possible delays will be determined for the different arrangements under consideration.

11.13.1 THE MIXER STAGE

Fig. 11.9 shows a schematic diagram of the tuner and the first I.F. stage. The signal levels at the various points and under the various conditions of interest for our analyses are given for the starting point of the AGC (i.e. at the point at which the output voltage of the video detector becomes 3.5 V). The voltage levels quoted apply to the strictly nominal case and are related to the picture carrier of the television signal. According to sub-section 11.1.1 the permissible signal level at the mixer transistor is assumed to be 20 mV at the picture carrier. With reference to Fig. 11.9 it then follows that the possible

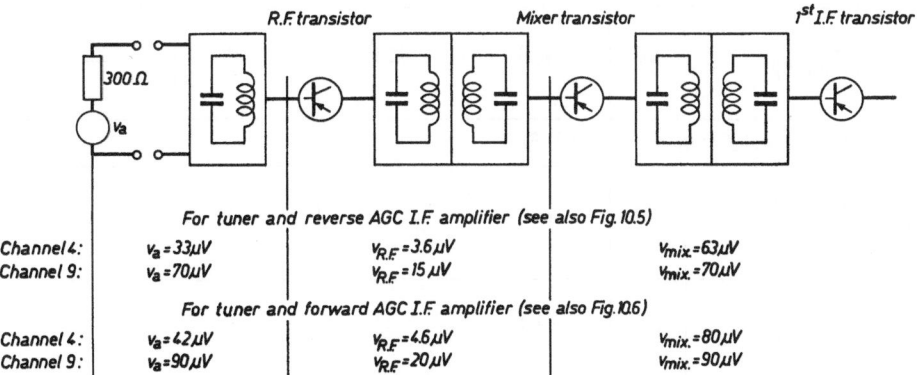

Fig. 11.9. Nominal voltage levels in the tuner at the starting point of the AGC on the I.F. amplifier.

delay of the AGC cross-over point is smallest in Channel 9 and that it amounts to:
— for the reverse AGC I.F. amplifier : 49 dB, and
— for the forward AGC I.F. amplifier: 47 dB,
provided no spreads occur.
These figures must be decreased with the spreads which occur in the input voltage to the mixer transistor. With the aid of Tables 11.8 and 11.12 these spreads can be determined. The possible delay at the mixer transistor of the AGC cross-over point, including an allowance of 2 dB for contrast expansion, is given in Table 11.17 for cases *without* and with (dynamic) adjustment of the cross-over point.

11.13.2 THE R.F. STAGE

Considering the signal levels at the aerial terminals at the starting point of the AGC as given in Fig. 11.9 and the signal levels that are assumed to be

TABLE 11.17

Possible Delay at the Mixer Transistor of AGC Cross-over Point.

I.F. amplifier with		reverse AGC		forward AGC	
resistance tolerance		$\Delta_R = 5\%$	$\Delta_R = 10\%$	$\Delta_R = 5\%$	$\Delta_R = 10\%$
without adjustment of cross-over point [1]):					
non-stabilized power supply	dB	41.4	40.7	34.0	30.3
stabilized power-supply	dB	42.0	42.1	35.0	32.2
with adjustment of cross-over point [1]) [2]):					
non-stabilized power supply	dB	43.5		40.9	
stabilized power supply	dB	44.9		42.9	

[1]) With an allowance of 2dB for contrast expansion.
[2]) These figures include an allowance of 1 dB for possible maladjustment.

TABLE 11.18

Possible Delay at the Aerial Terminals of the AGC Cross-Over Point

I.F. amplifier with:		reverse AGC		forward AGC	
Resistance tolerances:		$\Delta_R = 5\%$	$\Delta_R = 10\%$	$\Delta_R = 5\%$	$\Delta_R = 10\%$
with a fixed cross-over point [1]:					
non-stabilized power sypply	dB	46.0	45.3	39.2	36.5
stabilized power supply	dB	48.4	47.7	40.7	38.0
with an adjusted cross-over point [1] [2]:					
non-stabilized power supply	dB	48.7		47.4	
stabilized power supply	dB	51.1		48.9	

[1] With an allowance of 2 dB for contrast expansion.

[2] These figures include an allowance of 1 dB for a possible maladjustment.

permissible (see subsection 11.3.1), it follows that the possible delay of the AGC cross-over point is smaller in Band III than in Band I. It is therefore sufficient to consider Band III only. In the strictly nominal case the possible delay of the cross-over point would be:

— with the I.F. amplifier with reverse AGC : 55.2 dB, and

— with the I.F. amplifier with forward AGC: 53.0 dB.

These figures have to be reduced with the spread in signal voltage, which can be determined by combining Tables 11.8 and 11.12. The possible delay at the aerial terminals of the AGC cross-over point then becomes as set out in Table 11.18.

11.14 The Choice of the Nominal Cross-Over Point

After having evaluated the effects of deviations from nominal conditions in a combination of tuner and I.F. amplifier on the signal levels that can be handled in practice, we are in the position to make an optimum choice of the AGC cross-over point. Comparing the figures for the possible AGC range of the I.F. amplifier itself in Table 11.14 and those for the possible delay of the cross-over point in view of overloading in the mixer stage or R.F. stage of the tuner, given in Tables 11.17 and 11.18 respectively, the following conclusions can be drawn:

In the combination of tuner and I.F. amplifier with reverse gain control, the delay of the cross-over point is determined by the signal handling capability of the first I.F. transistor. The AGC amplifier must therefore be designed in such a way that cross-over takes place at the relevant amount of I.F. gain control given in Table 11.14.

In the combination of tuner and I.F. amplifier with forward gain control the signal handling capability of the mixer stage determines the delay of the cross-over point. The various figures given in Table 11.17 must therefore be taken in relevant cases as the amount of gain control of the *I.F. amplifier* at which cross-over of the AGC should occur. The above conclusions are also valid in cases in which the cross-over point is adjusted in each receiver by means of a measurement of overall gain. It follows from the various tables that, especially in the case of an I.F. amplifier with forward gain control, a considerable improvement of the AGC range can be obtained by such an adjustment.

CHAPTER 12

THE EFFECTS OF SPREADS IN TRANSISTOR ADMITTANCE PARAMETERS

In this Chapter an analysis of the extent to which spreads in transistor admittance parameters affect the performance of the amplifier with reverse AGC in the non controlled condition will be developed, and may be regarded as an example of how, in general, such an analysis can be carried out.

The transimpedance, the amplitude response curve and the envelope delay curve of the amplifier will be determined with transistors of the type AF179, selected at random, inserted in the amplifier designed for nominal transistors (see Chapters 10 and 11).

It will be assumed that spreads occur only in the admittance parameters of the transistors. This means that no spreads in transistor biasing points and no spreads or tolerances in the high-frequency properties of the bandpass filters are assumed to be present. The effects of adjacent channel and own sound wave-traps of the amplifier are not taken into account in this investigation.

It will, moreover, be assumed that the amplifier is aligned for each set of transistors inserted in the amplifier. This conforms with normal practice during the manufacturing process of such amplifiers.

12.1 Amplifier Determinants with Spreads in Transistor Parameters

The method of incorporating spreads in the analysis of bandpass amplifiers consists of an extension of the amplifier determinants considered in the various chapters of book I.

The amplifier determinants contain normalized admittances of the various tuned circuits in the main diagonal and regeneration coefficients of the transistor and coefficients of coupling of the double-tuned bandpass filters in an adjacent sub-diagonal. In the determinant below which applies to the amplifier configuration under consideration (III) the various factors are indicated by means of squares, triangles and circles respectively. We will now consider how spread in transistor parameters affects these factors. For this purpose the nominal value of a parameter will be denoted by a suffix N. Furthermore, spread factors for the transistor input and output dampings and the tap at the input side of the transistor are defined.

$$\delta(x) = \begin{vmatrix} \boxed{7} & ③ & 0 & 0 & 0 & 0 & 0 \\ 1 & \boxed{6} & \triangle{3} & 0 & 0 & 0 & 0 \\ 0 & 1 & \boxed{5} & ② & 0 & 0 & 0 \\ 0 & 0 & 1 & \boxed{4} & \triangle{2} & 0 & 0 \\ 0 & 0 & 0 & 1 & \boxed{3} & ① & 0 \\ 0 & 0 & 0 & 0 & 1 & \boxed{2} & \triangle{1} \\ 0 & 0 & 0 & 0 & 0 & 1 & \boxed{1} \end{vmatrix} \qquad (12.1.1)$$

$$-- M_2(x)$$

12.1.1 INPUT-DAMPING SPREAD FACTOR m_i

An input-damping spread factor is defined for each transistor in the amplifier as:

$$m_i = \frac{g_{11}}{g_{11\,N}} . \qquad (12.1.2)$$

A second index will be added to the symbol m_i to denote to which transistor in the amplifier it belongs. The symbol $_2m_i$, for instance, denotes the input-damping spread factor of the second transistor in the amplifier. The numbering of the transistors will be consecutive, starting at the output side.

12.1.2 OUTPUT-DAMPING SPREAD FACTOR m_o

Similarly, an output-damping spread factor is defined as:

$$m_o = \frac{g_{22}}{g_{22\,N}} . \qquad (12.1.3)$$

12.1.3 INPUT-TAP SPREAD FACTOR m_n

In the amplifier under consideration a capacitive tap is used at the input terminals of the various transistors, see Fig. 12.1. For determining the effects of spreads on the tapping ratio we assume that the effect of the residual feedback of the neutralized transistors on the respective input admittances is negligibly small.

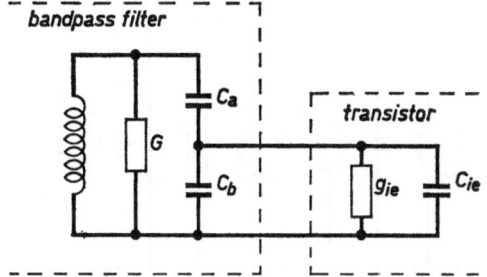

Fig. 12.1. Schematic diagram of part of an I.F. amplifier showing the capacitive taps on the circuit at the input terminals of a transistor.

The nominal value of the tapping ratio equals (disregarding effects due to the heavy loading):

$$n_N = \frac{C_a}{C_a + C_b + C_{11\,N}}.$$
(12.1.4)

The quantities C_a and C_b are defined in Fig. 12.1.

The damping due to g_{11} at the top of the tuned circuit then becomes for a nominal transistor:

$$n^2_N \cdot g_{11N},$$
(12.1.5)

and for an arbitrary transistor:

$$n^2 \cdot g_{11}.$$
(12.1.6)

In the last expression:

$$n = \frac{C_a}{C_a + C_b + C_{11}}.$$
(12.1.7)

Now an input-tap spread factor is defined as:

$$m_n = \frac{n}{n_N}.$$
(12.1.8)

Substitution of the above expressions gives:

$$m_n = \frac{1 + \dfrac{C_{11\,N}}{C_a + C_b}}{1 + \dfrac{C_{11}}{C_a + C_b}}.$$
(12.1.9)

As appears from the complete circuit diagram shown in Fig. 10.7, part of the damping G^* on the secondary of a double-tuned bandpass filter is obtained by connecting a resistor across the tap. Before the foregoing equations are applied, the damping of this resistor must be transformed to the top of the

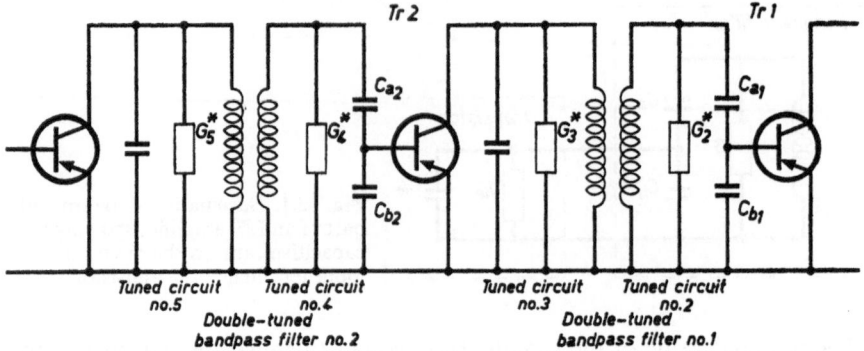

Fig. 12.2. Schematic diagram of part of an I.F. amplifier showing the numbering of the various elements employed.

respective tuned circuits with the nominal tapping ratio and added to the damping of the coil to obtain G^*. Obviously, expressions can be written down which include this transformation, but these expressions would be less general.

12.2 Transistor Regeneration Coefficients

In Fig. 12.2 a part of the simplified equivalent circuit diagram of the amplifier under consideration is given. The regeneration coefficient of transistor 2 is defined as:

$$T_2 = \frac{|_2 y_{12}| \cdot |_2 y_{21}|}{G_3 \cdot G_4} \tag{12.2.1}$$

Now: $\quad G_3 = g_{22} + G_3^*,$

$$= \frac{_2 g_{22\,N}}{\Phi_{p1}} - _2 g_{22\,N} + _2 g_{22}.$$

With Eq. (12.1.3) this becomes:

$$G_3 = _2 g_{22\,N} \cdot \left(\frac{1}{\Phi_{p1}} + _2 m_0 - 1\right). \tag{12.2.2}$$

Furthermore: $\quad G_4 = _2 g_{11} + \dfrac{G_4^*}{n^2}.$

The value of G_4^* (which remains nominal in all cases) follows from:

$$\frac{G_4^*}{n^2_N} = \frac{g_{11\,N}}{\Phi_{s2}} - g_{11\,N}.$$

With Eqs. (12.1.2) and (12.2.2) we then obtain:

$$G_4 = {}_2g_{11\,N}\Big\{ {}_2m_i + ({}_2m_n)^2\,(\frac{1}{\varPhi_{s2}} - 1)\Big\}. \qquad (12.2.3)$$

Also we have for the intrinsic regeneration coefficient of the transistor:

$$t_2 = \frac{|{}_2y_{12}| \cdot |{}_2y_{21}|}{{}_2g_{11} \cdot {}_2g_{22}}. \qquad (12.2.4)$$

Combining the various expressions derived in this section we find for the regeneration coefficient of the transistor in the second stage of the amplifier:

$$T_2 = t_2 \frac{{}_2m_i \cdot {}_2m_o}{({}_2m_o + \frac{1}{\varPhi_{p1}} - 1)\big\{ {}_2m_i + ({}_2m_n)^2\,(\frac{1}{\varPhi_{s2}} - 1)\big\}}. \qquad (12.2.5)$$

At the position indicated by $\triangle\!\!\!2$ in the determinant of Eq. (12.1.1) we thus obtain:

$$\triangle\!\!\!2 \equiv T_2 \cdot \exp(\mathrm{j}\,\varTheta_2), \qquad (12.2.6)$$

in which T_2 follows from Eq. (12.2.5).

12.3 The Coefficient of Coupling of the Double-Tuned Bandpass Filters

The coefficient of coupling of double-tuned bandpass filters no. 1 in the circuit given in Fig. 12.2 is:

$$q_1{}^2 \equiv k_1{}^2 \cdot Q_2\,Q_3. \qquad (12.3.1)$$

In this expression the quantity $k_1{}^2$ denotes the coupling factor of primary and secondary of bandpass filter no. 1.

The quality factors Q_2 and Q_3 are inversely proportional to the dampings G_2 and G_3 respectively.

If $q_1{}^2{}_N$ denotes the nominal value of the coefficient of coupling, we obtain with arbitrary transistors in stages 1 and 2:

$$q_1{}^2 = q_1{}^2{}_N \cdot \frac{G_{2N}}{G_2} \cdot \frac{G_{3N} \cdot n_N{}^2}{G_3 \cdot n^2}. \qquad (12.3.2)$$

With Eqs. (12.1.8), (12.2.2) and (12.2.3) we obtain:

$$q_1{}^2 = \frac{q_1{}^2{}_N \cdot {}_1m_n{}^2}{\varPhi_{p1}\,\varPhi_{s1} \cdot (\frac{1}{\varPhi_{p1}} + {}_2m_o - 1)\,\big\}{}_1m_i + {}_1m_n{}^2\,(\frac{1}{\varPhi_{s1}} - 1)\big\}}. \qquad (12.3.3)$$

At the position ① in Eq. (12.1.1) we then find:

$$① \equiv -q_1^2, \tag{12.3.4}$$

in which q_1 is given by Eq. (12.3.3).

12.4 The Normalized Admittances of the Tuned Circuits

The main diagonal of the determinant of Eq. (12.1.1) contains the normalized admittances of the various tuned circuits in the amplifier. These normalized admittances are also subject to spreads in transistor parameters.

Consider circuit no. 3 of the amplifier represented in Fig. 12.2. For an arbitrary transistor T_{r2} the normalized admittance can be written as:

$$y_3 = 1 + j \, x_{2N} \cdot \frac{G_{3N}}{G_3} \cdot \tag{12.4.1}$$

With Eq. (12.2.2) this becomes:

$$y_3 = 1 + j \, x_{2N} \cdot \frac{1}{1 + (_2m_o - 1) \, \Phi_{p1}} \cdot \tag{12.4.2}$$

For circuit no. 4 we obtain with Eq. (12.2.3):

$$y_4 = 1 + j \, x_{4N} \cdot \frac{1}{\Phi_{s_2} \left\{ _2m_1 + _2m_n^2 \left(\frac{1}{\Phi_{s_2}} - 1 \right) \right\}} \tag{12.4.3}$$

To these normalized admittances, tuning correction terms should be added corresponding with the tuning method employed. In our analysis only tuning method B will be considered. Then tuning correction terms appear only for the tuned circuits connected to the input terminals of the various transistors (i.e. for tuned circuits nos. 2, 4 and 6 in the amplifier under consideration). For tuned circuit no. 4 this tuning correction term equals:

$$x_4' = T_2 \sin \Theta_2 \cdot \frac{M_2 \, (x = 0)}{M_3 \, (x = 0)}, \tag{12.4.4}$$

in which M_2 and M_3 are minor determinants of the reduced amplifier determinant of Eq. (12.1.1).

Eq. (12.4.3) then becomes:

$$y_4 = 1 + j \left[x_{4N} \cdot \frac{1}{\Phi_{s2} \left\{ _2m_t + _2m_n^2 \left(\frac{1}{\Phi_{s2}} - 1 \right) \right\}} + \right. $$
$$\left. + T_2 \sin \Theta_2 \frac{M_2 \, (x = 0)}{M_3 \, (x = 0)} \right] \cdot \tag{12.4.5}$$

12.5 Survey of Terms of the Interchangeability Determinant

In Table 12.1 the various terms of the interchangeability determinant presented in Eq. (12.1.1) are set out. These terms are derived by analogy with the considerations of the preceding sections.

In deriving the various expressions the following assumptions are made:

TABLE 12.1 Terms of the Interchangeability Determinant

Normalized admittances:

$$\boxed{1} \equiv 1 + j\,0.75x \cdot \frac{1}{1 + (_1m_o - 1) \cdot {}_1\Phi_{11}}$$

$$\boxed{2} \equiv 1 + j\left[x\sqrt{r} \cdot \frac{1}{\Phi_{s1}\left\{ {}_1m_i + {}_1m_n^2 \left(\frac{1}{\Phi_{s1}} - 1 \right) \right\}} + T_1 \sin\Theta \right],$$

$$\boxed{3} \equiv 1 + j\left[x\,\frac{1}{\sqrt{r}} \cdot \frac{1}{1 + (_2m_o - 1)\,\Phi_{p1}} \right],$$

$$\boxed{4} \equiv 1 + j\left[x\sqrt{r} \cdot \frac{1}{\Phi_{s2}\left\{ {}_2m_i + {}_2m_n^2 \left(\frac{1}{\Phi_{s2}} - 1 \right) \right\}} + T_2 \sin\Theta_2 \cdot \frac{M_2\,(x = 0)}{M_3\,(x = 0)} \right],$$

$$\boxed{5} \equiv 1 + j\left[x \cdot \frac{1}{\sqrt{r}} \cdot \frac{1}{1 + (_3m_o - 1)\,\Phi_{p2}} \right],$$

$$\boxed{6} \equiv 1 + j\left[x\sqrt{r} \cdot \frac{1}{\Phi_{s3}\left\{ {}_3m_i + {}_3m_n^2 \left(\frac{1}{\Phi_{s3}} - 1 \right) \right\}} + T_3 \sin\Theta_3 \cdot \frac{M_4\,(x = 0)}{M_5\,(x = 0)} \right],$$

$$\boxed{7} \equiv 1 + j\left[x \cdot \frac{1}{\sqrt{r}} \right].$$

Coefficients of coupling of the double-tuned bandpass filters:

$$\textcircled{1} \equiv \frac{-q\,\overset{2}{N} \cdot {}_1m_n^2}{\Phi_{p1}\,\Phi_{s1}\left(\frac{1}{\Phi_{p1}} + {}_2m_o - 1 \right)\left\{ {}_1m_i + {}_1m_n^2 \left(\frac{1}{\Phi_{s1}} - 1 \right) \right\}},$$

$$\boxed{2} = \frac{-q \, {}_2N^2 \cdot {}_2m_n^2}{\Phi_{p2} \, \Phi_{s2} \left(\dfrac{1}{\Phi_{p2}} + {}_2m_0 - 1 \right) \left\{ {}_2m_i + {}_2m_n^2 \left(\dfrac{1}{\Phi_{s2}} - 1 \right) \right\}},$$

$$\boxed{3} = \frac{-q \, {}_3N^2 \cdot {}_3m_n^2}{\Phi_{p3} \, \Phi_{s3} \left(\dfrac{1}{\Phi_{p3}} + {}_3m_0 - 1 \right) \left\{ {}_3m_i + {}_3m_n^2 \left(\dfrac{1}{\Phi_{s3}} - 1 \right) \right\}}.$$

Transistor regeneration.coefficients:

$$\boxed{\triangle 1} = t_1 \exp(j\Theta_1) \cdot \frac{{}_1m_i \cdot {}_1m_0}{\left({}_1m_0 + \dfrac{1}{{}_1\Phi_{11}} - 1 \right) \left\{ {}_1m_i + {}_1m_n^2 \left(\dfrac{1}{\Phi_{s1}} - 1 \right) \right\}} = T_1 \exp(j\Theta_1),$$

$$\boxed{\triangle 2} = t_2 \exp(j\Theta_2) \cdot \frac{{}_2m_i \cdot {}_2m_0}{\left({}_2m_0 + \dfrac{1}{\Phi_{p1}} - 1 \right) \left\{ {}_2m_i + {}_2m_n^2 \left(\dfrac{1}{\Phi_{s2}} - 1 \right) \right\}} = T_2 \exp(j\Theta_2),$$

$$\boxed{\triangle 3} = t_3 \exp(j\Theta_3) \cdot \frac{{}_3m_i \cdot {}_3m_0}{\left({}_3m_0 + \dfrac{1}{\Phi_{p2}} - 1 \right) \left\{ {}_3m_i + {}_3m_n^2 \left(\dfrac{1}{\Phi_{s3}} - 1 \right) \right\}} = T_3 \exp(j\Theta_3).$$

${}_1\Phi_{11}$ = dampingratio of the last tuned circuit of the amplifier.

— the tuning of the amplifier is carried out according to method B,

— the double-tuned bandpass filters are identical,

— the ratio between the secondary and primary quality factors of the double-tuned bandpass filters is given by $r = Q_s/Q_p$

— the quality factor of the last (single-) tuned bandpass filter of the amplifier is given by:

$$Q_1 = 0.75 \sqrt{Q_p \, O_s}.$$

12.6 Results of an Interchangeability Analysis

A vision I.F. amplifier with three transistors AF179 was investigated by the method described in the preceding sections. The design of this amplifier taking into account average values of the admittance parameters of the transistors, was described in Chapter 10.

The investigation was based on measurements of the admittance parameters of a large number of transistors AF179 at the proper biasing point and

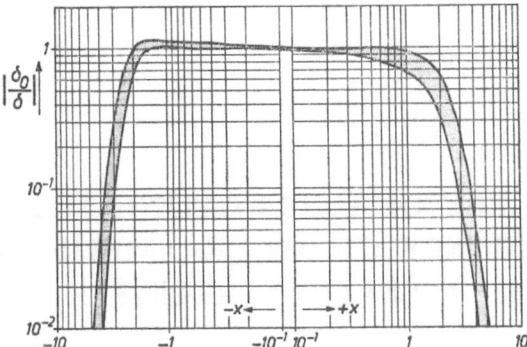

Fig. 12.3. Amplitude response curves of a three-stage vision I.F. amplifier equipped with transistors of the type AF179 which were selected at randon from a large batch. The area between the most extreme curves is indicated by shading.

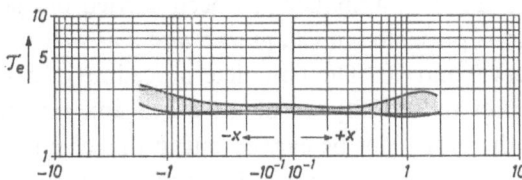

Fig. 12.4. Envelope delay characteristic for the same amplifier and for the same conditions as in Fig. 12.3.

signal frequency. From the measured parameters the interchangeability coefficients m_i, m_o and m_n, as required for the various terms in the inter-changeability determinant, are calculated.

It is assumed that the neutralizing network of the transistors in the prac-tical amplifier exactly cancels the feedback of transistors with an average value for this parameter. In the analysis the difference between the measured feedback admittance of each transistor and that of the "average" transistor has been taken into account.

For the damping ratios Φ_p, Φ_s and $_1\Phi_{11}$ appearing in the various expres-sions in Table 12.1, nominal values of these quantities as determined in Chapter 10 are taken into account.

The transistors are divided into groups of three in an arbitrary way and each group is assumed to be inserted in the amplifier. In each case the ampli-tude response and envelope delay curves of the amplifier are calculated. In Figs. 12.3 and 12.4 the various curves are drawn. The area between the most extreme curves obtained is indicated by shading.

CHAPTER 13

EXAMPLE OF CALCULATING THE GAIN PERFORMANCE OF AN I.F. AMPLIFIER FOR A TELEVISION RECEIVER

As described in Chapter 8, the gain performance of an I.F. amplifier can best be calculated by making use of the transimpedance concept. In this chapter this will be illustrated by calculating the performance of a vision I.F. amplifier equipped with the silicon planar transistors BF 167 and BF173.

The design of this I.F. amplifier has been carried out according to the basic design criteria presented in Chapters 5 and 8 as well as in the examples given in Chapters 9 and 10. The complete procedure of designing this amplifier with respect to stability, response curve and interchangeability of transistors will, therefore, not be carried out in this chapter.

Apart from the gain calculation of this amplifier, only a short description of the construction and the measured performance will be given. An extensive description of the design and construction of vision I.F. amplifiers employing the transistors BF 167 and BF 173 has been published elsewhere*.

13.1 Description of the Amplifier

The amplifier consists of three stages and employs double-tuned bandpass filters as interstage coupling networks. In Fig. 13.1. a block diagram of the amplifier is shown. The first stage is equipped with a transistor BF167. This transistor is very well suitable for gain control in the so-called forward-mode (see Chapter 7). The second and third stages of the amplifier are equipped with transistors BF173.

In Fig. 13.2 a complete circuit diagram of the amplifier is shown. No neutralization has been employed in any of the stages.

Adjacent channel and own sound trap circuits have not been incorporated in this design.

---

*) W. TH. H. HETTERSCHEID and A. H. J. NIEVEEN VAN DIJKUM, *The Silicon Planar Transistors BF167 and BF173 in Vision I.F. Amplifiers,* Electronic Applications, Vol. 26, No. 2, 1965-1966

Additional material from *Designing Transistor I. F. Amplifiers,*
ISBN 978-3-662-38672-9 (978-3-662-38672-9_OSFO2)

is available at http://extras.springer.com

Additional material from Designing Fordism ... Competitive ...
ISBN 978-3-662-38972-9 (978-3-662-38972-9 USDF...
is available at http://extras.springer.com

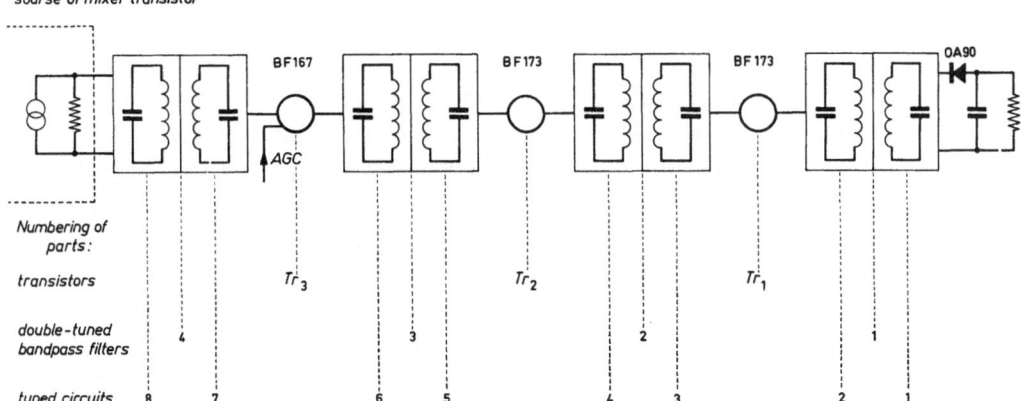

Fig. 13.1. Schematic representation of a three-stage I.F. amplifier showing the numbering of the constituting parts.

13.1.1 TRANSISTOR BIASING POINTS

The biasing points of the various transistors of the I.F. amplifier have been adjusted to such values that optimum performance of the respective stages is obtained. The transistor BF173 in the output stage is biased at $V_{CE} = 17$ V and $I_E = -6.8$ mA.

The nominal collector dissipation of 115 mW leads, when transistor biasing resistors and supply voltage spreads are taken into account, to a maximum dissipation of 175 mW. This dissipation is allowable for the BF173 without further cooling attachment up to ambient temperatures of 60°C.

The relatively high dissipation of this transistor has been chosen in order to deliver the maximum amount of power into the output double-tuned bandpass filter. This, in turn, gives the largest possible output voltage from the video detector. The current setting of 6.8 mA provides the maximum power gain in this output stage. The transistor BF173 in the middle stage is biased at $V_{CE} = 10$ V and $I_E = -7$ mA.

At this biasing point the best compromise between a high stage gain and a high safety factor against self-oscillations is found. The transistor BF167 in the control stage has a biasing point of $V_{CE} = 9$ V and $I_E = -4$ mA in the non-controlled condition. At 60 dB gain control the biasing point becomes approximately $V_{CE} = 1$ V and $I_E = -6$ mA.

13.1.2 THE OUTPUT STAGE

The transistor BF173 in the output stage drives a double-tuned bandpass filter with indirect inductive coupling. A diode OA 90 is used as video detector. The detector load impedance is 2700 Ω in parallel with 10 pF.

To prevent radiation of harmonics of the picture-carrier frequency generated by the detector circuit, the complete output bandpass filter is contained in a screening can. The indirect inductive coupling of this bandpass filter together with the collector decoupling network R_{17}, C_{21}, prevents these harmonics from reaching the supply rail of the amplifier. The network C_{24}, C_{27}, L_{10}, L_{11} and L_{12} suppresses signals of picture-carrier frequency and its harmonics before they reach the detector output terminals.

At the input terminals of the output stage transistor a double-tuned bandpass filter with capacitive top coupling is employed. The secondary of the bandpass filter is capacitively tapped to the base of the output transistor. The tap capacitors and the various dampings at input and output terminals of the transistors have been given such values that the stability factor of this stage is very large. For a discussion on these stability aspects reference is made to Book I, Chapter 12.

13.1.3 THE MIDDLE STAGE

The BF173 in the middle stage of the amplifier is biased at $V_{CE} = 10$ V and $I_E = -7$ mA. No gain control is applied. The circuitry of the middle stage is rather conventional. Capacitive top coupling is used in the bandpass filters at input and output terminals of the second transistor. In series with the coupling capacitor a resistor has been added to improve the symmetry of the response curve.

The primary and secondary quality factors of the collector bandpass filter of the second transistor as well as its coupling factor have been chosen such as to make an optimum compromise between response curve requirements and gain.

The quality factors and the coupling factor of the bandpass filter between the control stage and the middle stage of the amplifier are largely governed by response curve requirements during gain control. This will be discussed in the following sub-section.

Regarding the method of achieving sufficient stability of this stage the same remark can be made as with respect to the output stage; see the preceding sub-section.

13.1.4 THE INPUT STAGE

In the input stage of the amplifier, gain control is achieved by reducing the collector-emitter voltage from 9 V to approximately 1 V. The emitter current of the BF167 increases during this gain control from 4 mA to approximately 6 mA. The gain control range is then in excess of 60 dB. During gain control the input and output admittances of the BF167 vary considerably. Special measures have, therefore, been taken to keep the variations in response curve within narrow limits. These measures consist of:

— A type of current drive at the base of the control transistor. This current drive is achieved by employing a capacitor of suitably small value between the base of the transistor and the top of the parallel-tuned circuit which forms the secondary of the input double-tuned bandpass filter. In the circuit diagram of Fig. 13.2. this capacitor is denoted as C_4.

— A suitable choice of the ratio of primary and secondary quality factors of the double-tuned bandpass filter at the output side of the control transistor. An improvement in response curve constancy is obtained by making in the non-controlled condition, the quality factor of the tuned circuit adjacent to the transistor larger than the corresponding quality factor of the outer tuned circuit.

Obviously, the coupling factor of the double-tuned bandpass filter has been accomodated to the unequal quality factors.

By applying these measures it has been found that the response curve of the amplifier remains nearly constant over the entire control range. This will also become apparent from the results of measurements given in Section 13.3.

In the input double-tuned bandpass filter a capacitive bottom-end coupling has been employed. This type of coupling facilitates the use of a cable for connection of the bottom-end of the I.F. circuit located in the tuner section to the secondary of the input bandpass filter located in the I.F. amplifier. The capacitance of this cable then forms part of the total coupling capacitance. The value of the capacitor C_2 in Fig. 13.2 should therefore be adapted accordingly.

13.2 Gain Calculation

For calculating the gain performance of the amplifier it is required to know the admittance parameters of the transistors at their respective biasing points as well as the primary and secondary dampings of the double-tuned bandpass filters and their coefficients of coupling.

The admittance parameters of the transistors are given in Table 13.1. The values of these admittance parameters are derived from the tentative publication data of the transistors BF167 and BF173.

The primary and secondary dampings of the double-tuned bandpass filter

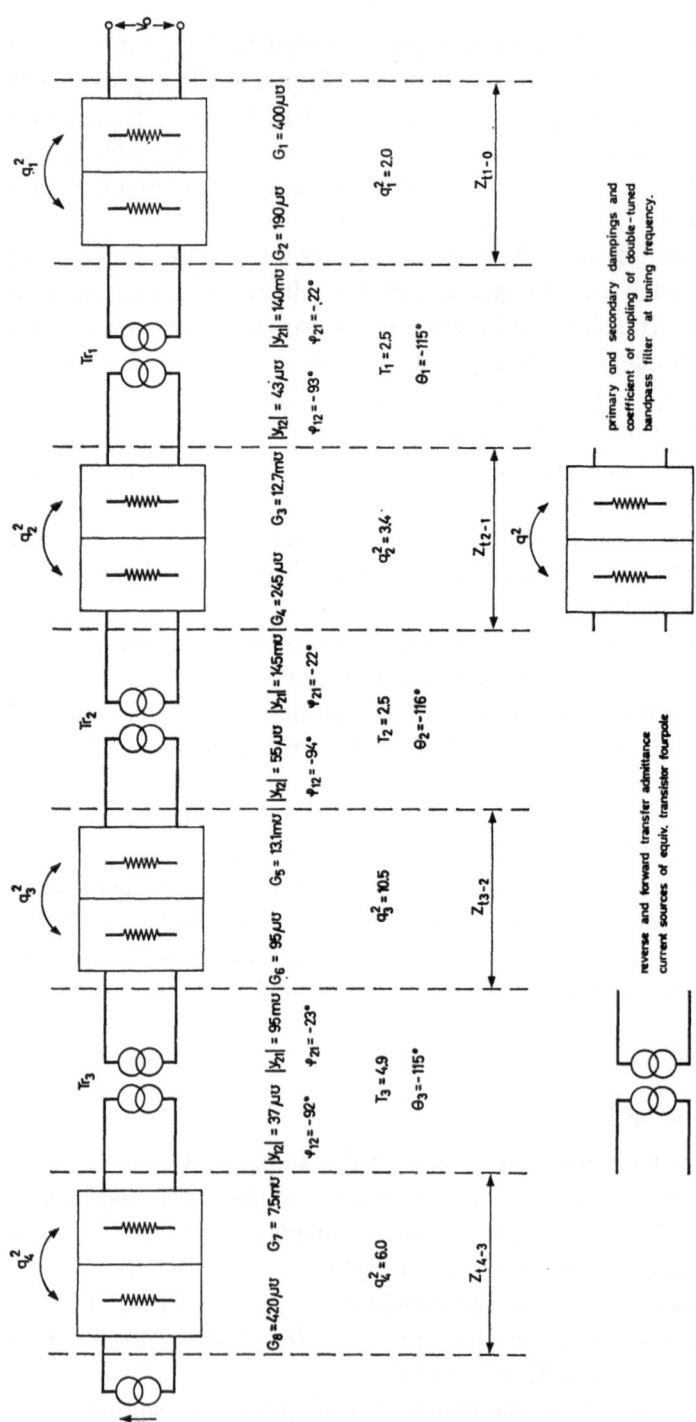

Fig. 13.3. Tabular representation of data required for calculating the gain performance of a three-stage vision I.F. amplifier.

TABLE 13.1 Admittance Parameters of the Transistors BF167 and BF173 at a Frequency
of 35 Mc/s

Transistor type		BF167	BF173	BF173
Biasing point				
V_{CE}	(V)	10	10	17
$-I_E$	(mA)	4	7	6.8
Admittance parameters				
g_{ie}	(m℧)	48	45	38
C_{ie}	(pF)	45	45	42
$\|y_{re}\|$	(μ℧)	37	55	43
φ_{re}	(<°)	—92	—94	—93
$\|y_{fe}\|$	(m℧)	95	145	140
φ_{fe}	(<°)	—23	—22	—22
g_{oe}	(μ℧)	30	65	50
C_{oe}	(pF)	1.2	2.1	2.0

can most easily be taken from the design procedure of the amplifier (not
described). They can, however, also be calculated directly from the circuit
diagram presented in Fig. 13.2. In the diagram shown in Fig. 13.3 the various damping values are entered. These dampings are expressed in terms of
the total damping present in the amplifier at the collector or base terminals
respectively of the various transistors. This means that these dampings
include the dampings g_{oe} and g_{ie} respectively of the transistors.
The various damping levels are determined at the midband frequency of the
amplifier (36.5 Mc/s).

Also the coefficients of coupling are entered in the diagram shown in
Fig. 13.3.

To increase accuracy, the various damping values entered in Fig. 13.3
are determined as follows:

— Firstly, the required values of damping on the various tuned circuits (of
the double-tuned bandpass filters) are determined from the design on
stability and transistor interchangeability.

— Secondly, double-tuned bandpass filters are constructed using capacitors
and resistors from the standard range of values (E 24).

— Thirdly, the various dampings are calculated from the constructed double-
tuned bandpass filters. These dampings should show a fairly close approximation to those obtained from the stability and interchangeability designs. The damping values determined from the actual bandpass filters
are used in the gain-calculation.

13.2.1 THE OUTPUT STAGE

The voltage gain of the last stage of the I.F. amplifier is given by the product of the transimpedance Z_{t1-0} of the output double-tuned bandpass filter, the transadmittance y_{21} of the output transistor and the power efficiency of the video detector circuit.

The Video Detector Circuit

Measurements have been carried out on the OA90 video detector circuit shown in Fig. 13.4 to determine its input damping and "voltage" efficiency at a frequency of 36.5 Mc/s. It was found that for a detector output voltage of 4.0 V (D.C.) an input voltage of 5.3 V (R.M.S.) is required. This gives a detector efficiency of $\eta = 0.53$. The detector input damping was measured as 360 $\mu\mho$ under these conditions.

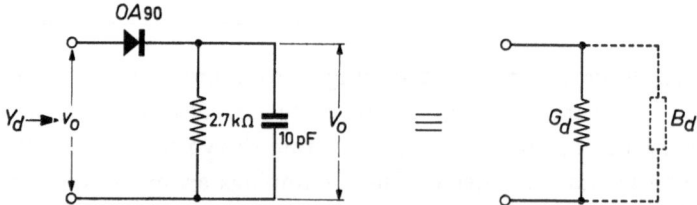

Fig. 13.4. Equivalent circuit for the input admittance of the video detector.

The Secondary of the Collector Bandpass Filter

In Fig. 13.5 a circuit diagram of the secondary of the output bandpass filter of the amplifier is shown. The video detector is represented by a damping G_d.
The damping of the coil is indictated by $G_o{}^*$ and the parasitic capacitance of the circuit by C_{par}.
Assuming that the quality factors of the coils is $Q_o{}^* = 60$ and that the parasitic capacitance is 2 pF, we obtain for the secondary damping at resonance:

$$G_1 = 395 \ \mu\mho$$

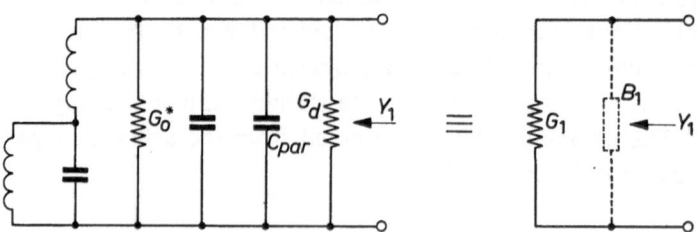

Fig. 13.5. Circuit for determining the equivalent damping at resonance of the secondary of the output bandpass filter.

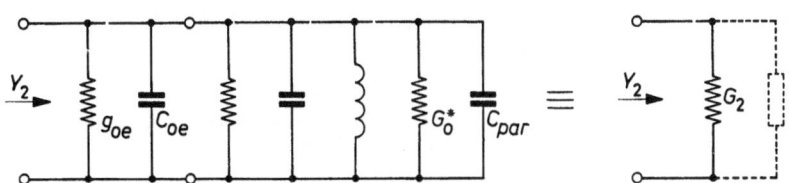

Fig. 13.6. As Fig. 13.5, but for the primary.

The Primary of the Output Double-Tuned Bandpass Filter

The primary of the output double-tuned bandpass filter of the amplifier is represented in the circuit diagram shown in Fig. 13.6. It can be calculated that:

$$G_2 = 190 \ \mu\mho$$

The Transimpedance of the Output Double-Tuned Bandpass Filter.

The coefficient of coupling of the output double-tuned bandpass filter is $q^2 = 2$. This coupling value can either be obtained from the initial design assumptions or by direct calculation of the circuit.

With the primary and secondary dampings obtained above, the transimpedance of the bandpass filter can be calculated. According to Chapter 8, Eq. (8.6.10):

$$Z_{t_{1-0}} = Z_{t_1}$$

From Table 8.5:

$$Z_{t_1} = \frac{q}{\sqrt{G_1 G_2}} \ \frac{1}{P_{2M}} \quad \text{and} \quad P_{2M} = 1 + q_1^2$$

This gives:

$$Z_{t_{1-0}} = \left(\frac{q}{1+q^2}\right)_1 \cdot \frac{1}{\sqrt{G_2 G_1}}$$

Substituting figures we find:

$$Z_{t_{1-0}} = 2120 \ \Omega$$

Gain of the Output Stage

The voltage gain of the output stage follows from:

$$\mu_1 = |\,_1 y_{21}\,| \ Z_{t_{(1-0)}} \quad \text{or} \quad \mu_1 = 295$$

The input damping of the output stage transistor, including feedback effects, equals:

$$g_{\text{in } 1} = g_{11} - G_3 \cdot T_1 \cos \Theta_1 \cdot \frac{1}{P_{2M}} \quad \text{(see Book I, p. 182).}$$

The damping G_3 can be calculated in the same way as has been shown above for the dampings G_2 and G_1. It is found that:

$$G_3 = 12.7 \text{ m}\mho$$

Substituting figures we obtain:

$$g_{in1} = 8.2 \text{ m}\mho$$

The power gain Φ_1 of the output stage can now be calculated from:

$$\Phi_1 = \mu_1^2 \, \frac{g_{in1}}{G_d} = 3800 \qquad \text{or} \qquad \Phi_1 = 35.9 \text{ dB}$$

The power losses Φ_3 of the detector stage follow from:

$$\Phi_d = 2\eta^2 \cdot \frac{G_d}{G_L} = 0.57 \qquad \text{or} \qquad \Phi_d = 2.4 \text{ dB}$$

13.2.2 THE MIDDLE STAGE

In the same way as for the output stage the gain performance of the middle stage can be calculated.

Here we find:

$$G_3 = 12.7 \text{ m}\mho, \ G_4 = 245 \text{ }\mu\mho, \ G_5 = 13.1 \text{ m}\mho$$
$$q_2^2 = 3.4$$

$$|\, _2y_{21}\,| = 145 \text{ m}\mho$$
$$T_2 = 2.5, \ \Theta = -116°$$
$$P_{1M} = 1, \ P_{2M} = 3, \ P_{3M} = 4, \ P_{4M} = 14$$

According to Table 8.6, the voltage gain of the middle stage is:

$$\mu_2 = |\, _2y_{21}\,| \cdot \frac{q^2}{\sqrt{G_3 G_4}} \cdot \frac{P_{2M}}{P_{4M}} \qquad \text{or} \qquad \mu_2 = 34.6$$

and the transimpedance is: $Z_{t_{2-1}} = 238 \ \Omega$.

The input damping including feedback becomes:

$$g_{in2} = 2g_{11} - G_5 \cdot T_2 \cos \Theta_2 \cdot \frac{P_{2M}}{P_{4M}}$$

or

$$g_{in2} = 9.3 \text{ m}\mho$$

and the power gain:

$$\Phi_2 = \mu_2^2 \; \frac{g_{in2}}{g_{in1}} = 1200$$

$$\Phi_2 = 30.8 \text{ dB}$$

13.2.3 THE INPUT STAGE

Now we have

$$G_5 = 13.1 \text{ m}\mho, \; G_6 = 95 \text{ μ}\mho, \; G_7 = 7.5 \text{ m}\mho$$
$$q_3^2 = 10.5$$

$$|\,{}_3y_{21}\,| = 95 \text{ m}\mho$$

$$T_3 = 4.9, \; P_{5M} = 18, \; P_{6M} = 165$$

Transimpedance: $Z_{t_{3-2}} = 238 \; \Omega$

Voltage gain: $\mu_3 = 22.6$

Input damping: $g_{in3} = 6.5 \text{ m}\mho$

Power gain: $\Phi_3 = 28.2 \text{ dB}$

13.2.4 THE INPUT DOUBLE-TUNED BADPASS FILTER

Here:

$$G_7 = 7.5 \text{ m}\mho, \; G_8 = 420 \text{ μ}\mho$$
$$q_4^2 = 6$$

$$P_{6M} = 165, \; P_{7M} = 200, \; P_{8M} = 1200$$

Transimpedance: $Z_{t_{4-3}} = 192 \; \Omega$

Transducer gain to base of input transistor:

$$\Phi_t = 4G_5 \cdot g_{in3} \cdot |\,Z_t\,|^2 = \frac{1}{2.5} \qquad \text{or} \qquad \Phi_t = -4 \text{ dB}$$

13.2.5 OVERALL GAIN

The voltage gains of the different stages calculated above enable us to calculate the voltages at the bases of the three transistors that should appear for a certain output voltage of the video detector. In Table 13.2 these voltages are tabulated for an output voltage of the video detector of 4 V.

TABLE 13.2 Signal Levels

	Calculated	Measured
Output voltage of video detector	4 V	4 V
Signal level at base of output transistor	19 mV	18 mV
Signal level at base of middle transistor	0.5 mV	0.52 mV
Signal level at base of control transistor	24 μV	23 μV

The power gain between the base of the input transistor and the output terminals of the video detector follows from:

$$\Phi = \Phi_3 + \Phi_2 + \Phi_1 + \Phi_d = 92.5 \text{ dB}$$

The transducer gain of the complete amplifier is:
$$\Phi_t = 92.5 - 4 = 88.5 \text{ dB}$$

The transimpedance of the complete amplifier can then be calcutated as:
$$Z_t = 32 \text{ M}\Omega$$

13.3 Measurements

The signal levels at the base terminals of the respective transistors in the amplifier have been measured for an d.c. output voltage of 4 V from the video detector. The measurements were carried out at a midband frequency of 36.5 Mc/s. The results are given in Table 13.2. The calculated and measured values are found to correspond very well.

It appears from Table 13.2 that the transimpedance of the amplifier including the video detector circuit amounts to 33 MƱ. This corresponds to a transimpedance of 43 MƱ when the video detector circuit is not included.

The power gain measured in an actual circuit between the base of the first transistor and the output terminals of the amplifier is 93 dB. This gain figure includes the losses in the video detector circuit and takes into account the effect of the feedback of the BF167 on its input damping.

CHAPTER 14

STEP-BY-STEP METHOD OF DESIGNING
I.F. AMPLIFIERS

In the preceding chapters the various aspects of the design of intermediate frequency amplifiers for radio and television receivers have been considered in detail. Moreover, a method of designing such amplifiers has been developed which lends itself very well to systematic procedure. The object of this chapter is to present this systematic procedure in the form of a step-by-step method of design.

On the following pages the step-by-step method is worked out. On the left-hand page descriptions are given of the successive steps. The right-hand page contains further details regarding the particular design step as well as references to the different sections of this book in which the relevant aspects of design are considered.

The fold-out page at the end of this book carries a schematic diagram which can advantageously be used for collecting the data and results of calculations obtained from the various steps in design.

The diagram has been constructed in such a way that it represents the step-by-step method in a schematic form. It is therefore possible, in many cases, to use the diagram in the design of I.F. amplifiers without need to consult the description of the various design steps.

Description of step in amplifier design	
1. COLLECT DATA REGARDING PERFORMANCE REQUIRED OF THE AMPLIFIER. The data should comprise information on: gain, 3dB bandwidth, adjacent channel selectivity, and envelope delay characteristic.	
2. DECIDE ON THE NUMBER OF STAGES OF THE AMPLIFIER AND THE TYPES OF TRANSISTORS TO BE USED.	
3. DETERMINE AMPLIFIER CONFIGURATION.	
4. CHOOSE BIASING POINTS OF THE TRANSISTORS TO BE USED IN THE AMPLIFIER.	

Further details	References
The data should specify the nominal performance as well as acceptable tolerances. The gain should be specified in terms either of transimpedance or transducer gain of the complete amplifier or of voltage gain or powergain per stage of the amplifier if the number of stages is already known. In some cases also a specification of the 0 dB bandwidth will be necessary. The adjacent channel selectivity should be specified in terms of separation in frequency and attenuation with respect to the relevant performance at the midband frequency of the amplifier. Instead of the separation in frequency, specific frequencies of adjacent channels may also be quoted.	Chapter 2 p. 5
In most cases the decision on which types of transistors are going to be used is greatly simplified if use is made of data supplied by the transistor manufacturer. The data usually comprise a recommendation of a particular type of transistor for the specific application under consideration. Also data are usually given on which types of transistors and how many of them should be used to reach a quoted performance.	Chapter 8, p. 72
The term amplifier configuration refers to the sequence of transistors and single and/or double-tuned bandpass filters in the amplifier. Three configurations are distinguished in this design procedure, viz.:	Chapter 8, p. 72
Configuration I : n-stage amplifier with $(n + 1)$ single-tuned bandpass filters, Configuration II : n-stage amplifier with $(n + 1)$ double-tuned bandpass filters, and Configuration III: n-stage amplifier with n double-tuned bandpass filters and one single-tuned bandpass filter.	Chapter 15, p. 228
In choice of the biasing points of the transistors to be used in the various stages of the amplifier use can again be made of the recommendations of the manufacturer(s) of the transistors.	

Description of step in amplifier design	
5. FROM THE ADMITTANCE PARAMETERS OF THE TRANSISTORS AT THE CHOSEN BIASING POINTS DETERMINE THE QUANTITIES Θ, Φ_{uM}, N, t AND T_g. Θ : regeneration phase angle Φ_{uM}: maximum unilateralized power gain N : transfer admittance ratio t : intrinsic regeneration coefficient T_g : approximate boundary of stability M : transfer admittance product	
6. SELECT THE DESIGN CHARTS APPLICABLE TO THE SPECIFIC AMPLIFIER DESIGN UNDER CONSIDERATION FROM TABLE 15.1 ON PAGE 230.	
7. FROM THE SERIES OF DESIGN CHARTS APPLICABLE DETERMINE THE VALUE OF THE TRANSISTOR REGENERATION COEFFICIENT (T) AND THE VALUE OF THE COEFFICIENT OF COUPLING OF THE DOUBLE-TUNED BANDPASS FILTERS (q^2) THE VALUE OF T, ASCERTAINED FROM RESPONSE CURVE, WILL BE DENOTED BY T_R. The number of the design chart for which T_R and q^2 are determined should be quoted on a schematic diagram as exemplified in the fold-out page at the back of the book.	

Further details	References
$\Theta = \varphi_{12} + \varphi_{21}$ $\Phi_{uM} = \dfrac{\lvert y_{21} \rvert^2}{4g_{11}\,g_{22}}$ $N = \dfrac{\lvert y_{21} \rvert}{\lvert y_{12} \rvert}$ $t = \dfrac{\lvert y_{12}\,y_{21} \rvert}{g_{11}\,g_{22}}$ $T_g = \dfrac{2}{1 + \cos\Theta}$ $M = \lvert y_{12}\,y_{21} \rvert$	Chapter 4, p. 17
From the number of stages of the amplifier, the amplifier configuration and the value of Θ as determined in steps 2, 3 and 5 respectively, the proper series of design charts can be selected. If the value of Θ is different for the various transistors in the amplifier, the averaged value should be taken. If the value of Θ obtained does not appear in the table on page 230, the nearest value must be taken.	Chapter 8, p. 72 Chapter 15, p. 230
For T and q^2 those values should be taken which, in the opinion of the designer, give an acceptable shape of amplitude response and envelope delay characteristics. The values for T and q^2 obtained should be regarded as average values for the respective transistors and double-tuned bandpass filters in the amplifier.	Chapter 8, p. 72 Chapter 15, p. 230

Description of step in amplifier design	
8. DETERMINE VALUES OF NORMALIZED DETUNING x AT 3dB ATTENUATION AND SPECIFIED ADJACENT CHANNEL ATTENUATION OF AMPLITUDE RESPONSE CURVE FROM DESIGN CHART FOR T AND q^2 AS OBTAINED IN STEP 7.	
9. CALCULATE THE QUALITY FACTORS Q OF THE TUNED CIRCUITS OF WHICH THE BANDPASS FILTERS ARE COMPOSED, FROM: $$Q = \frac{+x + -x}{2} \cdot \frac{f_0}{B_{3dB}},$$ AND/OR FROM: $$Q = \frac{+x + -x}{4} \cdot \frac{f_0}{|f_0 - f_{adj\cdot ch\cdot}|}$$	
10. FIND, IF NECESSARY, A COMPROMISE BETWEEN 3dB BANDWIDTH AND ADJACENT CHANNEL SELECTIVITY REQUIREMENTS AND THE QUALITY FACTOR OBTAINED IN STEP 9. THE FINAL VALUE OF Q SHOULD BE SUCH THAT THE REQUIREMENTS OF STEP 1 ARE FULFILLED AS CLOSELY AS POSSIBLE.	
11. CALCULATE FOR EACH TRANSISTOR, THE VALUE OF T, REQUIRED IN VIEW OF TRANSISTOR INTERCHANGEABILITY, FROM THE VALUES OF INPUT AND OUTPUT DAMPING RATIOS: $$T = \frac{t}{\Phi_{11}\,\Phi_{22}}$$ Φ_{11} = input damping ratio, and Φ_{22} = output damping ratio. THIS VALUE OF T, WHICH FOLLOWS FROM INTERCHANGEABILITY REQUIREMENTS, WILL BE DENOTED BY T_I	

Further details	References
Normalized detuning is expressed in terms of $+ x$ and $- x$ as specified on the relevant design chart	Chapter 8 p. 72
In the equations for determining Q fulfilling the 3dB bandwidth and adjacent channel selectivity requirements respectively, the relevant values for $+ x$ and $- x$ must be taken.	Chapter 8 p. 72
	Chapter 8 p. 72
In an amplifier designed for a particular type of transistor, all transistors of that type should be interchangeable without impairing the performance of the amplifier. For that reason the contribution of the input and output dampings of the transistor to the total damping of the tuned circuits to which they are connected should not exceed a certain percentage.	Chapter 8, p. 72
These "damping ratios" as they are referred to depend on the type of amplifier and on the tolerances in the response curve that are acceptable. The "damping ratio" is defined as: $$\frac{\text{transistor damping } (g_{11} \text{ or } g_{22} \text{ resp.})}{\text{total tuned circuit damping}}$$	Chapter 3, p. 12

Description of step in amplifier design	
12. WITH THE MINIMUM PRACTICABLE VALUES OF THE CAPACITANCES OF THE TUNED CIRCUITS AT INPUT AND OUTPUT TERMINALS OF THE TRANSISTORS AND THE VALUE OF Q FROM STEP 10, DETERMINE: $$G_1 = \frac{\omega_0 C_{1\min}}{Q} \text{ and } G_2 = \frac{\omega_0 C_{2\min}}{n^2 Q},$$ IN WHICH n^2 DENOTES THE TAPPING RATIO OF THE INPUT CIRCUIT. THEN CALCULATE FOR EACH TRANSISTOR: $$T = \frac{|y_{12}\, y_{21}|}{G_1\, G_2}.$$ THIS VALUE OF T WHICH FOLLOWS FROM MINIMUM TUNING CAPACITANCE WILL BE DENOTED BY T_C	
13. CALCULATE THE VALUE OF T REQUIRED FOR A SUFFICIENTLY LARGE STABILITY FACTOR s FOR EACH STAGE OF THE AMPLIFIER, FROM $$T = \frac{T_g}{s} \cdot u_n.$$ u_n = stability reduction factor in multi-stage amplifiers with single-tuned bandpass filters THE VALUE OF T WHICH FOLLOWS FROM STABILITY CONSIDERATIONS WILL BE DENOTED BY T_S.	
14. DETERMINE WHICH OF THE T VALUES OBTAINED IN STEPS 7, 11, 12 AND 13 IS THE SMALLEST. WHEN NO NEUTRALIZATION IS EMPLOYED, THE DESIGN	

Further details	References
Practical estimates are: 50 % or − 3dB for transistors with a fixed biasing point, 30 % or − 5dB for transistors with a variable biasing point (as in the gain control stages).	
The minimum acceptable values of the tuning capacitances depend on: a) the spreads in transistor output and/or input capacitances, b) the spreads in the values of the components of which the bandpass filters are composed, c) in case already existing bandpass filters are used, the tuning capacitance of these filters. In calculation of the damping which corresponds with the minimum tuning capacitance of the input circuit the tapping ratio n is also a variable (it is assumed that no tap is present on the output circuit).	Chapter 8, p. 72
The stability factor s is usually taken as $s = 4$. In all cases $u_n = 1$ *except* in multi-stage amplifiers with single-tuned bandpass filters. For the latter amplifier types: single-stage: $u_n = 1$, two-stage : $u_n = 0.5$, three-stage : $u_n = 0.38$, four-stage : $u_n = 0.33$. T may also be obtained from the relevant graph on the design chart used in step 7.	Chapter 5, p. 48
The condition that yields the smallest value for T governs the further design of the amplifier (if no neutralization is employed).	Chapter 8, p. 72

Description of step in amplifier design	
SHOULD BE CONTINUED WITH THE SMALLEST VALUE OF THE REGENERATION COEFFICIENT (STEP 16). THE VALUE OF THE REGENERATION COEFFICIENT, OBTAINED HERE, WILL BE DENOTED BY T (WITHOUT SUBSCRIPT).	
15. IF COMPARISON OF THE T VALUES IN STEP 14 REVEALS THAT THE VALUE OF T_S (BASED ON STABILITY) OBTAINED IN STEP 13 IS CONSIDERABLY SMALLER THAN THE OTHER VALUES, NEUTRALIZATION OF THE VARIOUS TRANSISTORS IN THE AMPLIFIER IS ADVISABLE. IF NO NEUTRALIZATION IS EMPLOYED: PROCEED WITH STEP 16, IF NEUTRALIZATION IS EMPLOYED: PROCEED WITH STEP 19.	
16. IF NO NEUTRALIZATION IS EMPLOYED: DETERMINE TRANSDUCER GAIN Φ_t OF THE COMPLETE AMPLIFIER FROM: $$\Phi_t = 4 \frac{G_S}{G_{\text{total input}}} \cdot \frac{G_L}{G_{\text{total output}}} \cdot T_1 N_1 q_1^2 \cdot T_2 N_2 q_2^2 \ldots$$ $$\ldots T_n N_n q_n^2 \cdot \frac{1}{|\delta_0|^2}$$ FURTHER GAIN EXPRESSIONS ARE REFERRED TO IN TABLE 15.2 ON PAGE 231.	
17. CHECK WHETHER TRANSDUCER GAIN OR TRANSIMPEDANCE DETERMINED IN STEP 16 COMPLIES WITH THE REQUIREMENT OF STEP 1. IF "YES": PROCEED WITH STEP 18, IF "NO": MODIFY ASSUMPTION OF STEP 2 AND REPEAT DESIGN PROCEDURE.	

Further details	References
For obtaining the maximum gain of the amplifier the largest value of the regeneration coefficient T which is allowable in view of the various design requirements must be taken. All T values are therefore determined separately and the smallest one is chosen for further design. The design requirements which allow a larger value are then automatically met.	
G_S = source damping, G_L = load damping, $G_{\text{total input}}$ = total damping of input tuned circuit, $G_{\text{total output}}$ = total damping of output circuit, $\dfrac{1}{\lvert\delta_0\rvert^2}$ = factor expressing effect of feedback on gain. The transimpedance $\lvert Z_t\rvert$ of the amplifier may be calculated from: $\lvert Z_t\rvert^2 = \dfrac{\Phi_t}{4 G_S G_L}$	Chapter 8, p. 72 Chapter 15, p. 231
In some cases a slight increase in gain can be obtained by changing (lowering) the requirements imposed on the design in steps 11 and 13. If stability requirements are governing the value of T, more gain may be obtained by employing neutralizing networks.	

Description of step in amplifier design	
18. CARRY OUT DIMENSIONING OF THE BANDPASS FILTERS TAKING INTO ACCOUNT VARIOUS DESIGN REQUIREMENTS OF STEPS 14, 11, 12 AND 13.	
19. IF NEUTRALIZATION NETWORKS ARE EMPLOYED: DETERMINE COMPONENT VALUES OF FIXED NEUTRALIZATION NETWORK FROM DATA ON TRANSISTOR PARAMETER SPREADS IF NO DATA ON SPREADS ARE AVAILABLE CALCULATE C_N, R_N AND n_N FOR AVERAGE TRANSISTORS.	

Further details	References
1. Calculate total damping of the tuned circuits at the input and output terminals of each transistor from: $$G_{\text{tot in}} \cdot G_{\text{tot out}} = \frac{|y_{12}\,y_{21}|}{T}$$ Apply T value from step 14.	Chapter 8, p. 72
2. Take into account the exchangeability conditions of step 11 and the minimum tuning capacitance conditions of step 12 for determining the damping of each tuned circuit. For the tuned circuit at the input of the transistor the tapping ratio also must be determined from these criteria. All requirements need not necessarily exist.	
3. Determine tuning inductance and tuning capacitance of each tuned circuit from $$G = \frac{1}{\omega_0 L Q} = \frac{\omega_0 C}{Q}$$ 4. When double tuned bandpass filters are used, determine coupling reactance	
$$C_N = \frac{1}{n_N} \cdot \frac{1}{\omega} \cdot \frac{|y_{12N}|}{\sin \varphi_{12}},$$ $$R_N = n_N \cdot \frac{\cos \varphi_{12}}{|y_{12N}|}.$$ For C_N and R_N values from the standard range of component values must be chosen. For C_N take nearest component value lower than calculated one. Furthermore: $$|Y_{12N}| = \frac{2\,M_a + \Delta\,M \cos \Theta}{2\,|y_{21a}| + \Delta\,|y_{21}| \cos \Theta},$$ $$M = |y_{12}\,y_{21}|,$$ index a : average, index Δ : spread.	Chapter 6, p. 61

Description of step in amplifier design	
20. CALCULATE TRANSDUCER GAIN Φ_t OF COMPLETE AMPLIFIER FROM EXPRESSIONS REFERRED TO IN CHAPTER 15, TABLE 2	
21. CHECK WHETHER THE GAIN OF THE AMPLIFIER FULFILS DESIGN REQUIREMENTS, EMPLOYING THE METHODS OF STEP 17.	
22. CARRY OUT DIMENSIONING OF THE BANDPASS FILTERS TAKING INTO ACCOUNT REQUIREMENTS FROM STEPS 11 AND 15.	

Further details	References
The gain expression comprises the product of the maximum unilateralized gain: Φ_{uM} of the various transistors and the insertion losses Φ_i of the various bandpass filters.	Chapter 15, p. 231
See step 19, point 2.	Chapter 8, p. 72

CHAPTER 15

DESIGN CHARTS

This chapter gives various calculated amplifier design charts. These comprise amplitude response curves and envelope delay curves for various values of the regeneration coefficient T, and also graphs in which the value of the reduced determinant $\left|\dfrac{1}{\delta_0}\right|$ and the losses Φ_f are plotted against T. The design charts are given for three different amplifier configurations, termed "I", "II" and "III" respectively, see Fig. 15.1.

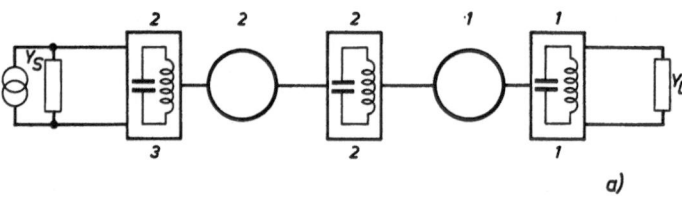

a)

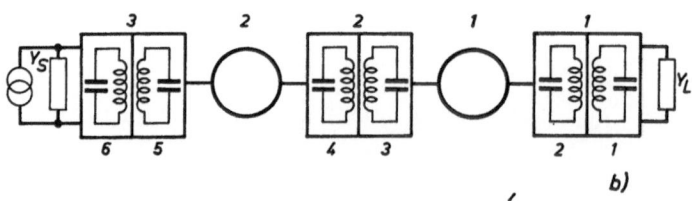

b)

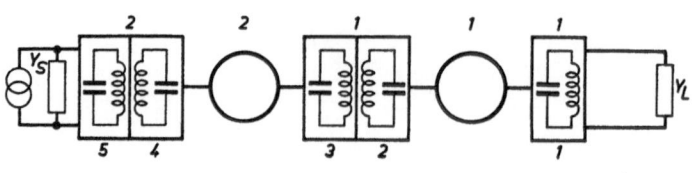

c)

Configuration I   : amplifier comprising n transistors and $(n + 1)$ *single-tuned* bandpass filters in the sequence bandpass filter-transistor-bandpass filter-transistor etc. This configuration is schematically shown in Fig. 15.1a for a two-stage amplifier.

Configuration II : amplifier comprising n transistors and $(n + 1)$ *double-tuned* bandpass filters in the same sequence as in configuration I.
Fig. 15.1b gives a schematic representation for a two-stage amplifier.

Configuration III: amplifier comprising n transistors, n *double-tuned* bandpass filters and one *single-tuned* bandpass filter, the latter being located at the output terminals of the amplifier.
Fig. 15.1c schematically shows a two-stage amplifier in thi: configuration.

15.1 Choice of Tuning Method

As already referred to, the various graphs in the Design Chart section are valid for tuning method B only.

This restriction to one of the three possible tuning methods is necessary because, otherwise, the number of design charts required would be unwieldy large.

Tuning method B has been chosen because it has been evidenced by experience that this method of tuning is most practical. The reason of its being so practical is that in this case the output stage of the amplifier is tackled first. Thinking in terms of radio or television receivers, this means that the part of the receiver following the output stage can be employed as output indicator.

Furthermore, when the other stages of the amplifier are aligned, starting at the output side, the same facilities can be used.

Another reason for the preference for tuning method B is that during alignment an impression is gained on a stage-by-stage basis of the operation of the complete amplifier. With tuning method C, this is not the case.

15.2 Survey of Design Charts

To facilitate the consultation of the relevant set of graphs, a survey is given in Table 15.1.

Table 15.1 Design chart parameters

Number of stages	Regeneration phase angle	Coefficient of coupling of double-tuned bandpass filters	Amplifier configuration					
			I		II		III	
n	Θ	q^2	Fig.	page	Fig.	page	Fig.	page
1	270	1.0			15.5	239		
1	270	1.5			15.6	240		
1	270	1.25			15.7	241		
1	270	1.875			15.8	242		
2	240	1.0			15.12	247		
2	240	1.5			15.13	248		
2	255	1.0			15.14	249		
2	270	1.0			15.15	250		
3	30		15.19	254				
3	30	1.0			15.35	272	15.63	304
3	60		15.20	255				
3	60	1.0			15.36	273	15.64	305
3	210		15.21	256				
3	210	2.0			15.37	274	15.65	306
3	210	3.0			15.38	275	15.66	307
3	225	1.0			15.39	276	15.67	308
3	225	2.0			15.40	277	15.68	309
3	240		15.22	257				
3	240	1.0			15.41	278	15.69	310
3	240	2.0			15.42	279	15.70	311
3	255	1.0			15.43	280	15.71	312
3	255	1.5			15.44	281	15.72	313
3	270		15.23	258				
3	270	1.0			15.45	282	15.73	314
4	30		15.27	262				
4	30	1.0			15.49	288	15.77	320
4	60		15.28	263				
4	60	1.0			15.50	289	15.78	321
4	210		15.29	264				
4	210	2.0			15.51	290	15.79	322
4	210	3.0			15.52	291	15.80	323
4	225	1.0			15.53	292	15.81	324
4	225	2.0			15.54	293	15.82	325
4	240		15.30	265				
4	240	1.0			15.55	294	15.83	326
4	240	2.0			15.56	295	15.84	327
4	255	1.0			15.57	296	15.85	328
4	255	1.5			15.58	297	15.86	329
4	270		15.31	266				
4	270	1.0			15.59	298	15.87	330

The symbol r occurring in the caption of several design charts denotes the ratio of the quality factors of secondary and primary of the double tuned bandpass filters:

$$r = Q_s/Q_p$$

The symbol m which occurs in the captions of the design charts for amplifiers in the configuration III denotes the ratio of the quality factor Q_1 of the single-tuned bandpass filter at the output to the normalized quality factor $Q = \sqrt{Q_p Q_s}$ of the double-tuned bandpass filters.

At the beginning of each sub-section in the series of design charts given on the following pages a short introduction is given to the relevant type of amplifier. These introductions contain a definition of the various symbols related to the particular type of amplifier, a reference to the reduced amplifier determinant from which the various curves on the design charts are calculated and a survey of expressions for calculating the gain of the amplifier. General forms of the reduced determinant are given below.

In the step-by-step design method of I.F. amplifiers presented in Chapter 14 reference is made to the expressions for calculating the gain of a particular type of amplifier. To facilitate consulting these expressions a survey of where these expressions can be found is given in Table 15.2.

Table 15.2   Location of expressions for calculating gain

Number of sta ges	Amplifier configu- ration	I	II	III
1			p. 236	
2			p. 244	
3		p. 252	p. 268 and 269	p. 300 and 301
4		p. 260	p. 284 and 285	p. 316 and 317

When designing I.F. amplifiers it is sometimes convenient to know the variation of $\left|\dfrac{1}{\delta_0}\right|$ (value of the reduced amplifier determinant at the turning frequency) and Φ_f (losses due to the real component of the feedback) as a function of T and Θ. In the design chart section graphs are therefore incorporated for $\left|\dfrac{1}{\delta_0}\right|$ and Φ_f versus $T\cos\Theta$. Table 15.3 gives a survey of where these graphs are located.

Table 15.3   Location of graphs for $\dfrac{1}{\delta_0}$ and Φ_f

Number of stages	Amplifier configuration	I	II	III
1			Fig. 15.3, p. 237	
			Fig. 15.4, p. 238	
2			Fig. 15.10, p. 245	
			Fig. 15.11, p. 246	
3		Fig. 15.17, p. 253	Fig. 15.33, p. 270	Fig. 15.61, p. 302
		Fig. 15.18, p. 253	Fig. 15.34, p. 271	Fig. 15.62, p. 303
4		Fig. 15.25, p. 261	Fig. 15.47, p. 286	Fig. 15.75, p. 318
		Fig. 15.26, p. 261	Fig. 15.48, p. 287	Fig. 15.76, p. 319

Amplifier Configuration I.

Reduced Determinant for an n-Stage Amplifier with $(n + 1)$ Single-Tuned Bandpass Filters

Determinant no. I.

$$
\delta_I = \begin{vmatrix}
1 + j(x + x'_n) & T_n \exp(j\Theta_n) & 0 & 0 & 0 & 0 \\
1 & - & - & - & - & - \\
0 & - & - & - & - & - \\
0 & - & - & - & - & - \\
0 & - & -1 + j(x + x'_2) & T_2 \exp(j\Theta_2) & 0 \\
0 & - & - & 1 & 1 + j(x + x'_1) & T_1 \exp(j\Theta)_1 \\
0 & - & - & 0 & 1 & 1 + jx
\end{vmatrix}
$$

Amplifier Configuration II.

Reduced Determinant for an n-Stage Amplifier with $(n + 1)$ Double-Tuned Bandpass Filters.

Determinant no. II.

$$\delta_{II} = \begin{vmatrix}
1+jx\sqrt{r} & -q_{n+1}^2 & 0 & - & 0 & 0 & 0 & 0 \\
1 & 1+j\left(\frac{x}{\sqrt{r}}+x_n'\right) & T_s\exp(j\Theta_n) & - & 0 & 0 & 0 & 0 \\
- & - & - & - & - & - & - & - \\
- & - & - & - & - & - & - & - \\
0 & 0 & 0 & - & 1+jx\sqrt{r} & -q_2^2 & 0 & 0 \\
0 & 0 & 0 & - & 1 & 1+j\left(\frac{x}{\sqrt{r}}+x_1'\right) & T_1\exp(j\Theta_1) & 0 \\
0 & 0 & 0 & - & 0 & 1 & 1+jx\sqrt{r} & -q_1^2 \\
0 & 0 & 0 & - & 0 & 0 & 0 & 1+j\frac{x}{\sqrt{r}}
\end{vmatrix}$$

Amplifier Configuration III.

Reduced Determinant for an n-Stage Amplifier with n Double-Tuned and one Single-Tuned Bandpass Filters.

Determinant no. III.

$$\delta_{III} = \begin{vmatrix}
1+jx\sqrt r & -q_{n+1}^2 & 0 & 0 & 0 & 0 & 0 & 0 \\[4pt]
1 & 1+j\!\left(\dfrac{x}{\sqrt r}+x_n'\right)T_4\exp(j\Theta_n) & 0 & - & 0 & 0 & 0 & 0 \\[4pt]
0 & 1 & 1+jx\sqrt r & -q_n^3 & - & 0 & 0 & 0 \\[4pt]
- & - & - & - & - & - & - & - \\[4pt]
- & - & - & - & - & - & - & - \\[4pt]
0 & 0 & 0 & 1 & 1+j\!\left(\dfrac{x}{\sqrt r}+x_2'\right)T_2\exp(j\Theta_2) & 0 & 0 & 0 \\[4pt]
0 & 0 & 0 & 0 & 1 & 1+jx\sqrt r & -q_1^2 & 0 \\[4pt]
0 & 0 & 0 & 0 & 0 & 1 & 1+j\!\left(\dfrac{x}{\sqrt r}+x_1'\right)T_1\exp(j\Theta_1) & 1 \\[4pt]
0 & 0 & 0 & 0 & 0 & 0 & 1 & 1+jxm
\end{vmatrix}$$

Single-Stage Amplifier with Two Double-Tuned Bandpass Filters

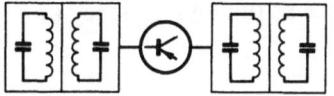

15.3   Single-stage amplifier with two double-tuned bandpass filters.

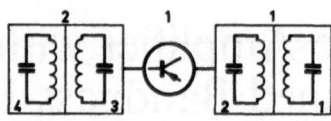

Fig. 15.2 Schematic representation of a single-stage amplifier with two double-tuned bandpass filters (Configuration II).

Reduced amplifier determinant:
See Determinant II on page 233 for $n = 1$.

Method of tuning: **B**
Reduced amplifier determinant at tuning frequency ($x = 0$):
$|_1\delta_0|$ $= P_{4M}$

Tuning correction term:
$x_1' = T_1 \sin\Theta_1$
Minor determinant values at tuning frequency ($x = 0$):
$P_{1M} = 1$

$P_{2M} = 1 + q_1^2$

$P_{3M} = 1 + q_1^2 - T_1 \cos\Theta_1$

$P_{4M} = (1 + q_1^2)(1 + q_2^2) - T_1 \cos\Theta_1$

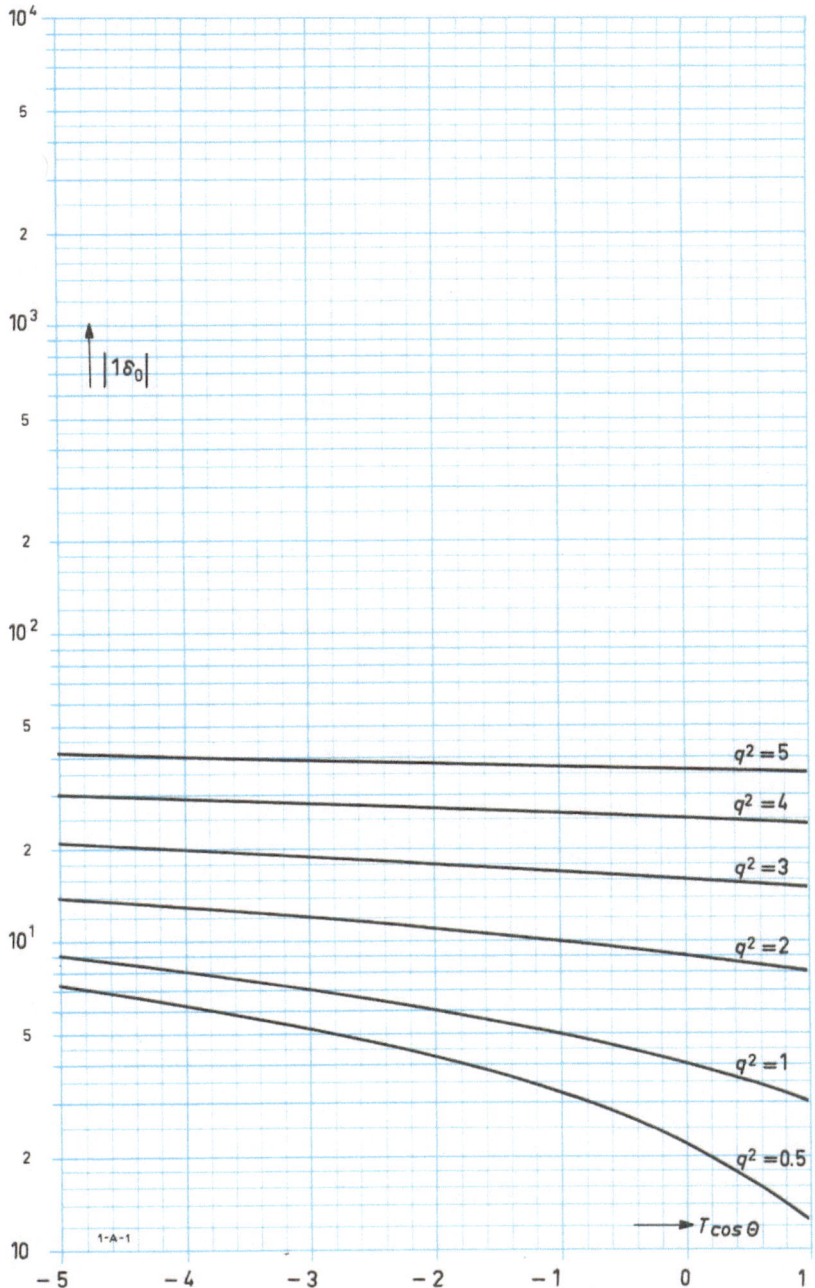

Fig. 15.3 Graph representing the value of the reduced determinant $|1/\delta_0|$ as a function of $T\cos\Theta$ for a single-stage amplifier with two double-tuned bandpass filters (Configuration II, tuning method B).

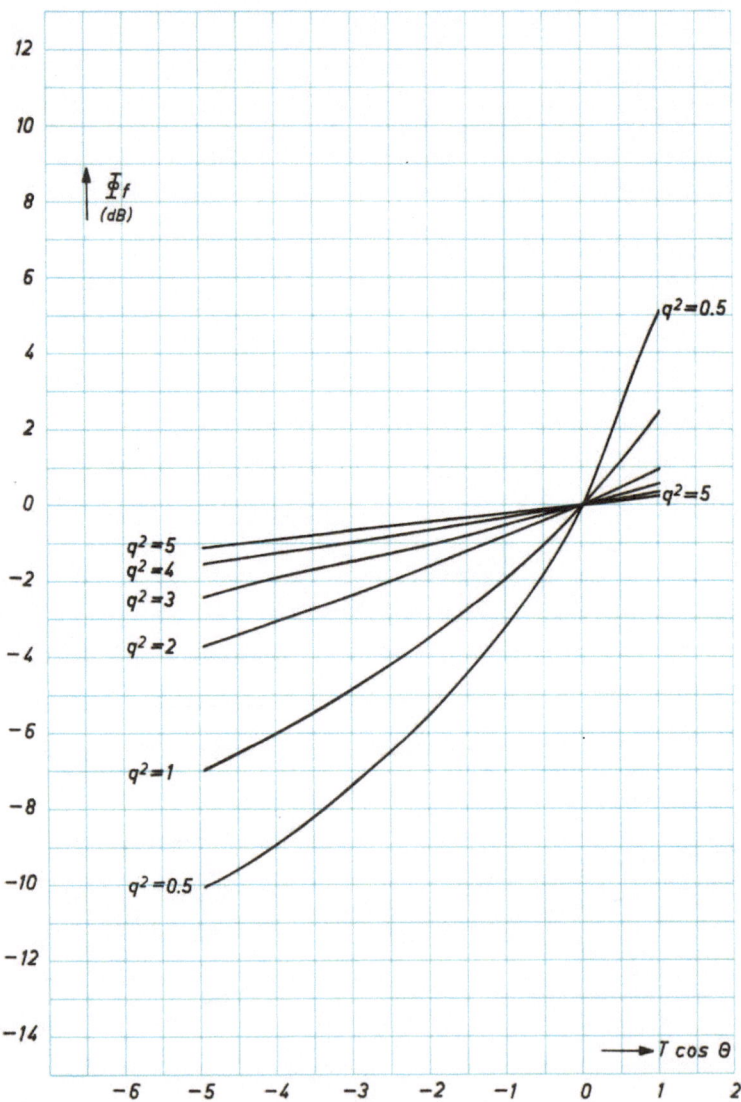

Fig. 15.4 Graph representing the feedback losses Φ_f as a function of $T\cos\theta$ for a single-stage amplifier with two double-tuned bandpass filters (Configuration II, tuning method B).

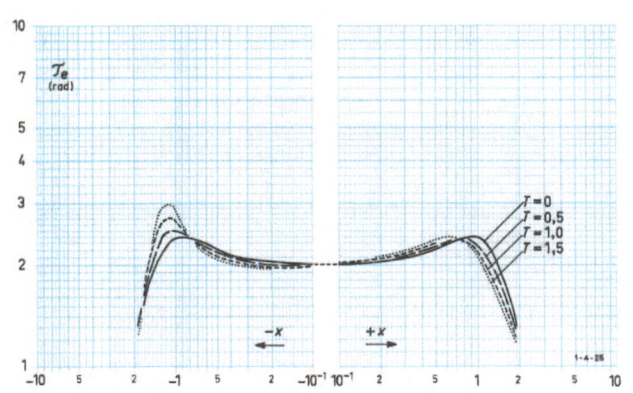

Fig. 15.5
Single-stage amplifier
Configuration II
Tuning method B
$\Theta = 270°$
$q^2 = 1.0, r = 1$

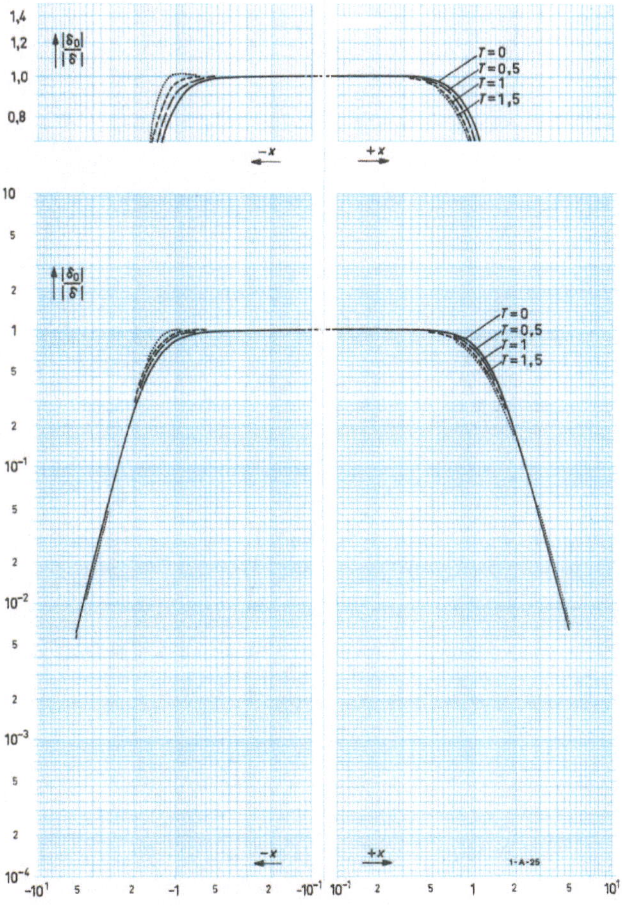

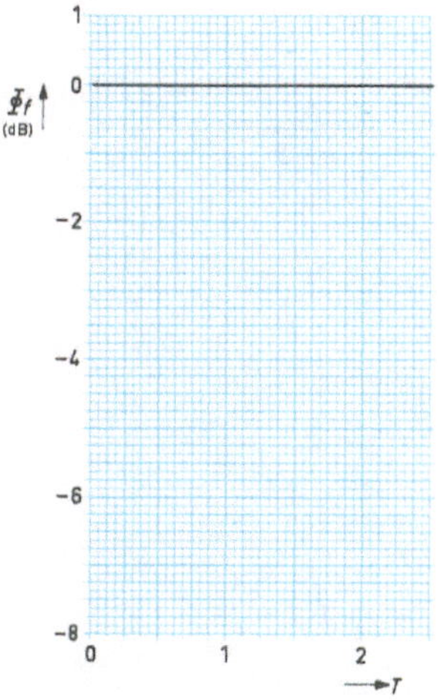

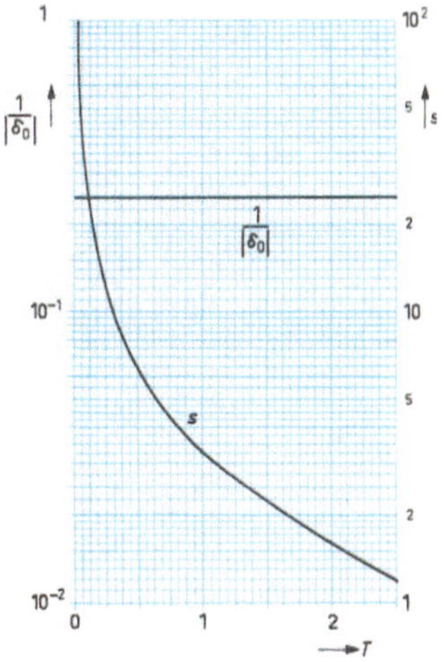

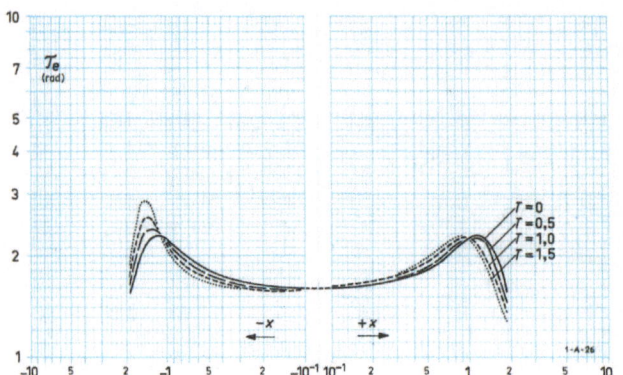

Fig. 15.6
Single-stage amplifier
Configuration II
Tuning method B
$\Theta = 270°$
$q^2 = 1.5, r = 1$

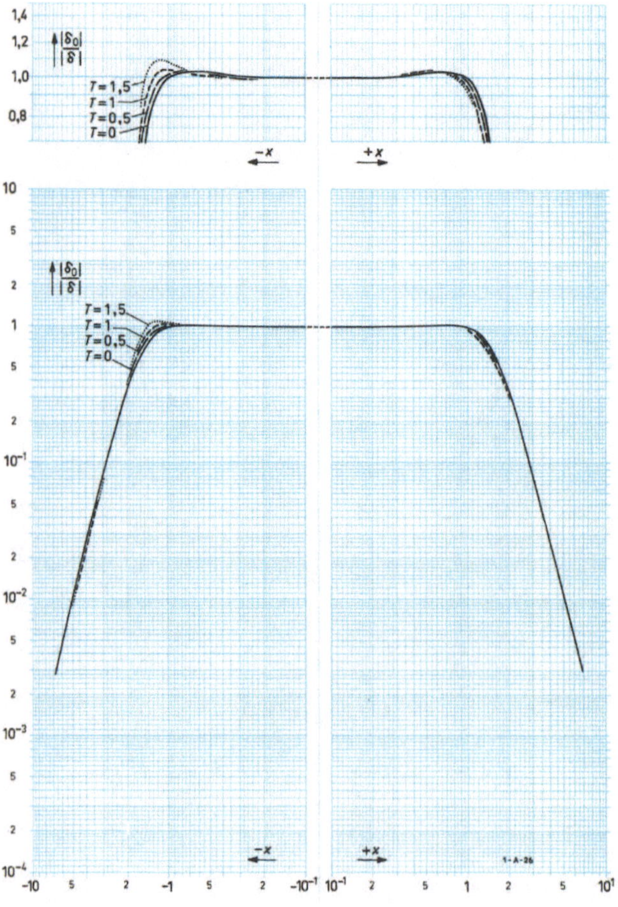

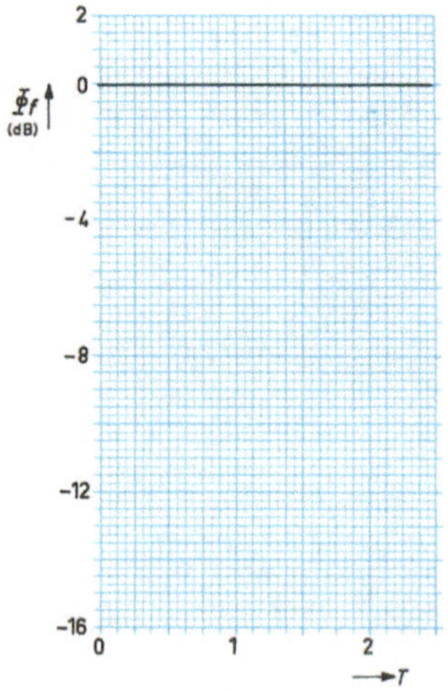

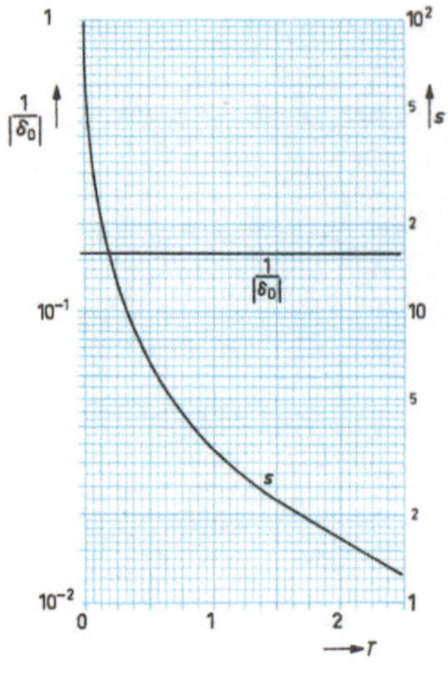

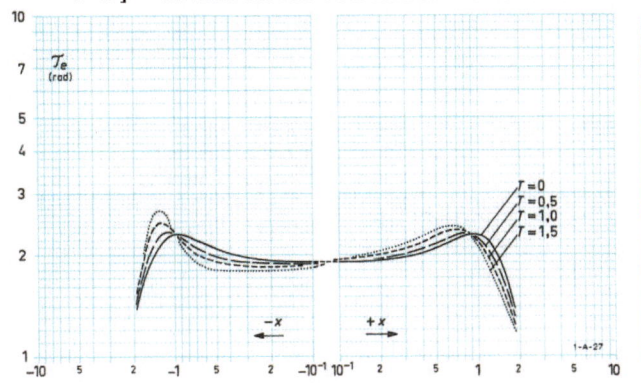

Fig. 15.7
Single-stage amplifier
Configuration II
Tuning method B
$\Theta = 270°$
$q^2 = 1.25, r = 2$

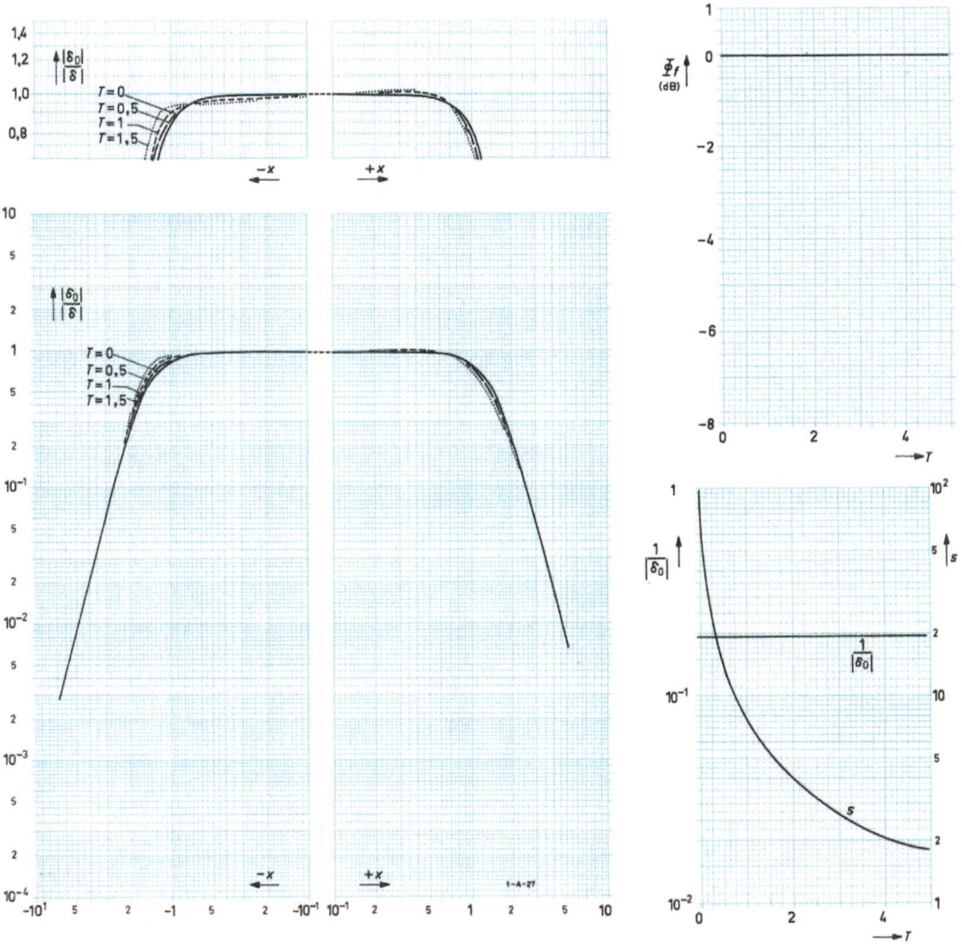

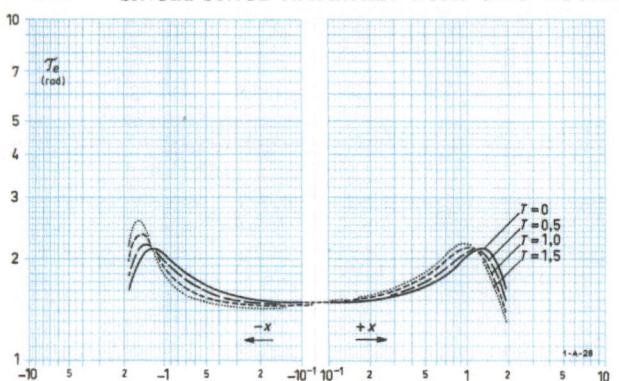

Fig. 15.8
Single-stage amplifier
Configuration II
Tuning method B
$\Theta = 270°$
$q^2 = 1.875, r = 2$

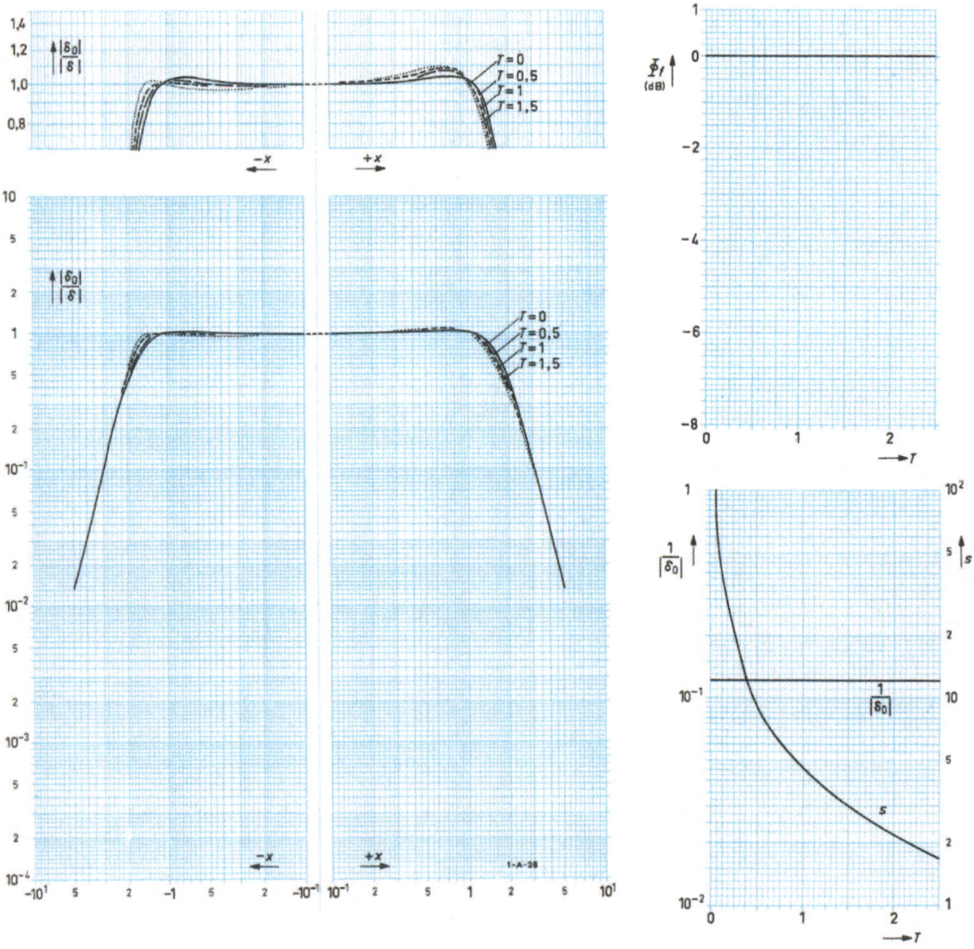

Two-Stage Amplifier with Three Double-Tuned Bandpass Filters

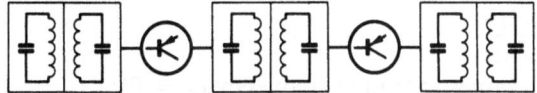

15.4   Two-stage amplifier with three double-tuned bandpass filters

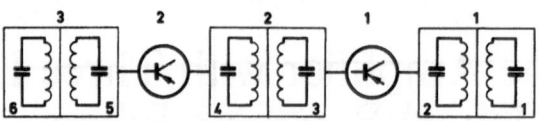

Fig. 15.9 Schematic representation of a two-stage amplifier with three double-tuned bandpass filters (Configuration II).

Reduced amplifier determinant:
See Determinant II on page 233 for $n = 2$.

Method of tuning: B
Reduced amplifier determinant at tuning frequency ($x = 0$):
$$|2\delta_0| \quad = P_{6M}$$
Tuning correction terms:

$$x_1' = T_1 \sin\Theta_1 \cdot \frac{P_{1M}}{P_{2M}}$$

$$x_2' = T_2 \sin\Theta_2 \cdot \frac{P_{3M}}{P_{4M}}$$

Minor determinant values at tuning frequency ($x = 0$):
$$P_{1M} = 1$$
$$P_{2M} = 1 + q_1^2$$

$$P_{3M} = 1 + q_1^2 - T_1 \cos\Theta_1$$

$$P_{4M} = (1 + q_1^2)(1 + q_2^2) - T_1 \cos\Theta_1$$

$$P_{5M} = (1 + q_1^2)(1 + q_2^2 - T_2 \cos\Theta_2) - T_1 \cos\Theta_1 (1 - T_2 \cos\Theta_2)$$

$$P_{6M} = (1 + q_1^2)\{(1 + q_2^2)(1 + q_3^2) - T_2 \cos\Theta_2\} - T_1 \cos\Theta_1 (1 + q_3^2 - T_2 \cos\Theta_2)$$

Minor determinant values at tuning frequency ($x = 0$) for identical stages:
$$P_{1M} = 1$$
$$P_{2M} = 1 + q^2$$
$$P_{3M} = 1 + q^2 - T\cos\Theta$$
$$P_{4M} = (1 + q^2)^2 - T\cos\Theta$$
$$P_{5M} = (1 + q^2)^2 - T\cos\Theta (2 + q^2 - T\cos\Theta)$$
$$P_{6M} = (1 + q^2)^3 - T\cos\Theta (2 + 2q^2 - T\cos\Theta)$$

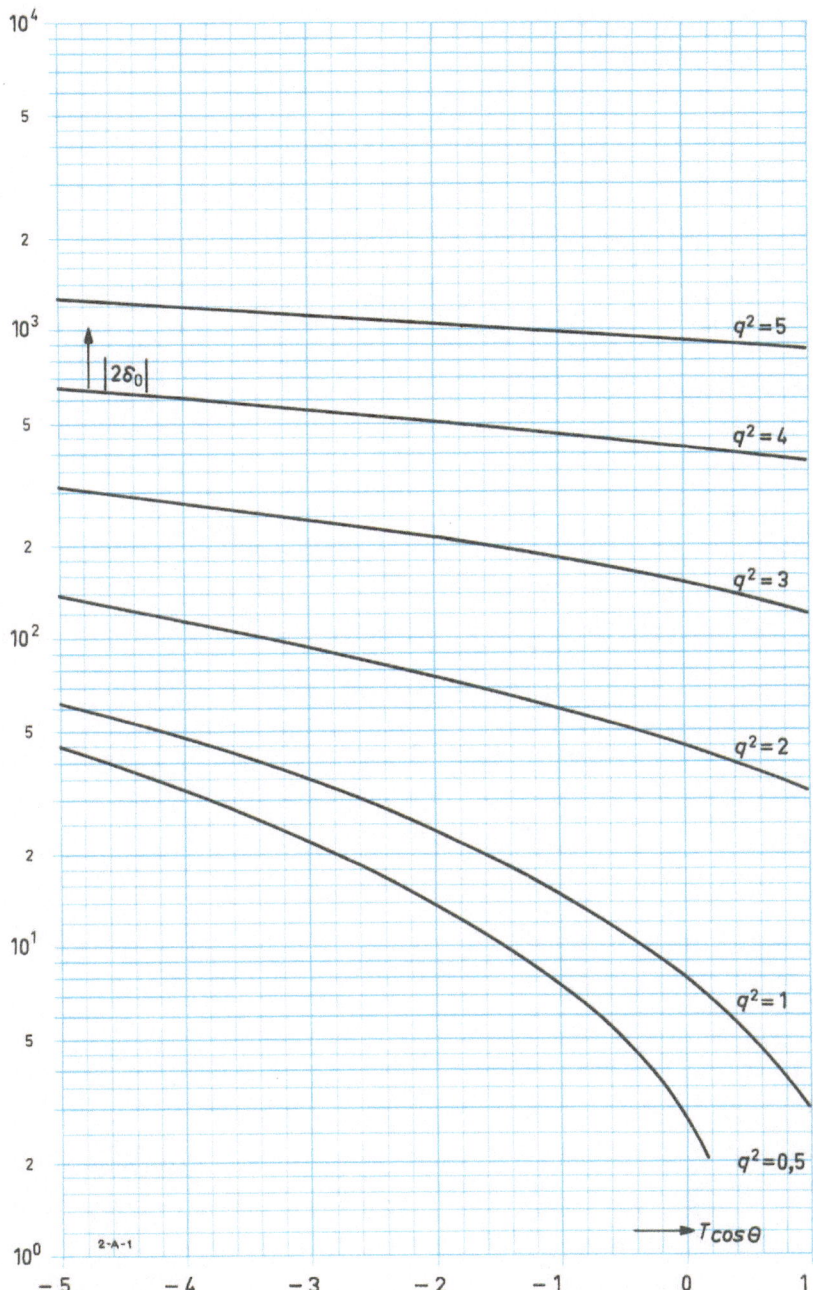

Fig. 15.10 Graph representing the value of the reduced determinant $\left|\,1/\delta_0\,\right|$ as a function of $T\cos\Theta$ for a two-stage amplifier with three double-tuned bandpass filters (Configuration II, tuning method B).

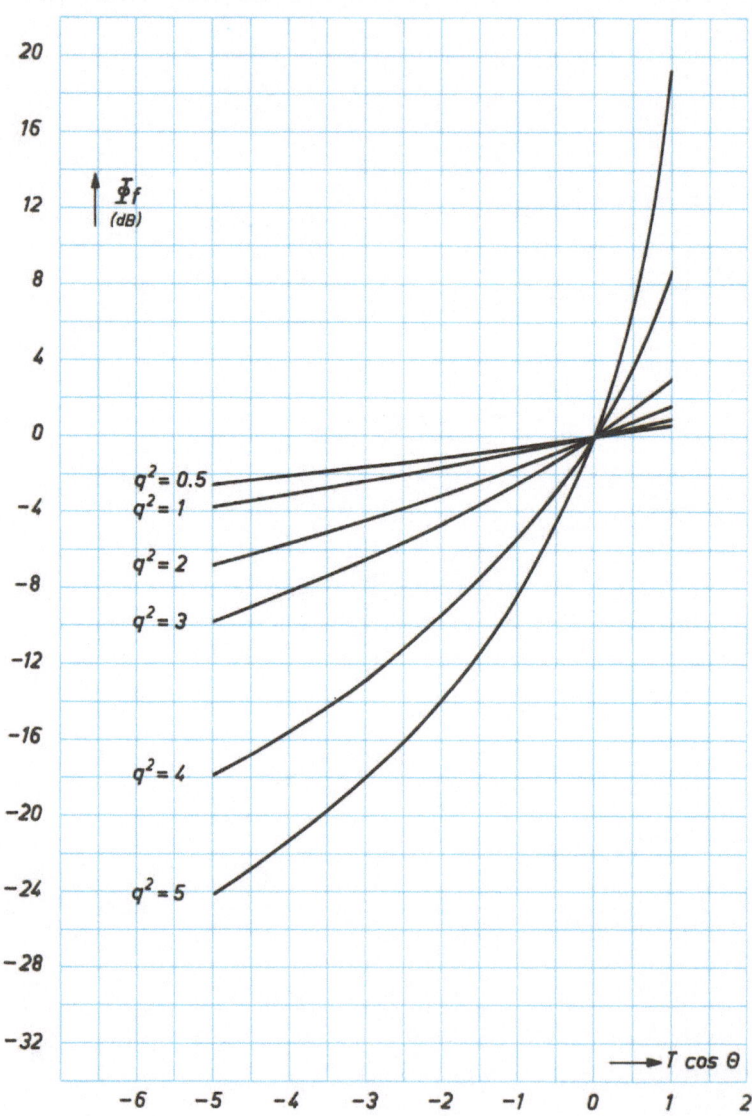

Fig. 15.11 Graph representing the feedback losses Φ_f as a function of $T\cos\theta$ for a two-stage amplifier with three double-tuned bandpass filters (Configuration II, tuning method B).

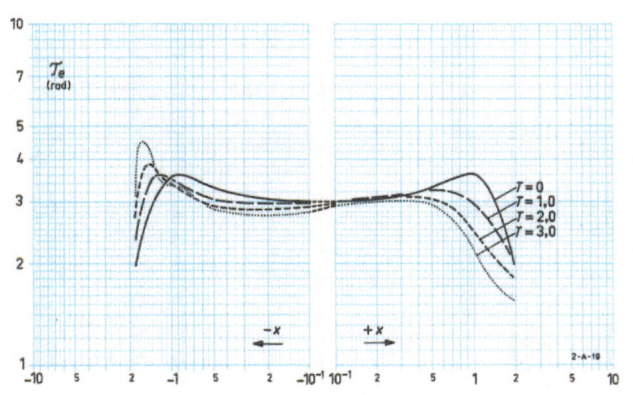

Fig. 15.12
Two-stage amplifier
Configuration II
Tuning method B
$\Theta = 240°$
$q^2 = 1.0, r = 1$

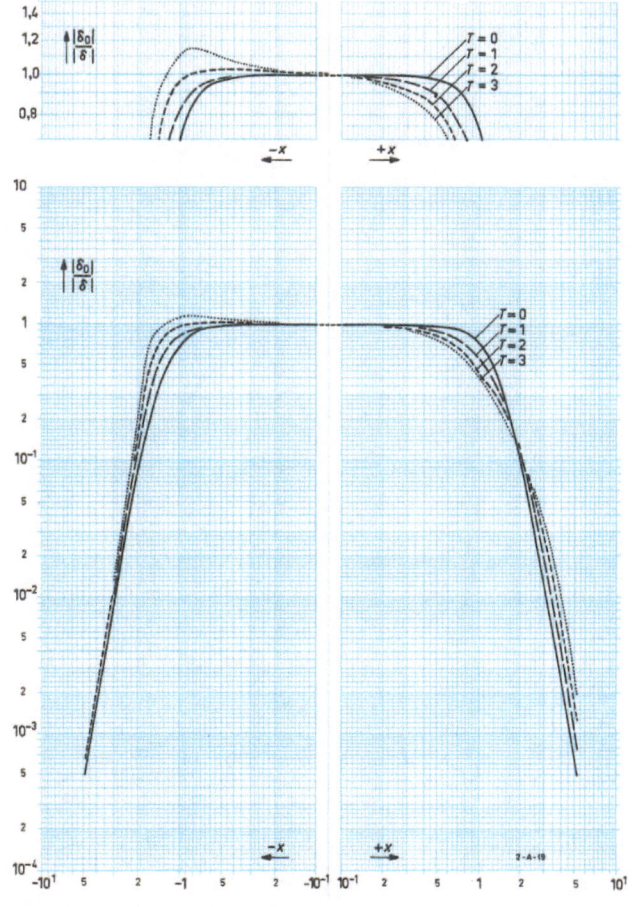

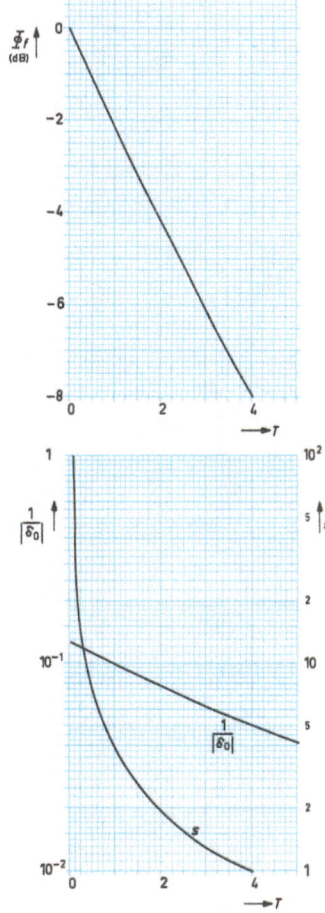

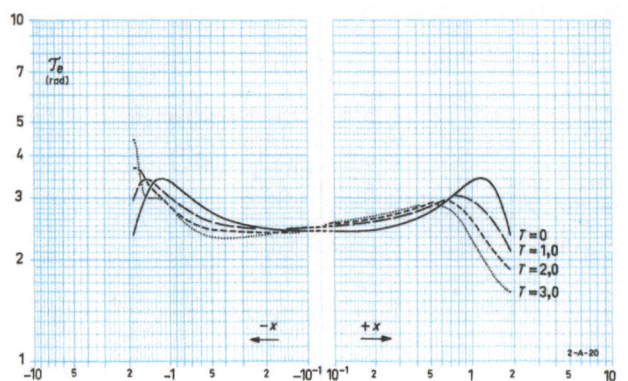

Fig. 15.13
Two-stage amplifier
Configuration II
Tuning method B
$\Theta = 240°$
$q^2 = 1.5, r = 1$

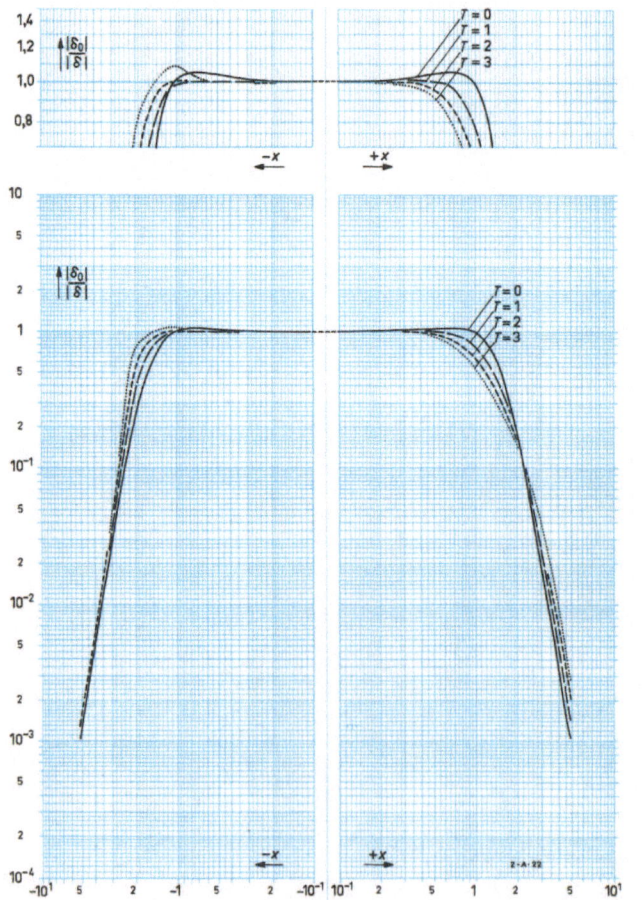

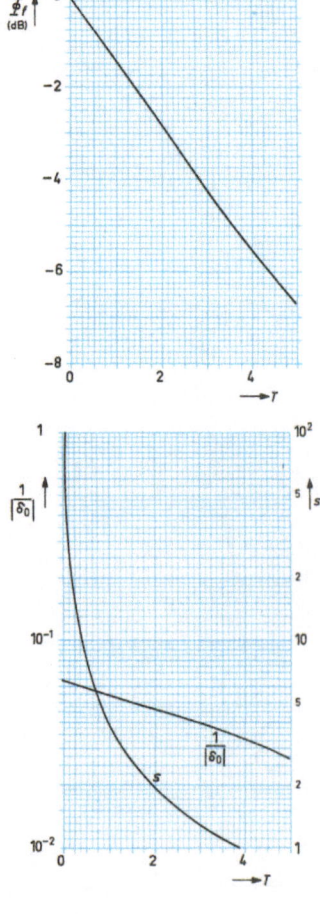

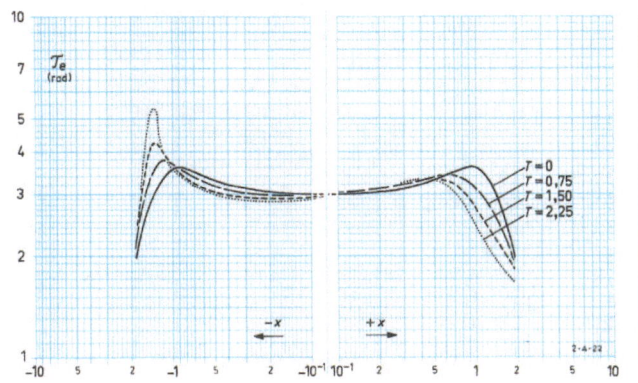

Fig. 15.14
Two-stage amplifier
Configuration II
Tuning method B
$\Theta = 255°$
$q^2 = 1.0, r = 1$

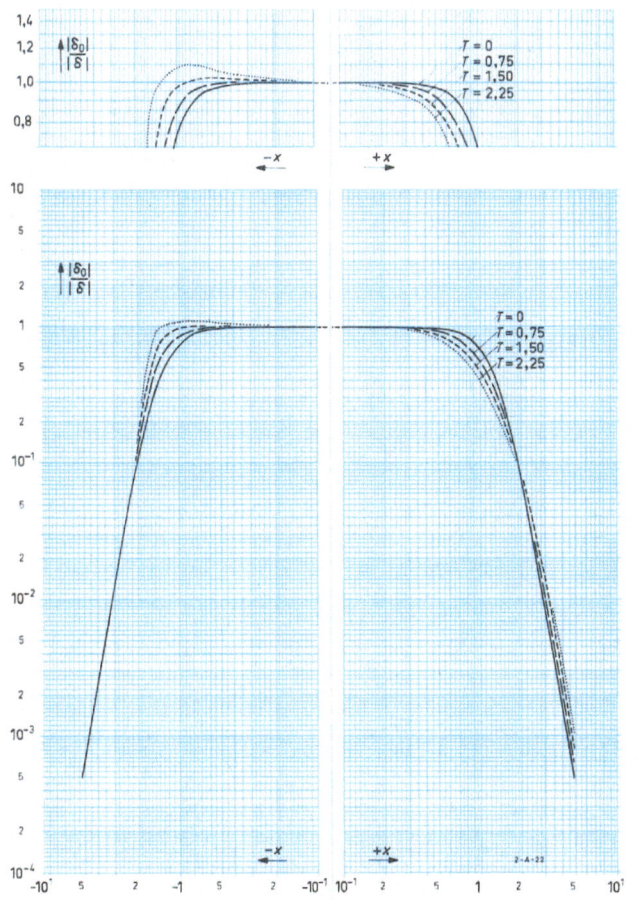

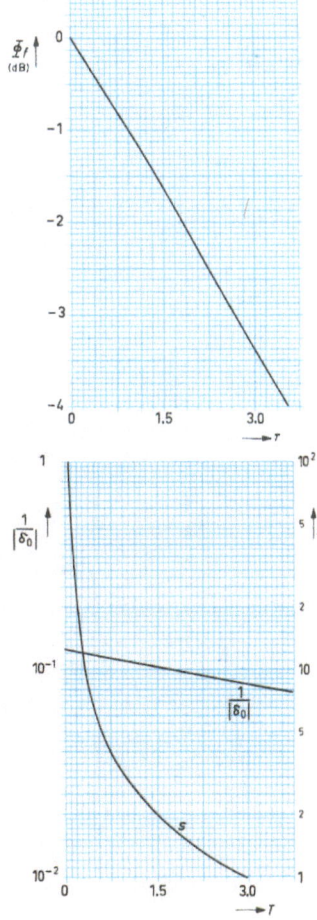

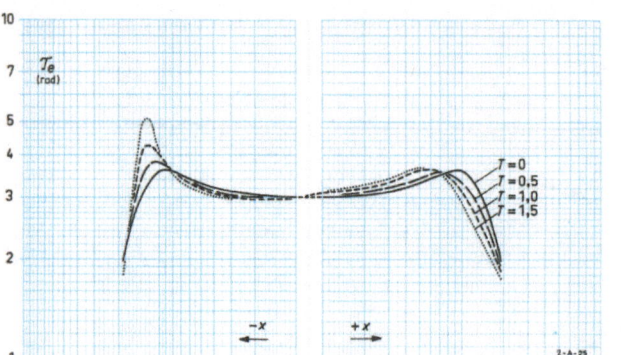

Fig. 15.15
Two-stage amplifier
Configuration II
Tuning method B
$\Theta = 270°$
$q^2 = 1.0, r = 1$

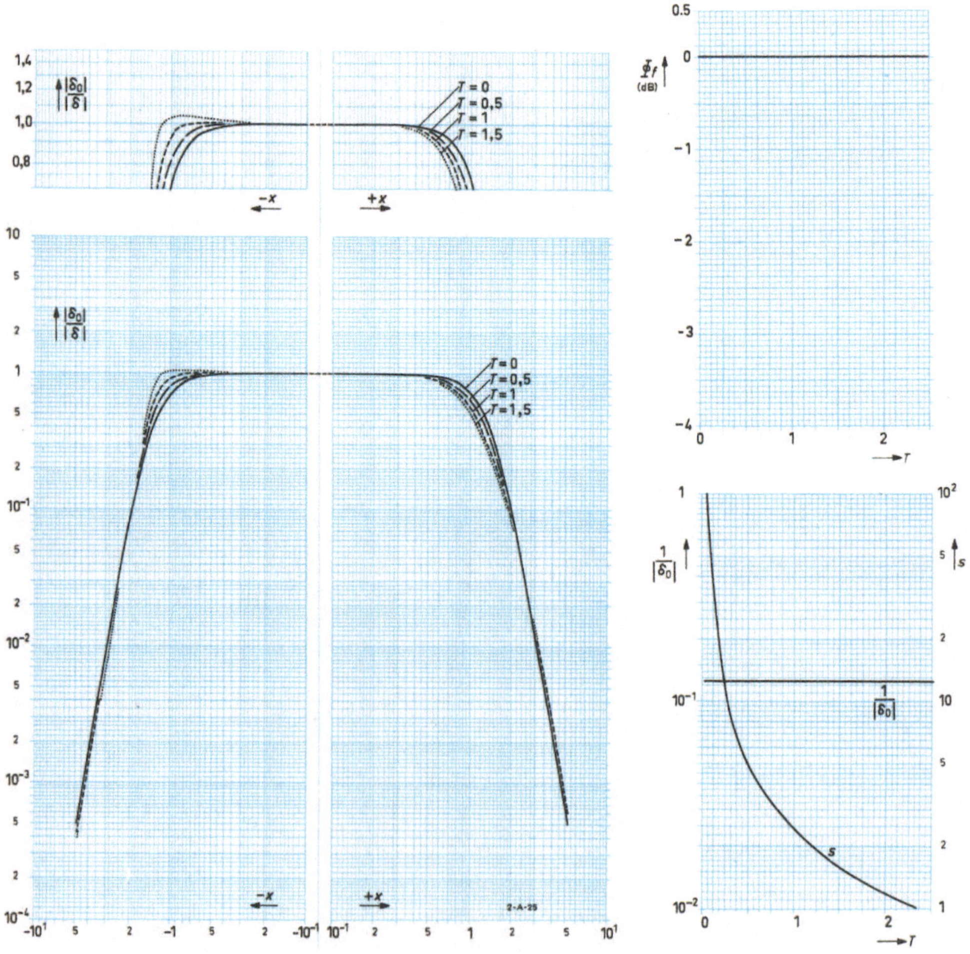

Three-Stage Amplifier with Four Single-Tuned Bandpass Filters

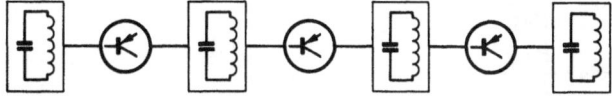

15.5   Three-stage amplifier with four single-tuned bandpass filters

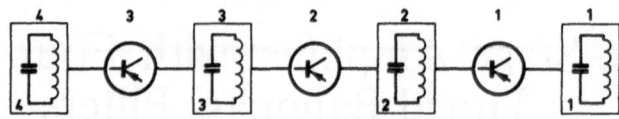

Fig. 15.16 Schematic representation of a three-stage amplifier with four single-tuned bandpass filters (Configuration I).

Reduced amplifier determinant:
See Determinant I on page 232 for $n = 3$.
Method of tuning: B
Reduced amplifier determinant at tuning frequency ($x = 0$):
$|_3\delta_0|$ $= P_{4M}$
Tuning correction terms:

$x'_1 = T_1 \sin\Theta_1$

$x'_2 = T_2 \sin\Theta_2 \cdot \dfrac{P_{1M}}{P_{2M}}$

$x'_3 = T_3 \sin\Theta_3 \cdot \dfrac{P_{2M}}{P_{3M}}$

Minor determinant values at tuning frequency ($x = 0$):
$P_{1M} = 1$
$P_{2M} = 1 - T_1 \cos\Theta_1$
$P_{3M} = 1 - T_1 \cos\Theta_1 - T_2 \cos\Theta_2$
$P_{4M} = 1 - T_1 \cos\Theta_1 (1 - T_3 \cos\Theta_3) - T_2 \cos\Theta_2 - T_3 \cos\Theta_3$

Minor determinant values at tuning frequency ($x = 0$) for identical stages:
$P_{1M} = 1$
$P_{2M} = 1 - T\cos\Theta$
$P_{3M} = 1 - 2T \cos \Theta$
$P_{4M} = 1 - 3T \cos\Theta + T^2 \cos^2\Theta$

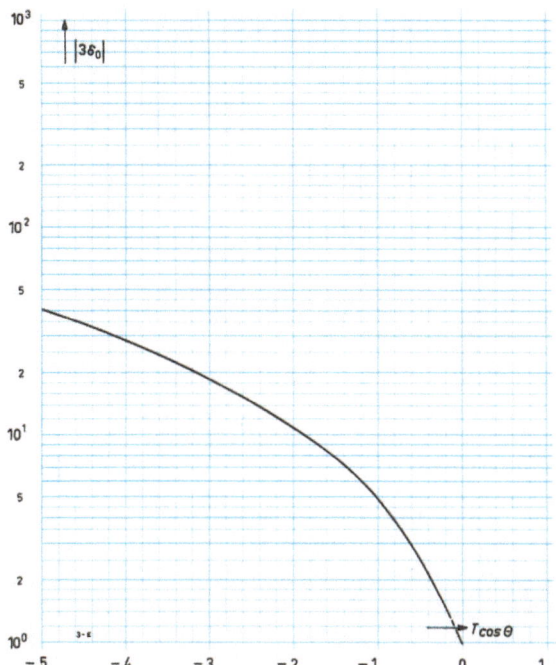

Fig. 15.17
Graph representing the value
of the reduced determinant
$|1/\delta_0|$ as a function of $T\cos\Theta$
for a three-stage amplifier
with four single-tuned bandpass filters (Configuration I,
tuning method B).

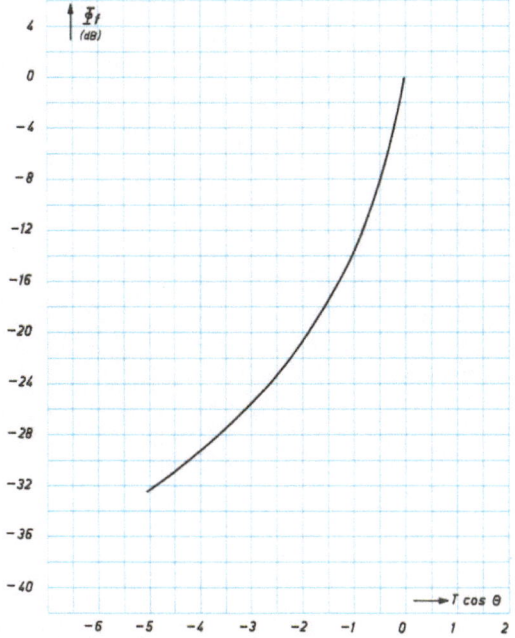

Fig. 15.18
Graph representing the feedback losses Φ_f as a function
of $T\cos\Theta$ for a three-stage
amplifier with four single-
tuned bandpass filters (Configuration I, tuning method
B).

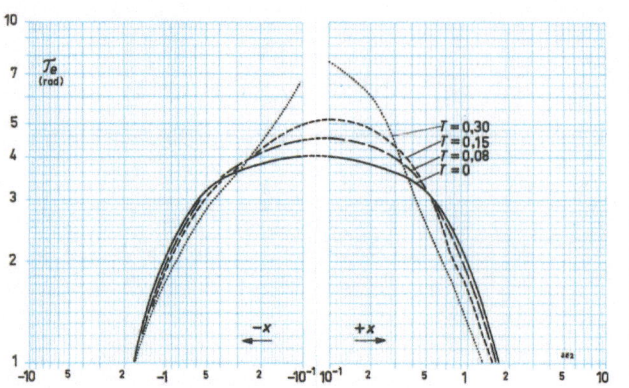

Fig. 15.19
Three-stage amplifier
Configuration I
Tuning method B
$\Theta = 30°$

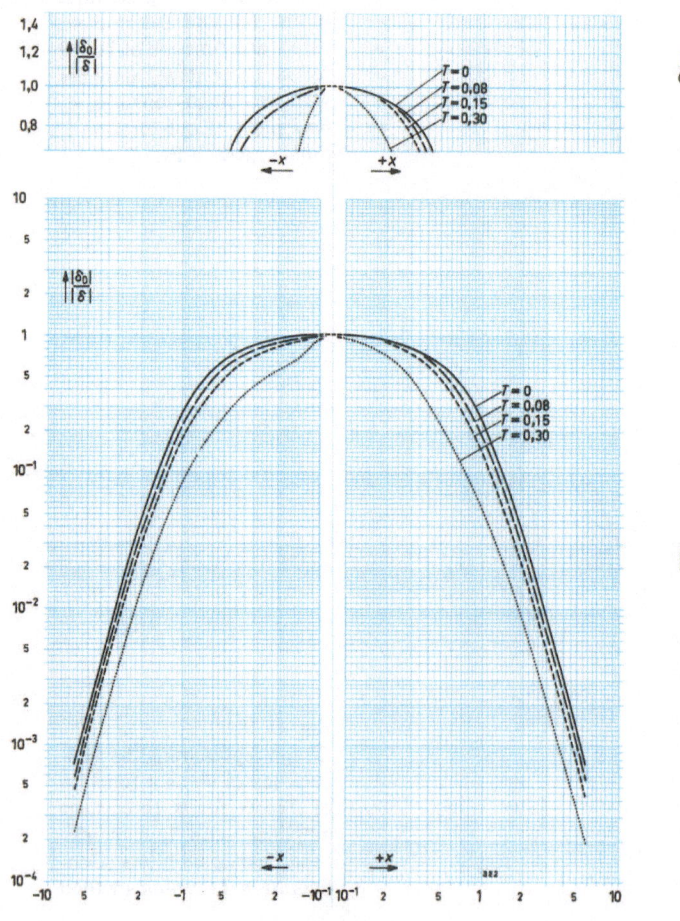

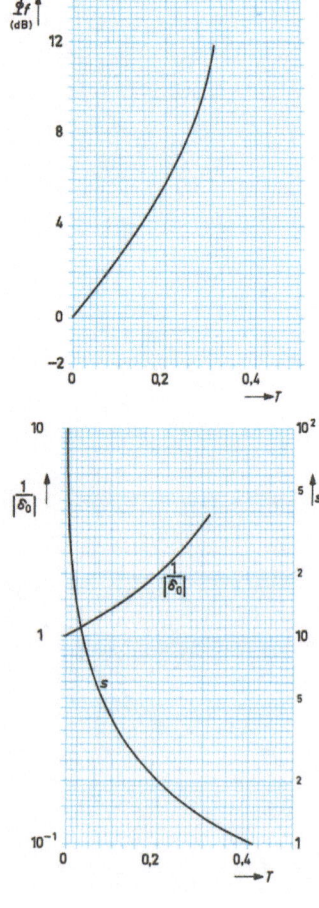

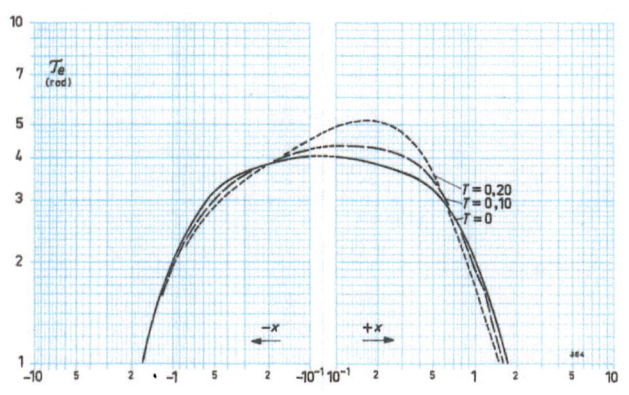

Fig. 15.20
Three-stage amplifier
Configuration I
Tuning method B
$\Theta = 60°$

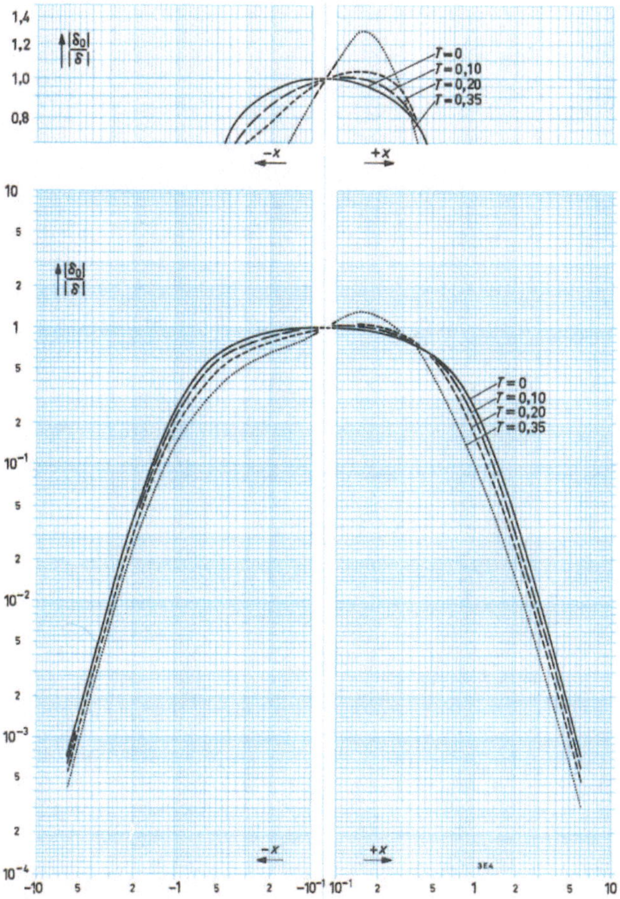

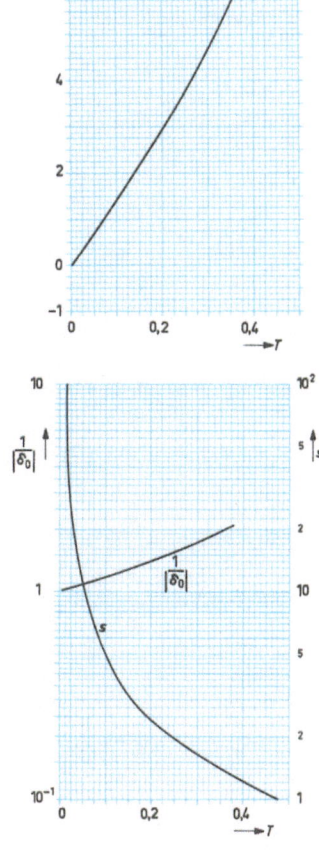

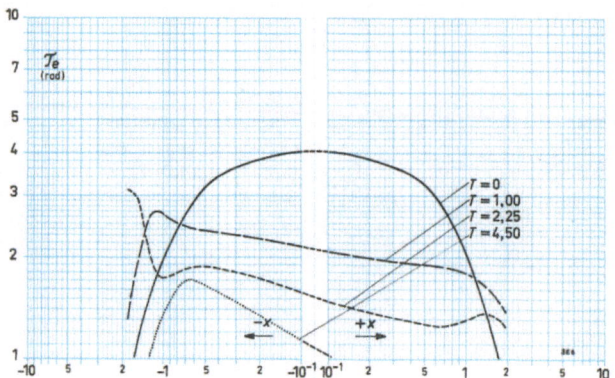

Fig. 15.21
Three-stage amplifier
Configuration I
Tuning method B
$\Theta = 210°$

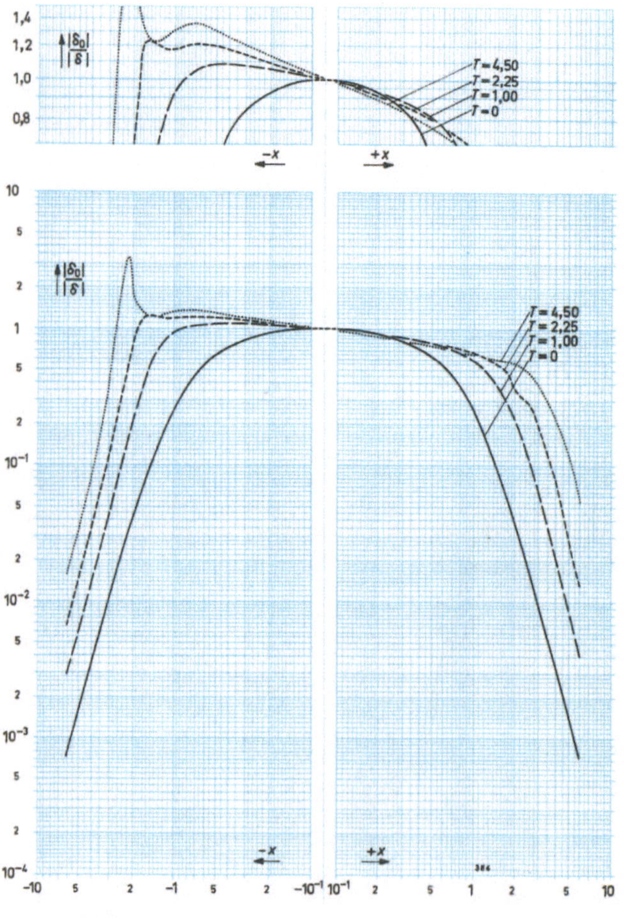

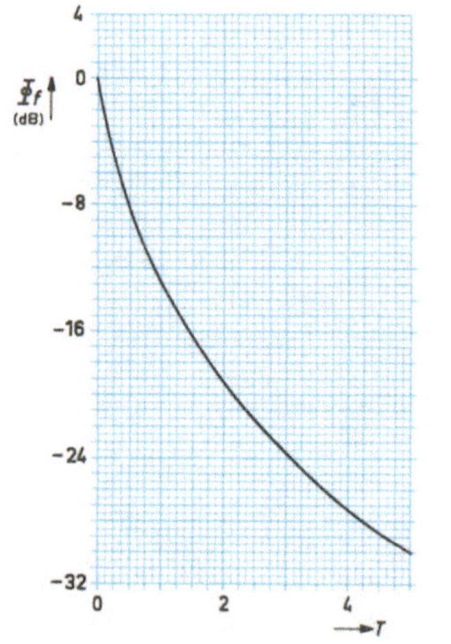

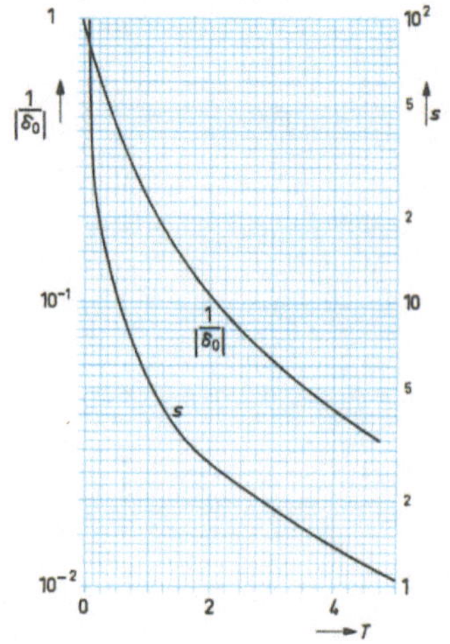

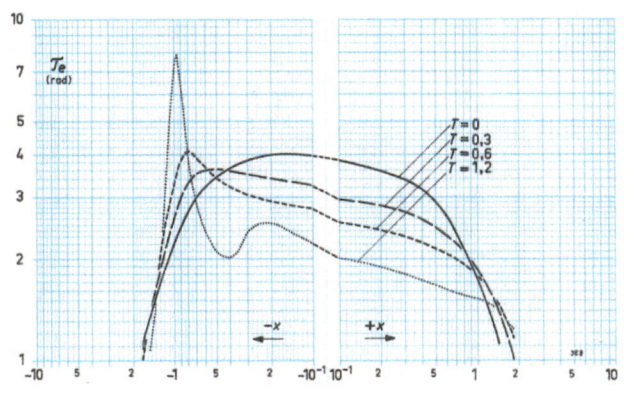

Fig. 15.22
Three-stage amplifier
Configuration I
Tuning method B
$\Theta = 240°$

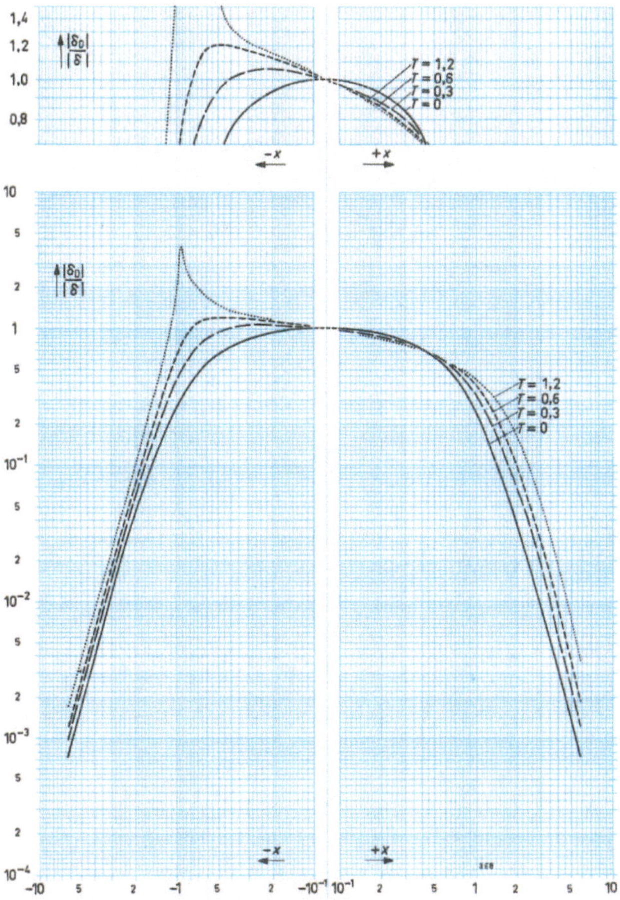

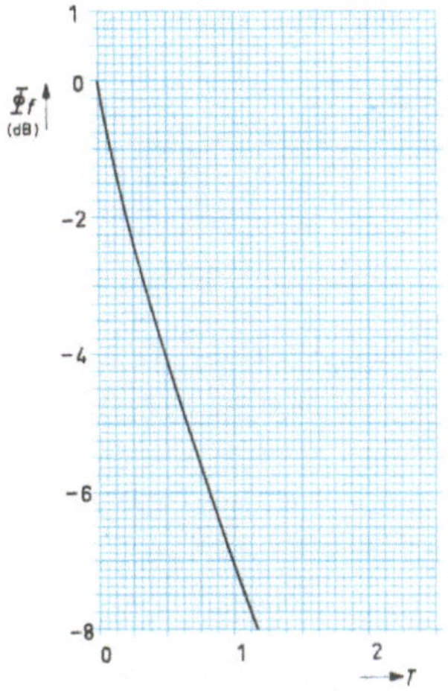

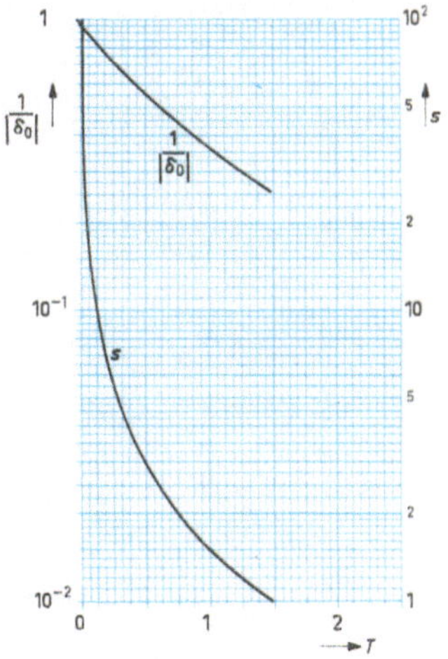

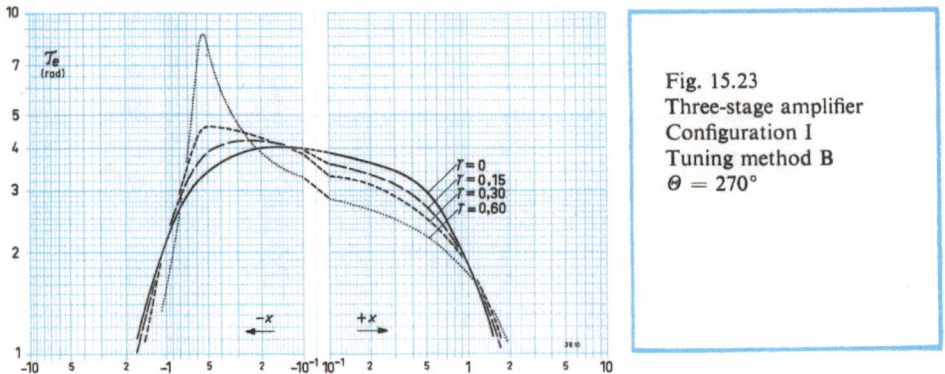

Fig. 15.23
Three-stage amplifier
Configuration I
Tuning method B
$\Theta = 270°$

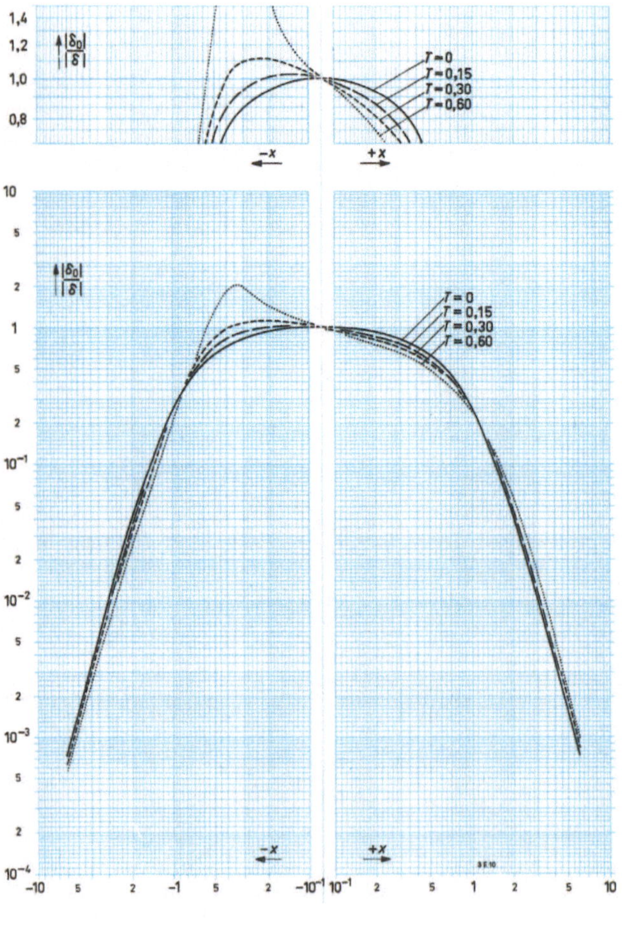

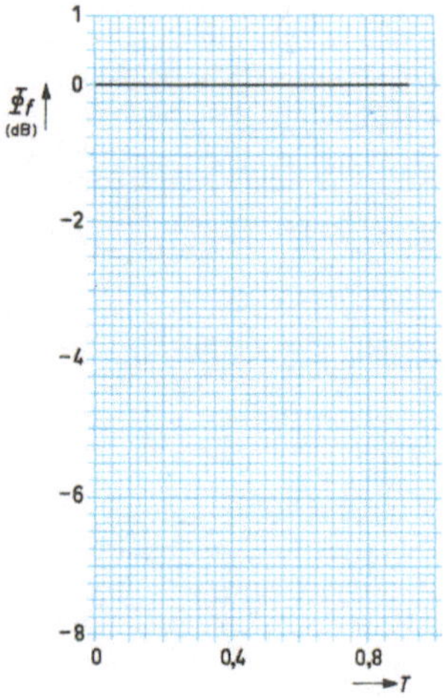

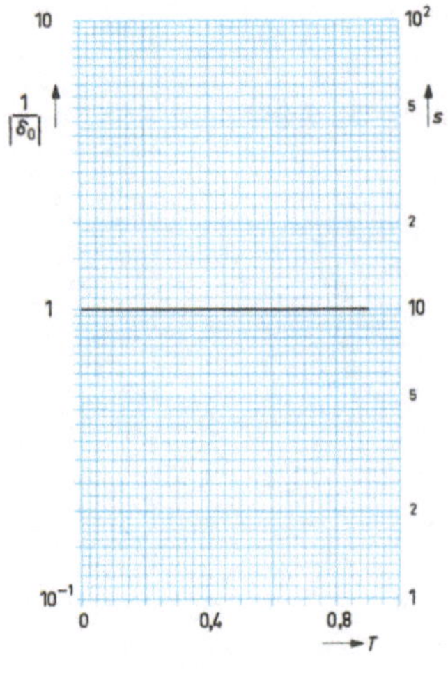

Four-Stage Amplifier with Five Single-Tuned Bandpass Filters

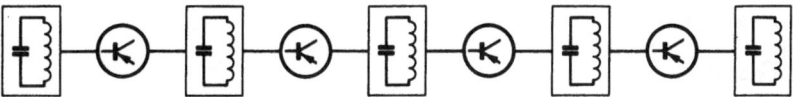

15.6   Four-stage amplifier with five single-tuned bandpass filters

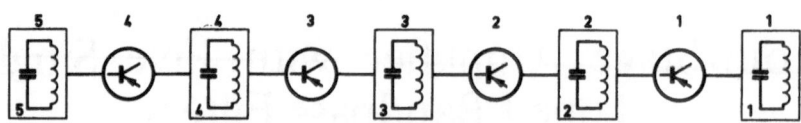

Fig. 15.24 Schematic representation of a four-stage amplifier with five single-tuned band pass filters (Configuration I).

Reduced amplifier determinant:
See Determinant I on page 232 for $n = 4$.

Method of tuning: B
Reduced amplifier determinant at tuning frequency ($x = 0$):
$|_4\delta_0| = P_{6M}$
Tuning correction terms:

$$x'_1 = T_1 \sin\Theta_1$$

$$x'_2 = T_2 \sin\Theta_2 \cdot \frac{P_{1M}}{P_{2M}}$$

$$x'_3 = T_3 \sin\Theta_3 \cdot \frac{P_{2M}}{P_{3M}}$$

$$x'_4 = T_4 \sin\Theta_4 \cdot \frac{P_{3M}}{P_{4M}}$$

Minor determinant values at tuning frequency ($x = 0$):
$P_{1M} = 1$
$P_{2M} = 1 - T_1 \cos\Theta_1$
$P_{3M} = 1 - T_1 \cos\Theta_1 - T_2 \cos\Theta_2$
$P_{4M} = 1 - T_1 \cos\Theta_1 (1 - T_3 \cos\Theta_3) - T_2 \cos\Theta_2 - T_3 \cos\Theta_3$
$P_{5M} = 1 - T_1 \cos\Theta_1 (1 - T_3 \cos\Theta_3 - T_4 \cos\Theta_4) - T_2 \cos\Theta_2 (1 - T_4 \cos\Theta_4)$
      $- T_3 \cos\Theta_3 - T_4 \cos\Theta_4$

Minor determinant values at tuning frequency ($x = 0$) for identical stages:
$P_{1M} = 1$
$P_{2M} = 1 - T \cos\Theta$
$P_{3M} = 1 - 2 T \cos\Theta$
$P_{4M} = 1 - 3 T \cos\Theta + T^2 \cos^2\Theta$
$P_{5M} = 1 - 4 T^2 \cos\Theta + 3 T^2 \cos^2\Theta$

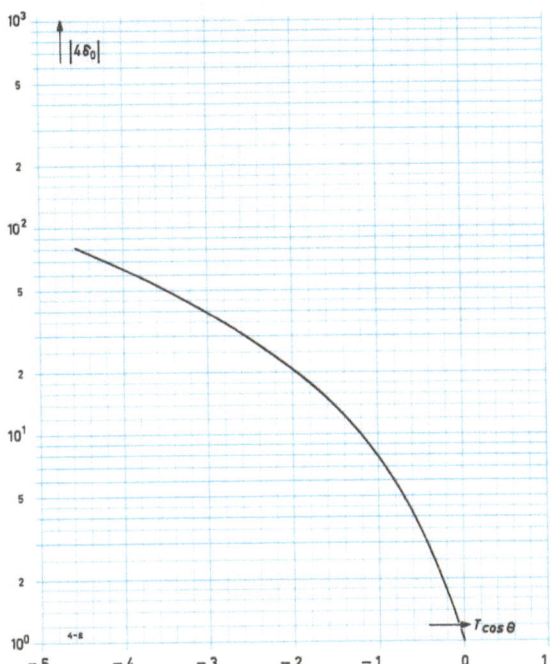

Fig. 15.25 Graph representing the value of the reduced determinant $|1/\delta_0|$ as a function of $T\cos\Theta$ for a four-stage amplifier with five single-tuned bandpass filters (Configuration I, tuning method B).

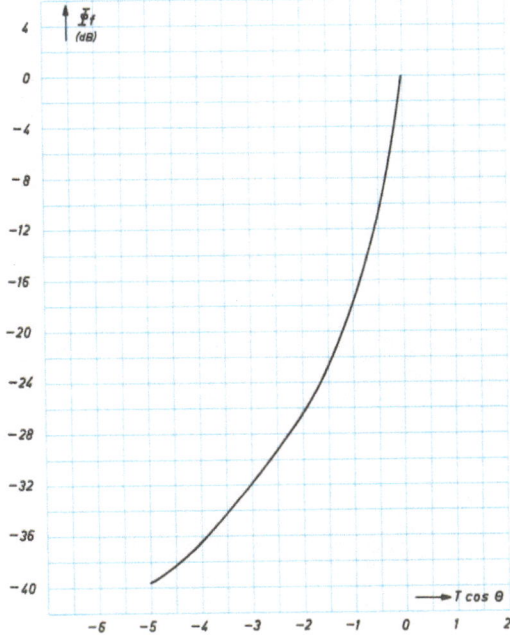

Fig. 15.26 Graph representing the feedback losses Φ_f as a function of $T\cos\Theta$ for a four-stage amplifier with five single-tuned bandpass filters (Configuration I, tuning method B).

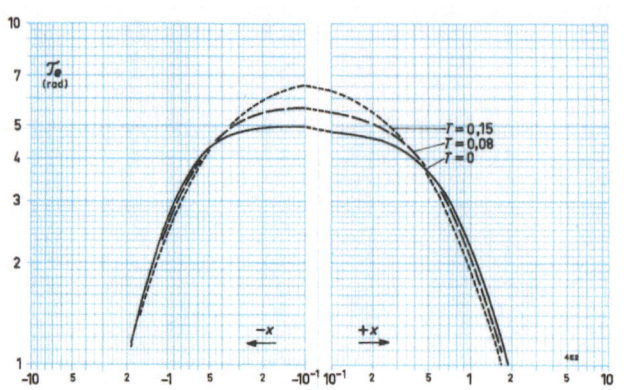

Fig. 15.27
Four-stage amplifier
Configuration I
Tuning method B
$\Theta = 30°$

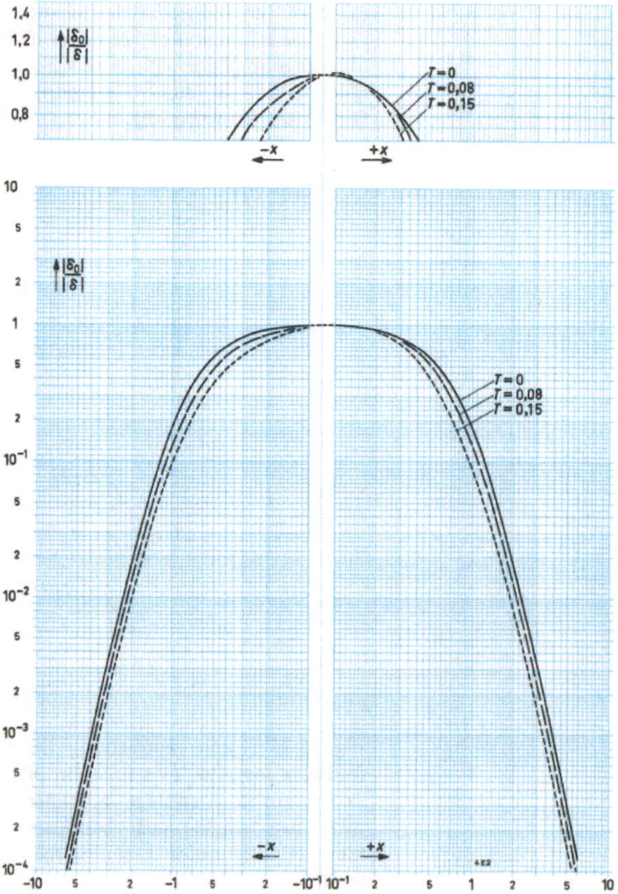

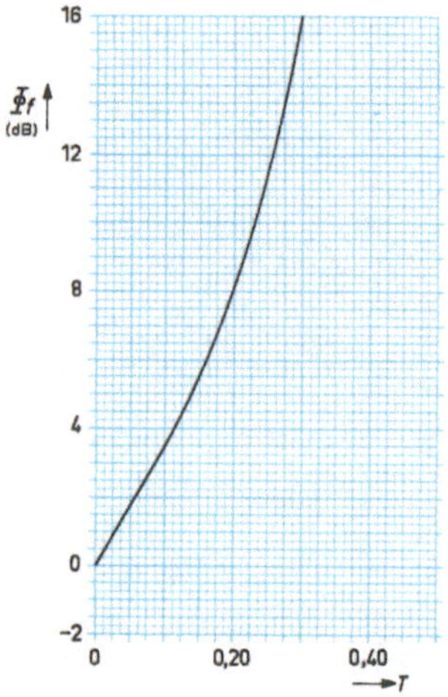

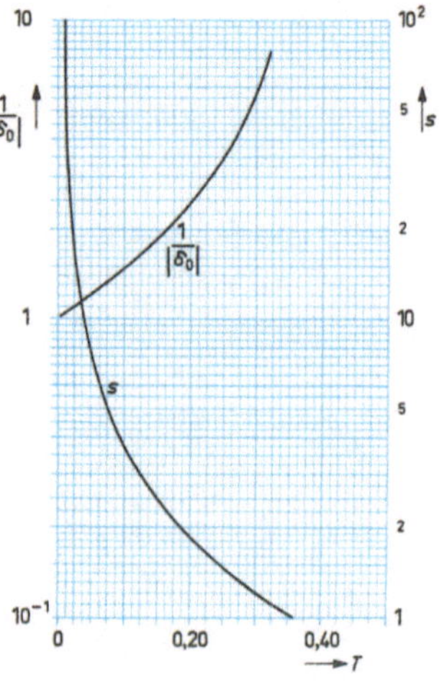

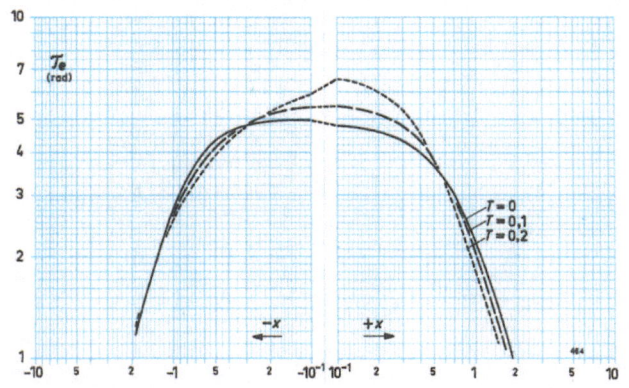

Fig. 15.28
Four-stage amplifier
Configuration I
Tuning method B
$\Theta = 60°$

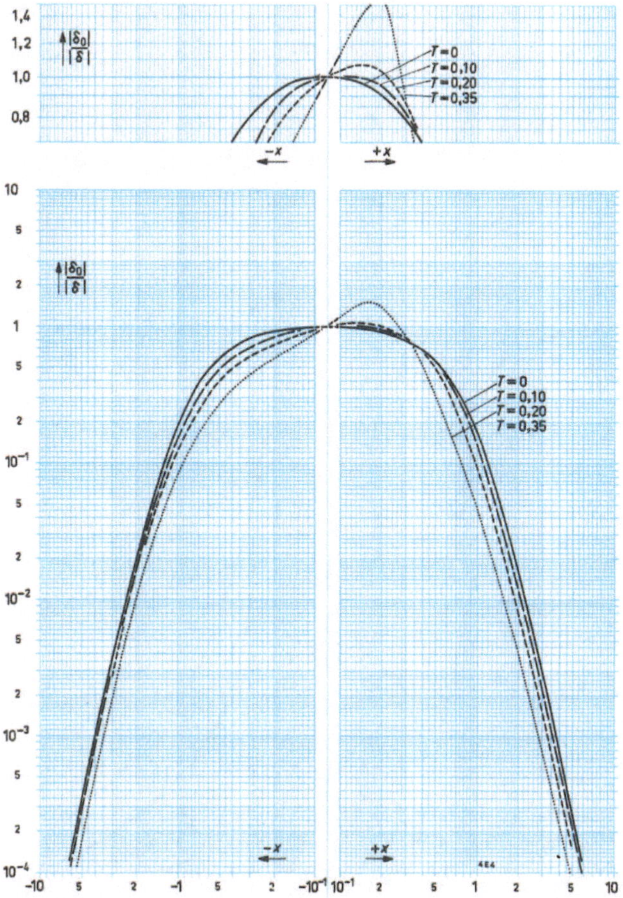

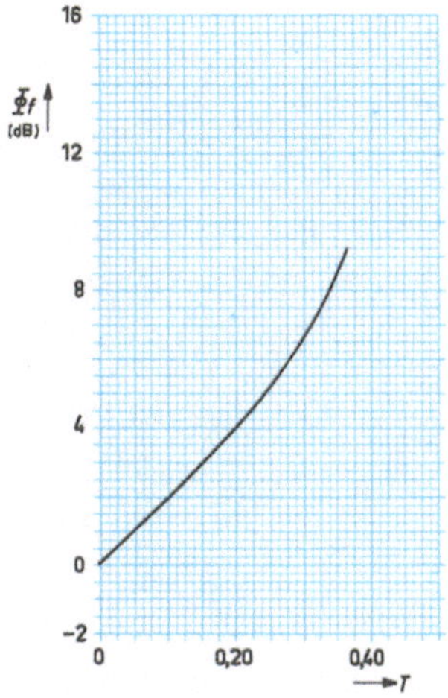

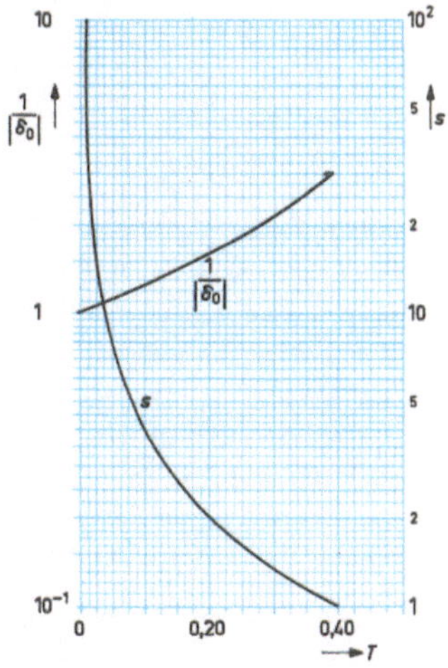

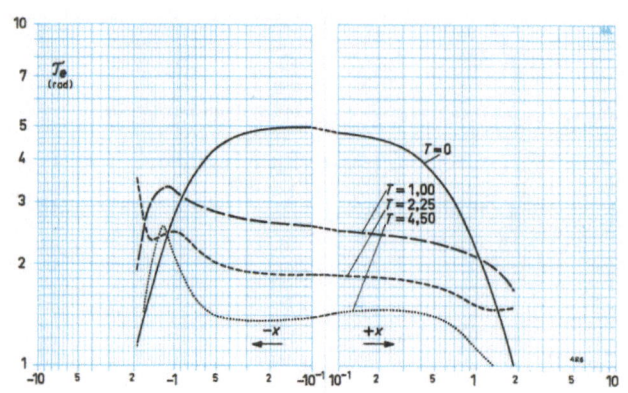

Fig. 15.29
Four-stage amplifier
Configuration I
Tuning method B
$\Theta = 210°$

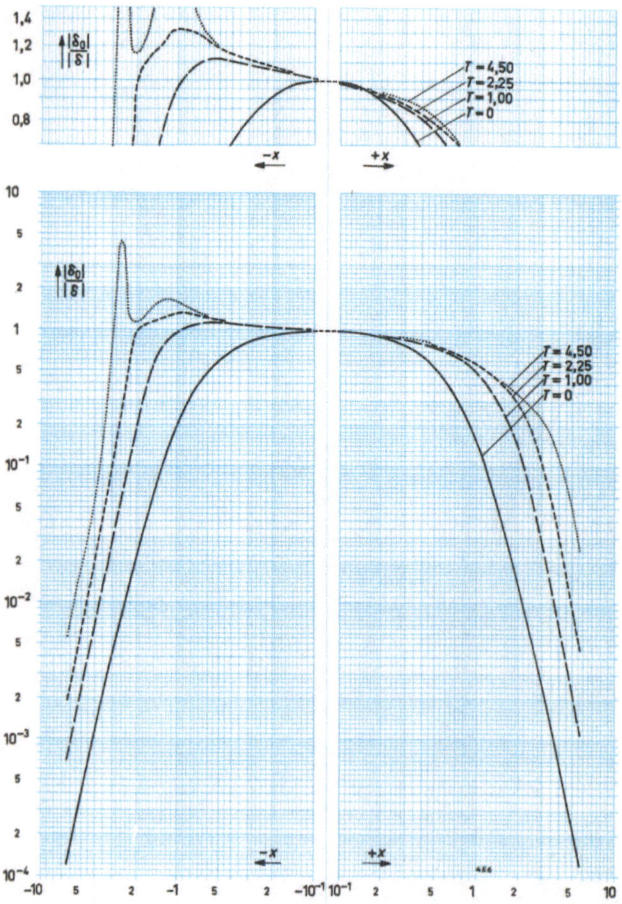

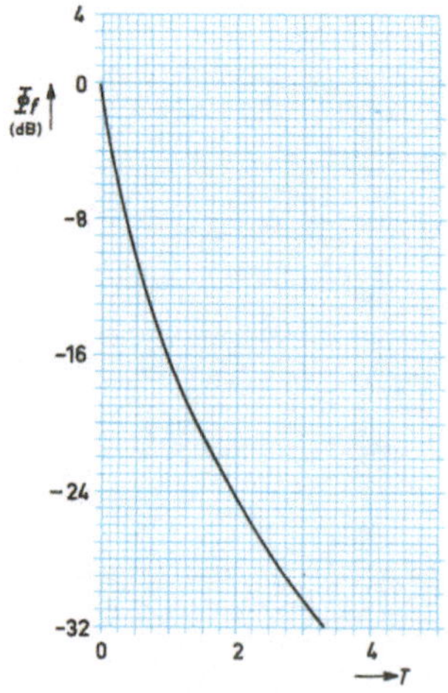

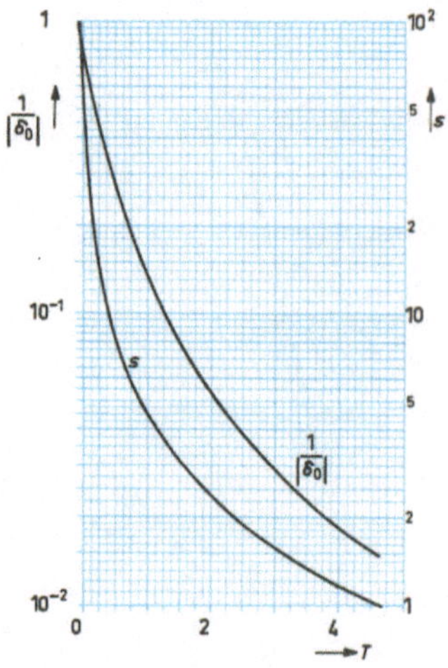

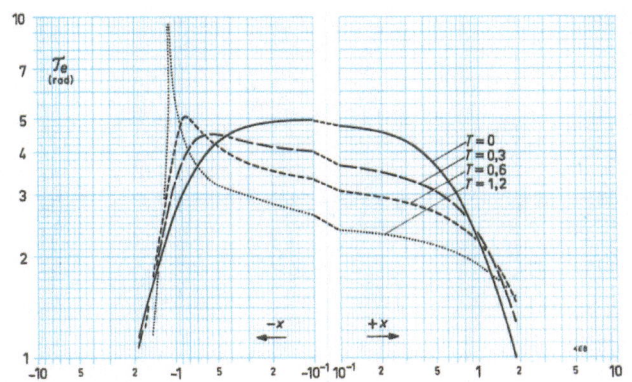

Fig. 15.30
Four-stage amplifier
Configuration I
Tuning method B
$\Theta = 240°$

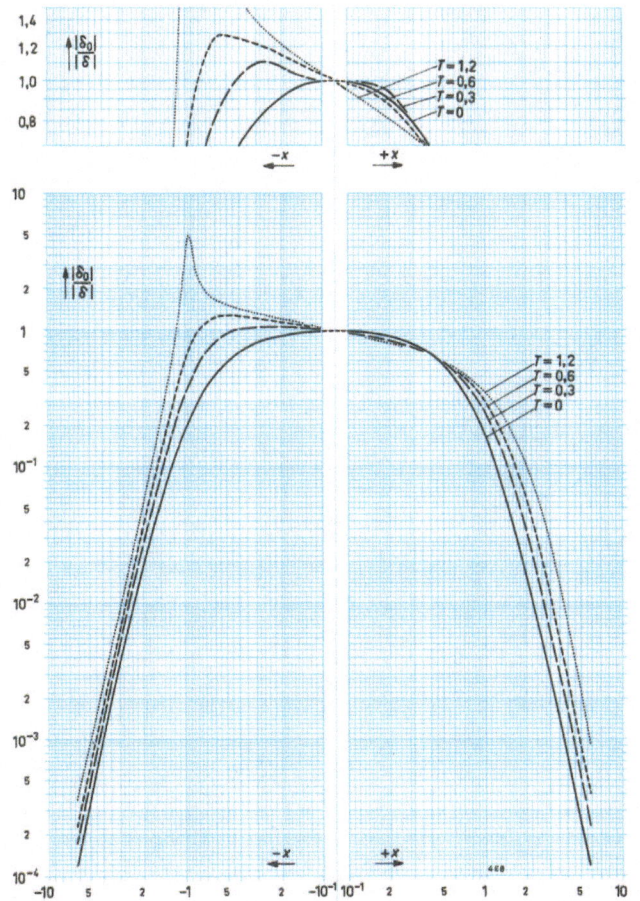

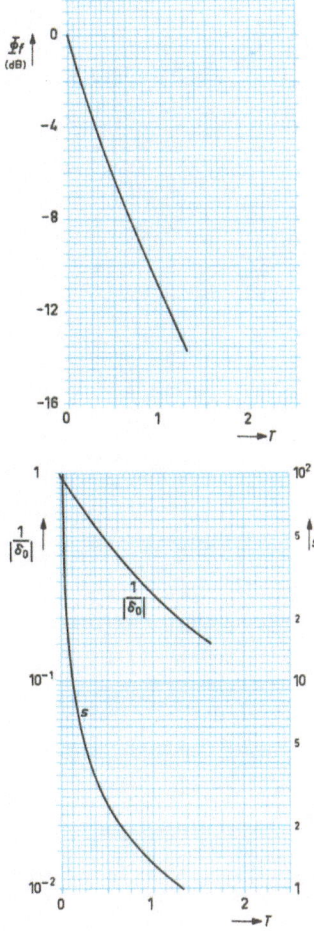

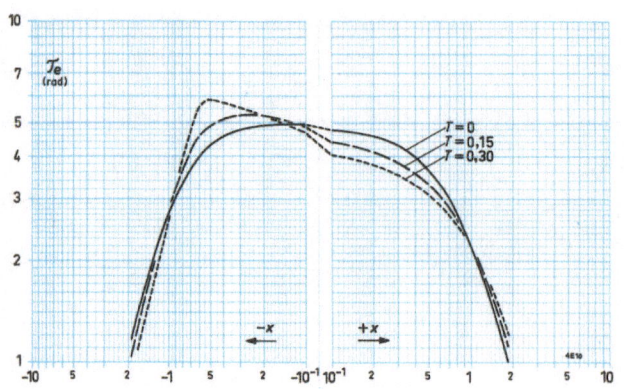

Fig. 15.31
Four-stage amplifier
Configuration I
Tuning method B
$\Theta = 270°$

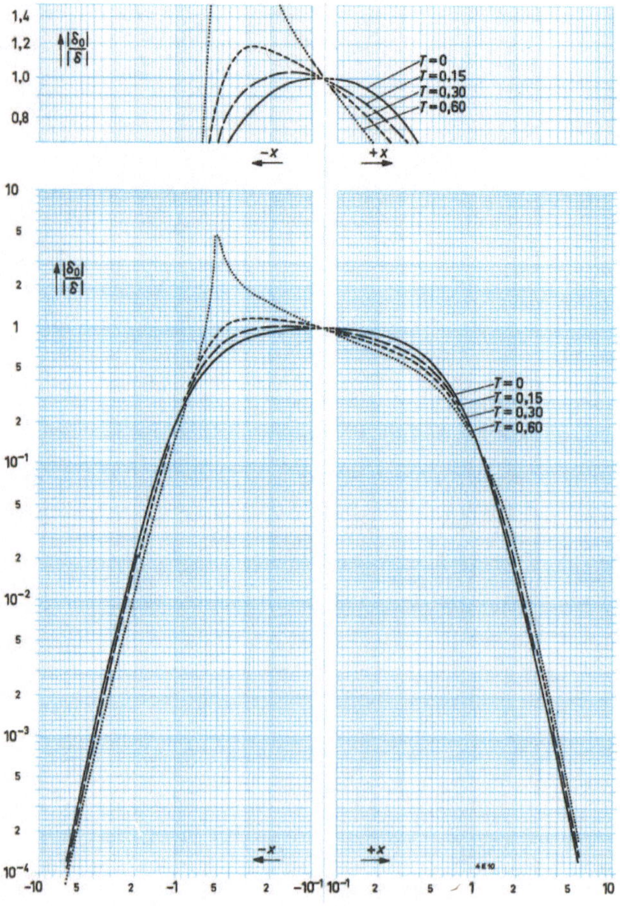

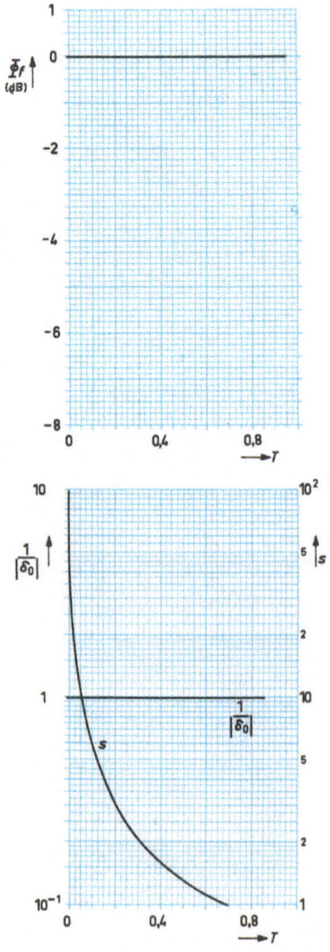

Three-Stage Amplifier with Four Double-Tuned Bandpass Filters

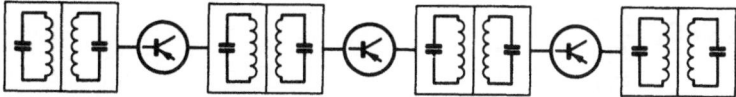

15.7   Three-Stage amplifier with four double-tuned bandpass filters

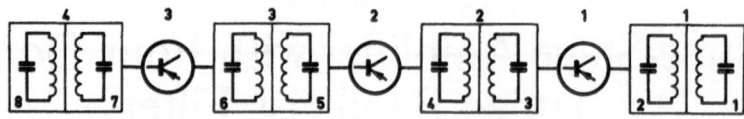

Fig. 15.32 Schematic representation of a three-stage amplifier with four double-tuned bandpass filters (Configuration II).

Reduced amplifier determinant:

See Determinant II on page 233 for $n = 3$.

Method of tuning: B

Reduced amplifier determinant at tuning frequency ($x = 0$):

$$|_3\delta_0| = P_{8M}$$

Tuning correction terms:

$$x_1' = T_1 \sin\Theta_1 \cdot \frac{P_{1M}}{P_{2M}}$$

$$x_2' = T_2 \sin\Theta_2 \cdot \frac{P_{3M}}{P_{4M}}$$

$$x_3' = T_3 \sin\Theta_3 \cdot \frac{P_{5M}}{P_{6M}}$$

Minor determinant values at tuning frequency ($x = 0$):

$P_{1M} = 1$

$P_{2M} = 1 + q_1^2$

$P_{3M} = 1 + q_1^2 - T_1 \cos\Theta_1$

$P_{4M} = (1 + q_1^2)(1 + q_2^2) - T_1 \cos\Theta_1$

$P_{5M} = (1 + q_1^2)(1 + q_2^2 - T_2 \cos\Theta_2) - T_1 \cos\Theta_1 (1 - T_2 \cos\Theta_2)$

$P_{6M} = (1 + q_1^2)\left\{(1 + q_2^2)(1 + q_3^2) - T_2 \cos\Theta_2\right\} - T_1 \cos\Theta_1 (1 + q_3^2 - T_2 \cos\Theta_2)$

$P_{7M} = (1 + q_1^2)\left\{(1 + q_2^2)(1 + q_3^2 - T_3 \cos\Theta_3) - T_2 \cos\Theta_2 (1 - T_3 \cos\Theta_3)\right\}$

$\qquad - T_1 \cos\Theta_1 \left\{- T_2 \cos\Theta_2 (1 - T_3 \cos\Theta_3) + 1 + q_3^2 - T_3 \cos\Theta_3\right\}$

$P_{8M} = (1 + q_1^2)\left[(1 + q_2^2)\left\{(1 + q_3^2)(1 + q_4^2) - T_3 \cos\Theta_3\right\}\right.$

$\qquad \left. - T_2 \cos\Theta_2 (1 + q_4^2 - T_3 \cos\Theta_3)\right]$

$\qquad - T_1 \cos\Theta_1 \left\{- T_2 \cos\Theta_2 (1 + q_4^2 - T_3 \cos\Theta_3) + 1 + q_3^2 - T_3 \cos\Theta_3\right.$

$\qquad \left. + q_4^2 (1 + q_3^2)\right\}$

Minor determinant values at tuning frequency ($x = 0$) for identical stages:

$P_{1M} = 1$

$P_{2M} = 1 + q^2$

$P_{3M} = 1 + q^2 - T\cos\Theta$

$P_{4M} = (1 + q^2)^2 - T \cos\Theta$

$P_{5M} = (1 + q^2)^2 - T \cos\Theta (2 + q^2 - T \cos\Theta)$

$P_{6M} = (1 + q^2)^3 - T \cos\Theta (2 + 2q^2 - T \cos\Theta)$

$P_{7M} = (1 + q^2)^3 - T \cos\Theta \left\{3 + 4q^2 + q^4 - (3 + q^2) T\cos\Theta + T^2 \cos^2\Theta\right\}$

$P_{8M} = (1 + q^2)^4 - T \cos\Theta \left\{3 + 6q^2 + 3q^4 - (3 + 2q^2) T \cos\Theta + T^2 \cos^2\Theta\right\}$

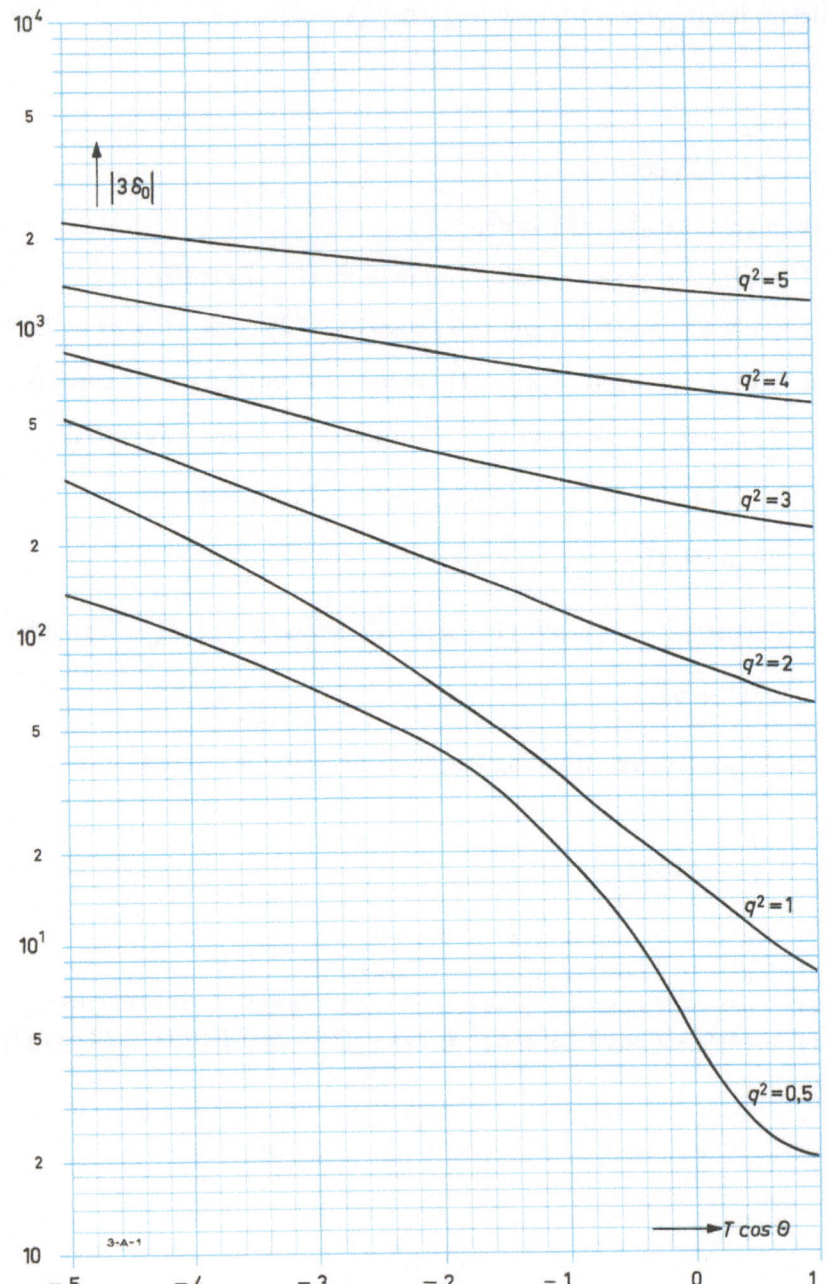

Fig. 15.33 Graph representing the value of the reduced determinant $\mid 1/\delta_0 \mid$ as a function of $T\cos\theta$ for a three-stage amplifier with four double-tuned bandpass filters (Configuration II, tuning method B).

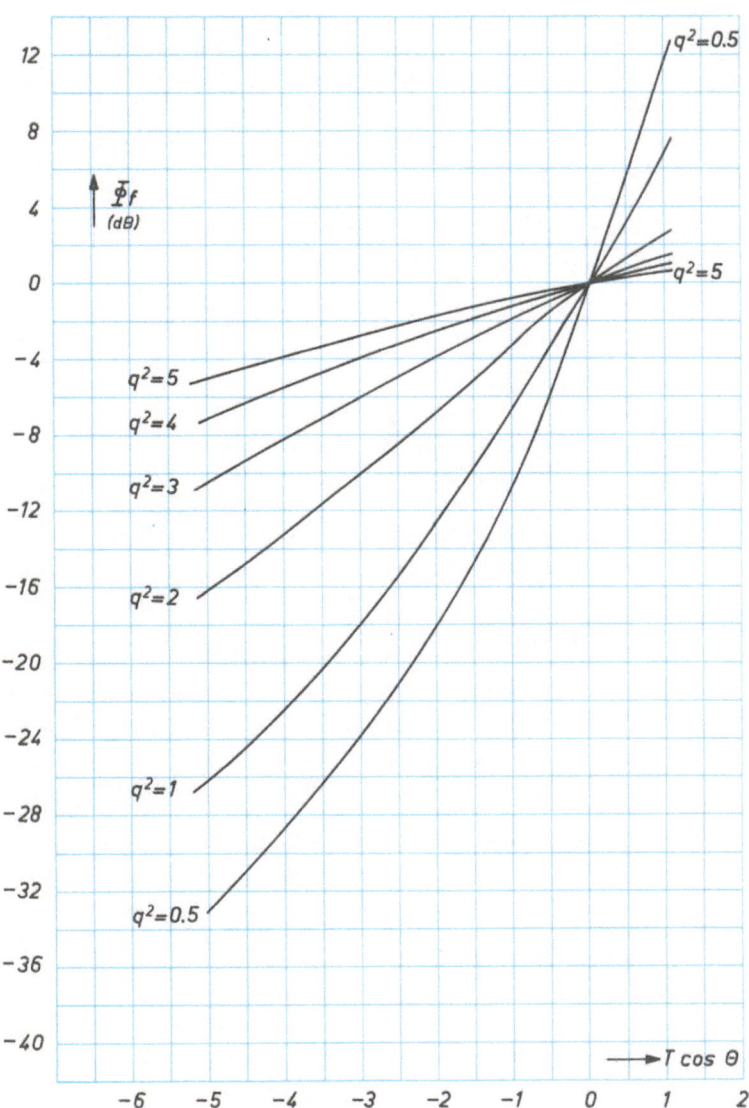

Fig. 15.34 Graph representing the feedback losses Φ_f as a function of $T\cos\Theta$ for a three-stage amplifier with four double-tuned bandpass filters (Configuration II, tuning method B).

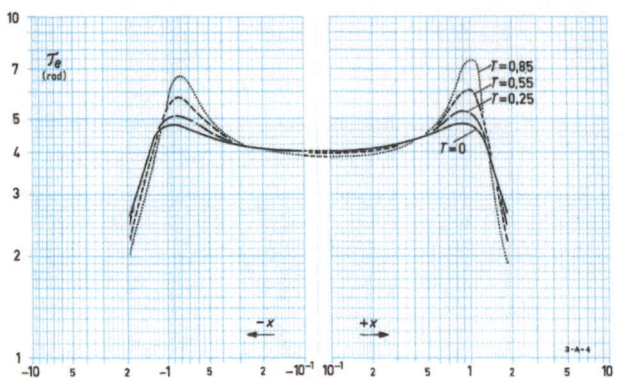

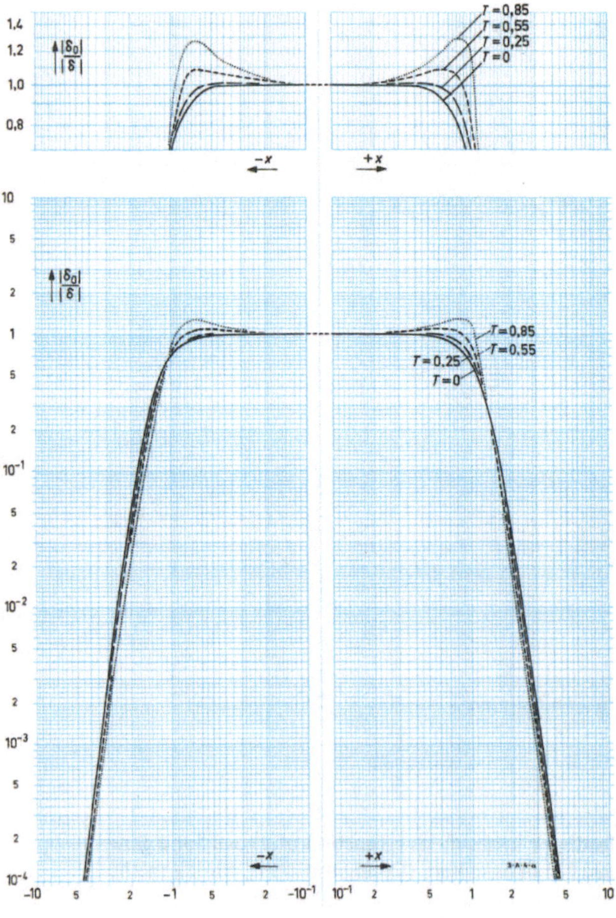

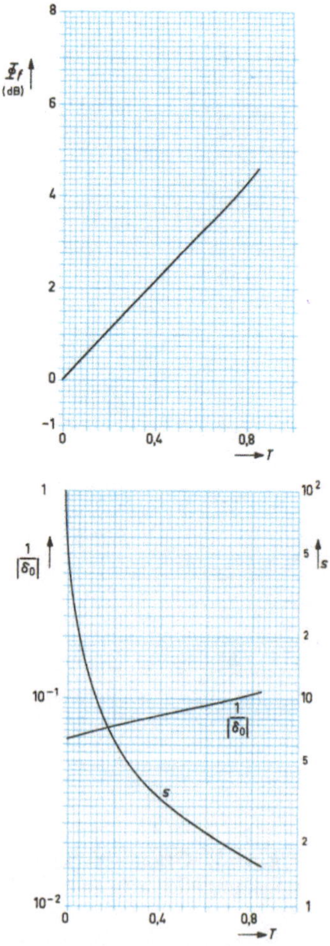

Fig. 15.35
Three-stage amplifier
Configuration II
Tuning method B
$\Theta = 30°$
$q^2 = 1, r = 1$

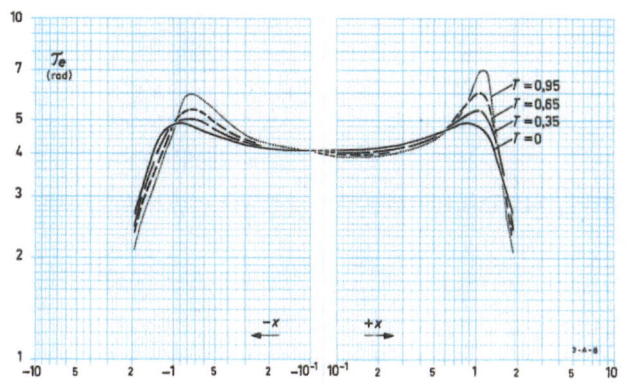

Fig. 15.36
Three-stage amplifier
Configuration II
Tuning method B
$\Theta = 60°$
$q^2 = 1, r = 1$

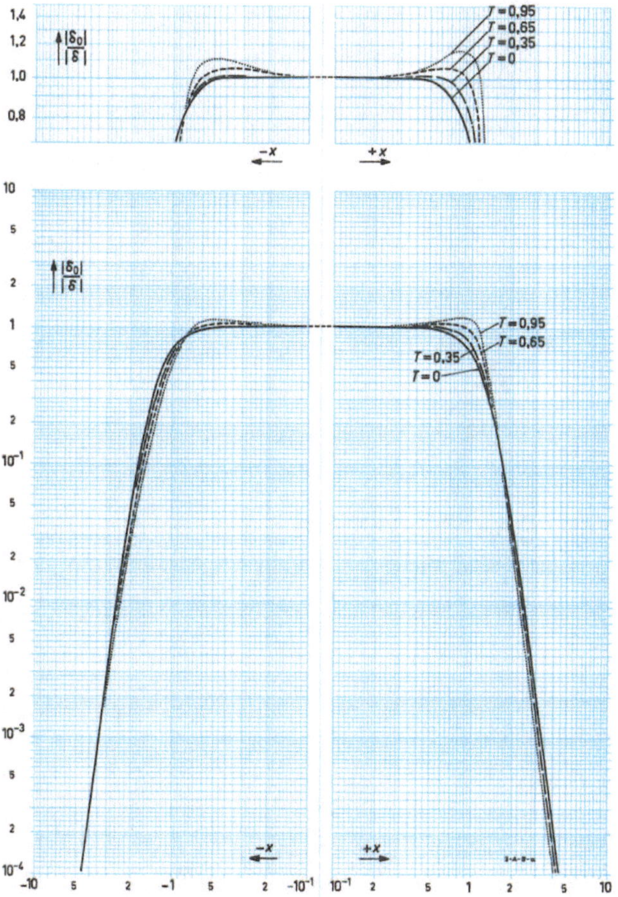

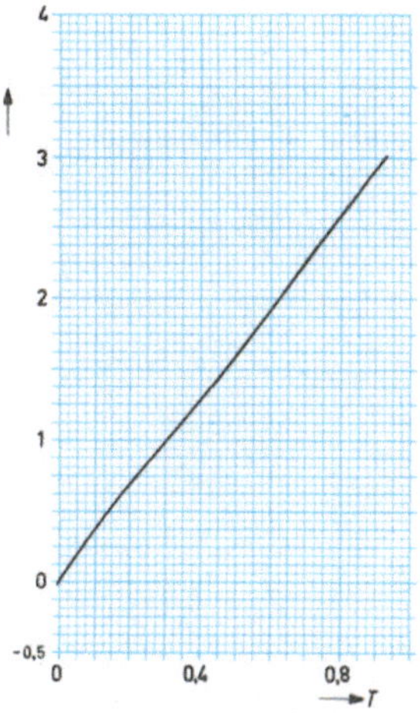

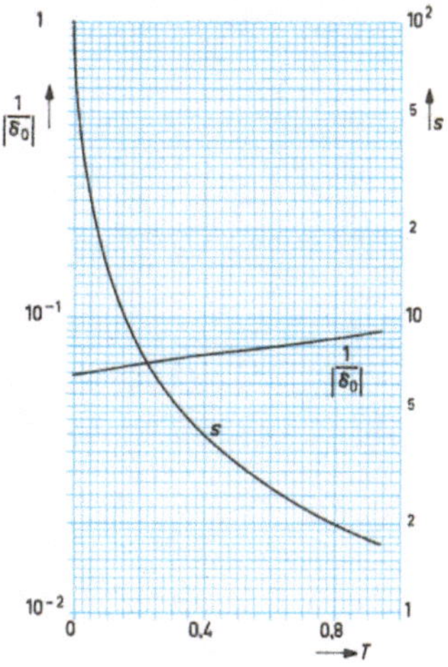

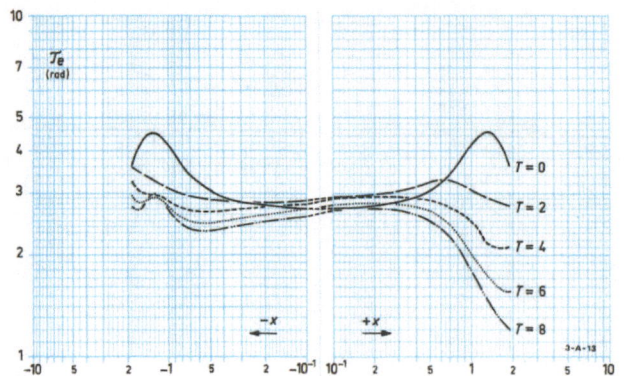

Fig. 15.37
Three-stage amplifier
Configuration II
Tuning method B
$\Theta = 210°$
$q^2 = 2, r = 1$

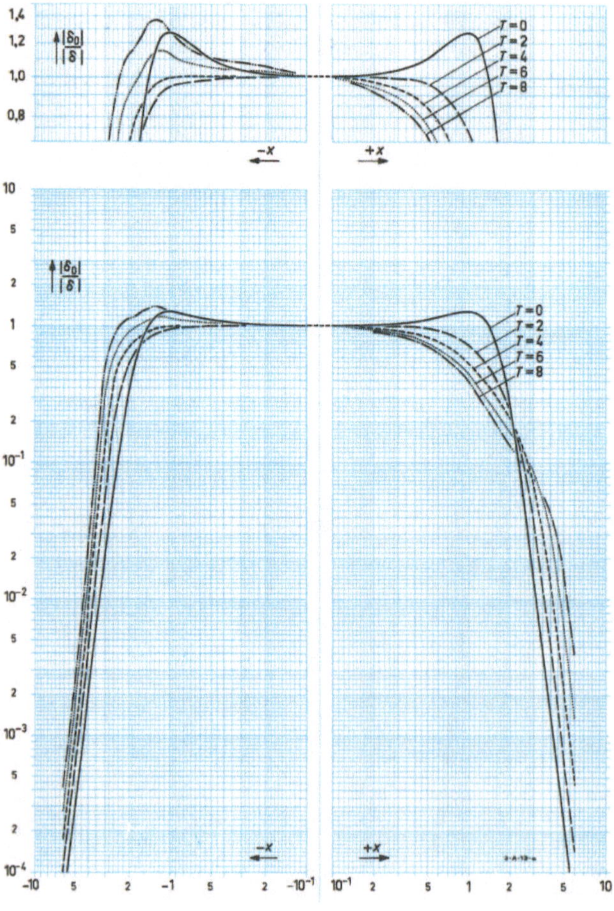

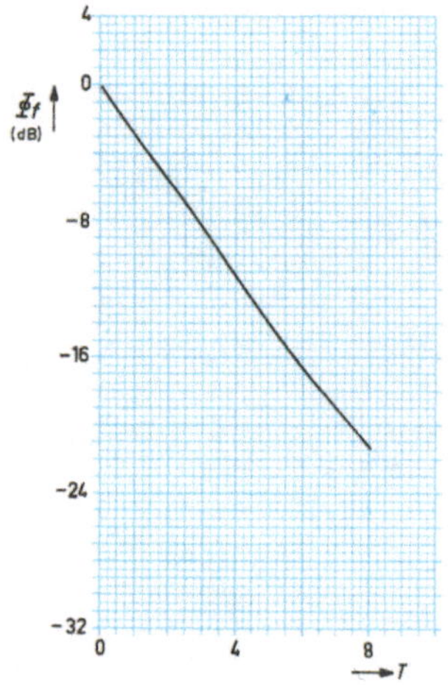

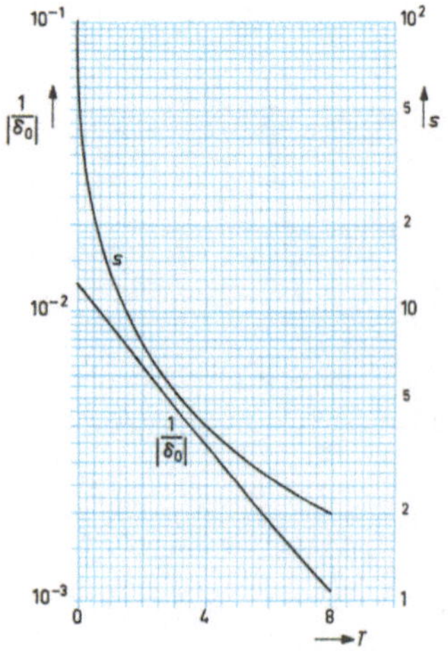

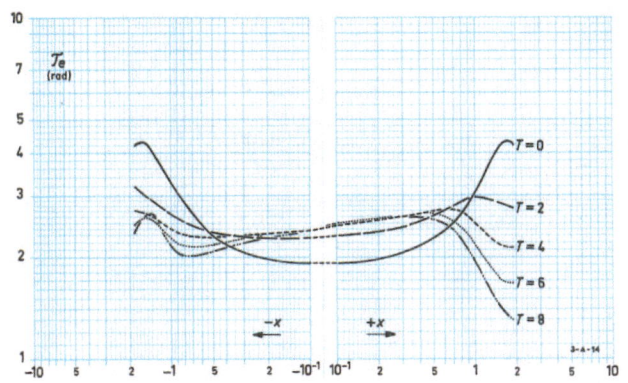

Fig. 15.38
Three-stage amplifier
Configuration II
Tuning method B
$\Theta = 210°$
$q^2 = 3, r = 1$

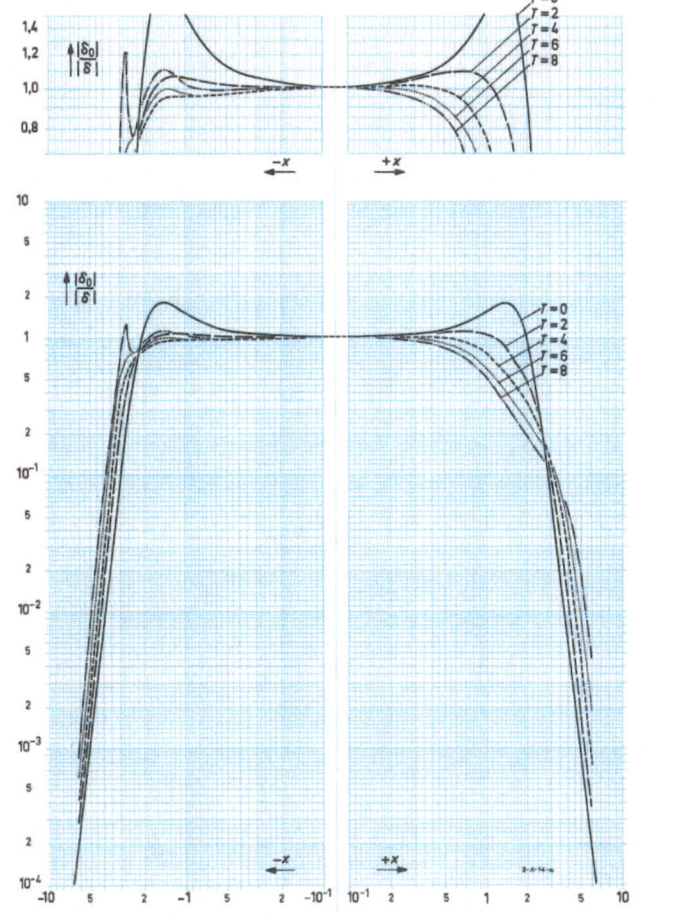

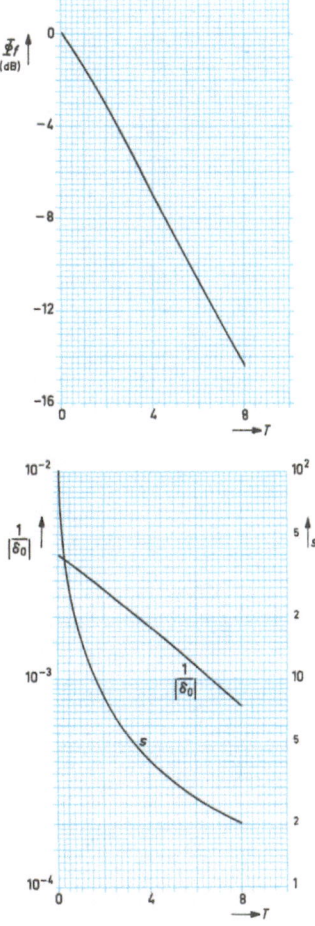

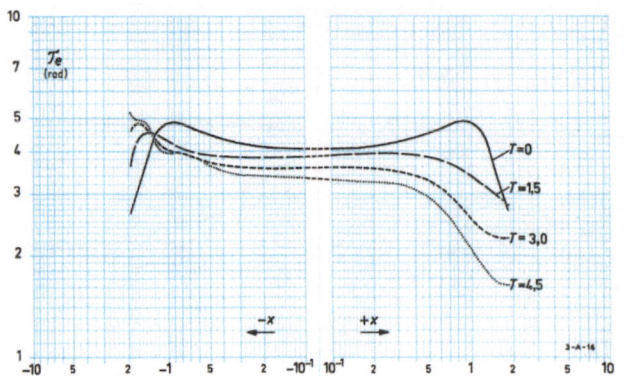

Fig. 15.39
Three-stage amplifier
Configuration II
Tuning method B
$\Theta = 225°$
$q^2 = 1, r = 1$

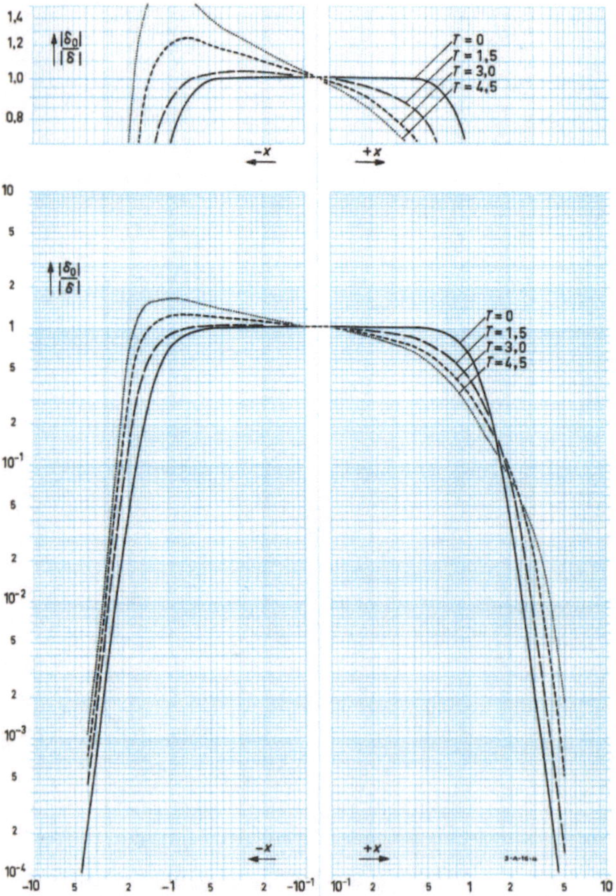

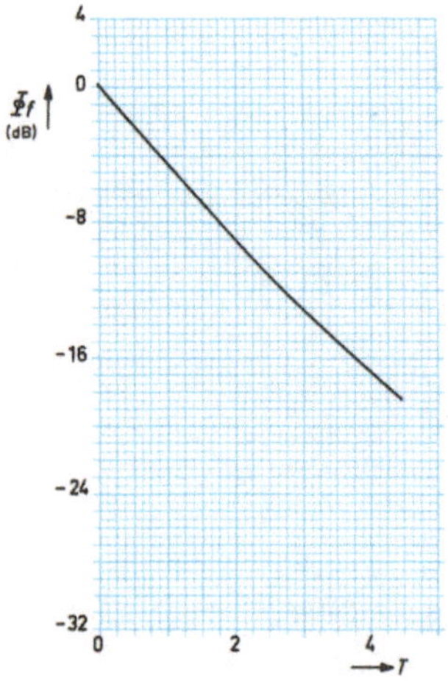

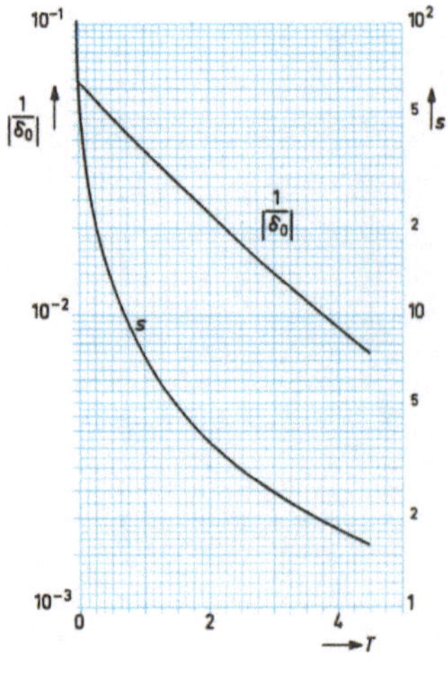

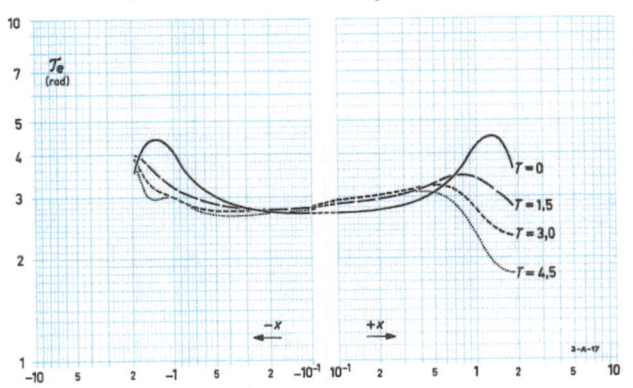

Fig 15.40
Three-stage amplifier
Configuration II
Tuning method B
$\Theta = 225°$
$q^2 = 2, r = 1$

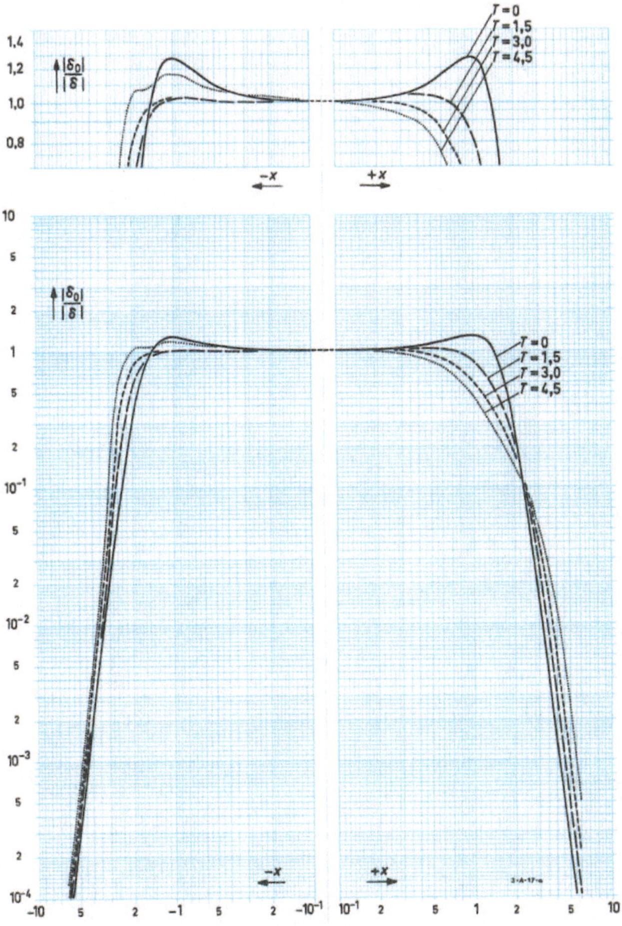

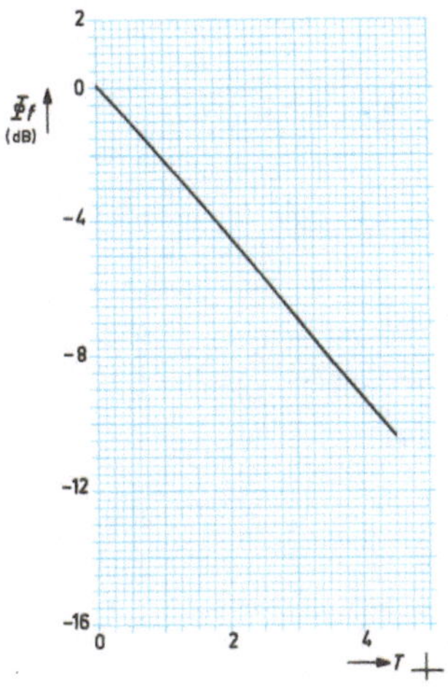

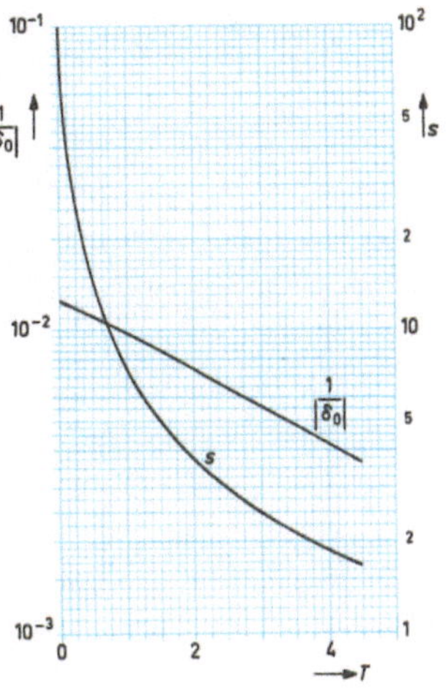

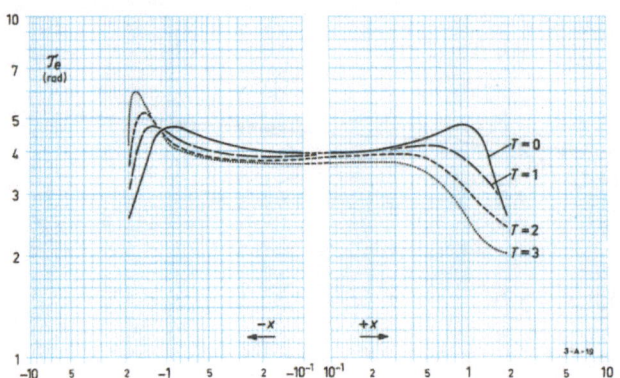

Fig. 15.41
Three-stage amplifier
Configuration II
Tuning method B
$\Theta = 240°$
$q^2 = 1, r = 1$

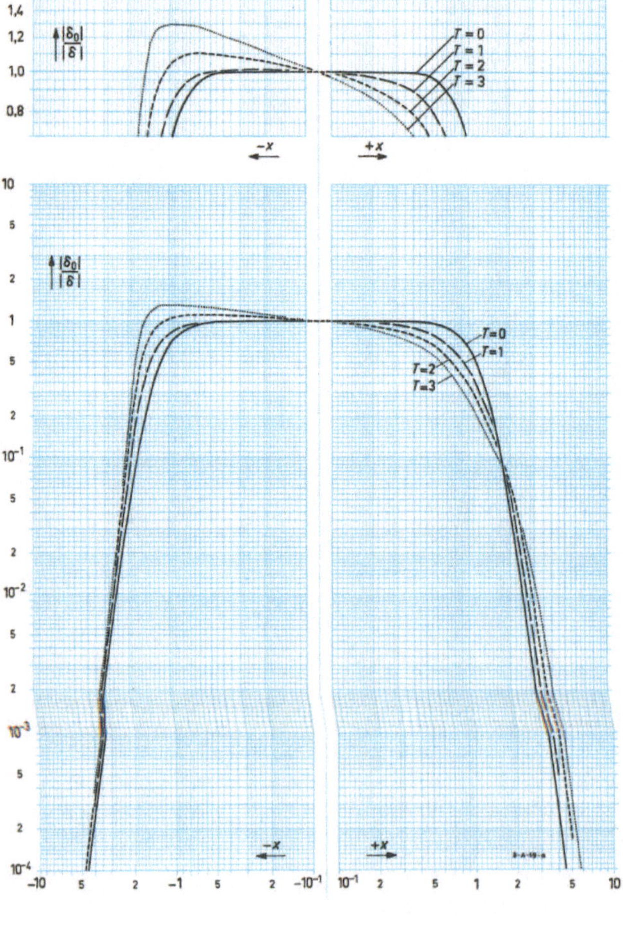

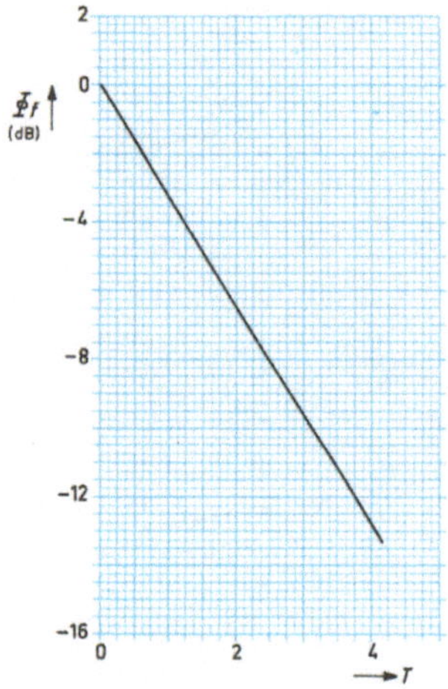

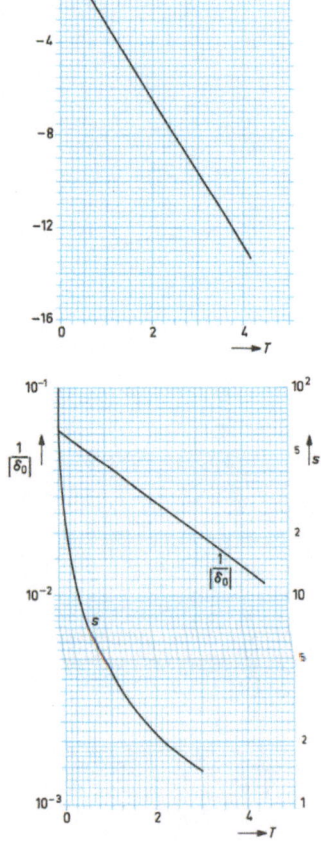

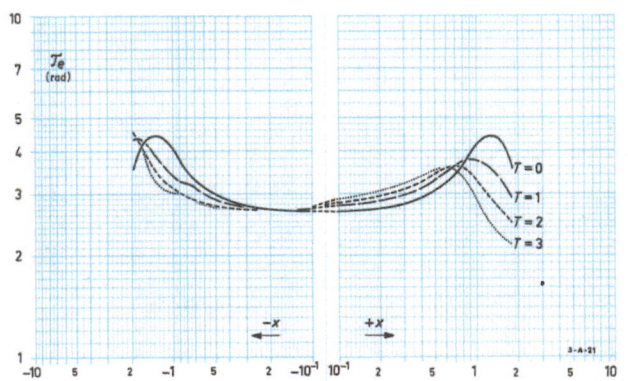

Fig. 15.42
Three-stage amplifier
Configuration II
Tuning method B
$\Theta = 240°$
$q^2 = 2, r = 1$

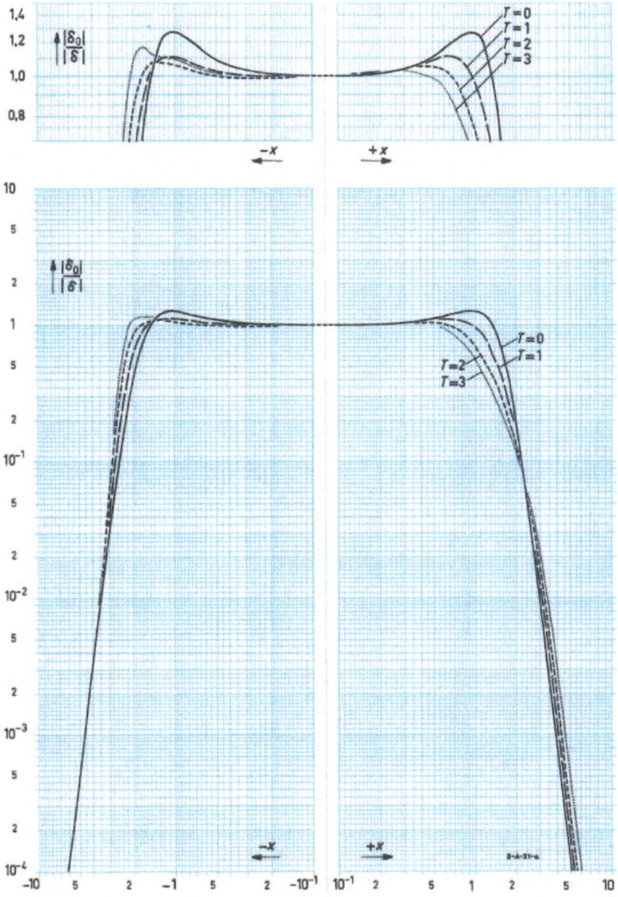

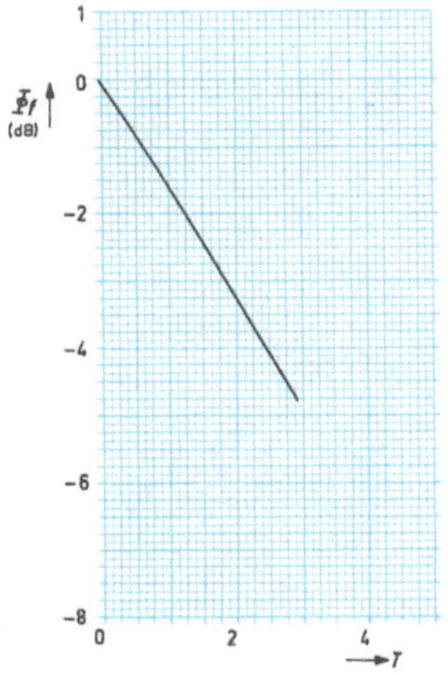

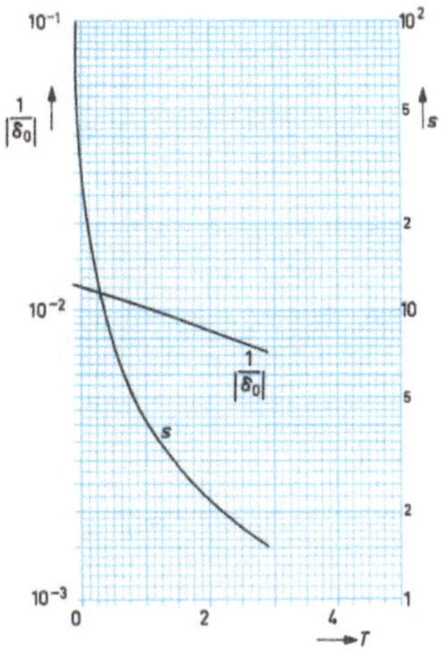

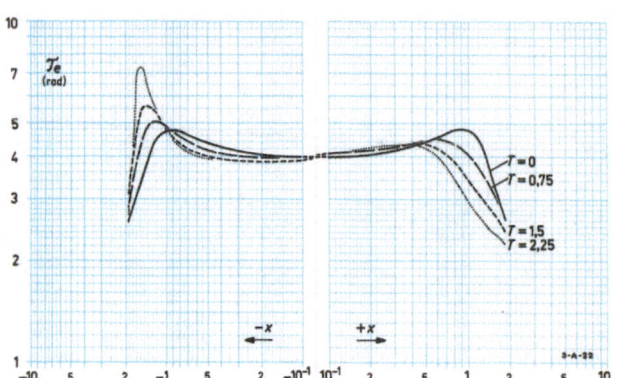

Fig. 15.43
Three-stage amplifier
Configuration II
Tuning method B
$\Theta = 255°$
$q^2 = 1, r = 1$

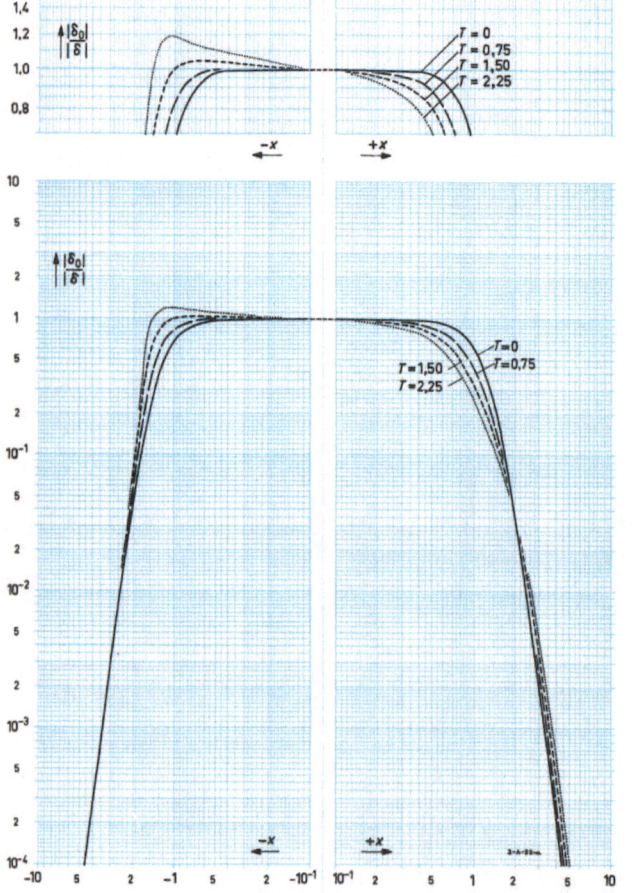

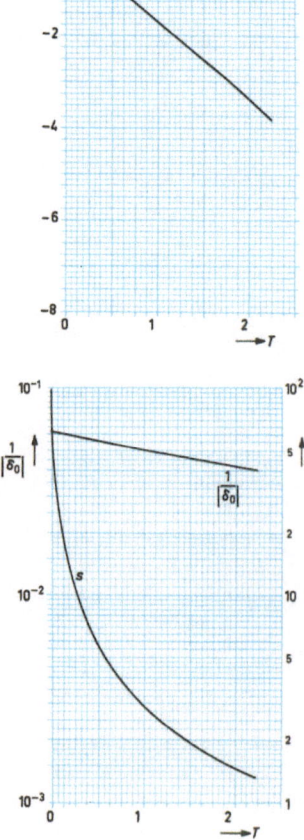

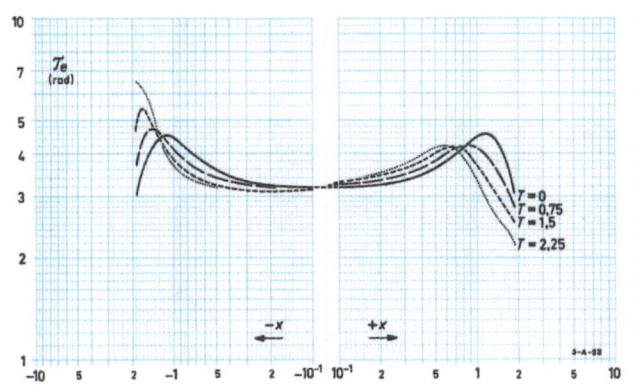

Fig. 15.44
Three-stage amplifier
Configuration II
Tuning method B
$\Theta = 255°$
$q^2 = 1.5$, $r = 1$

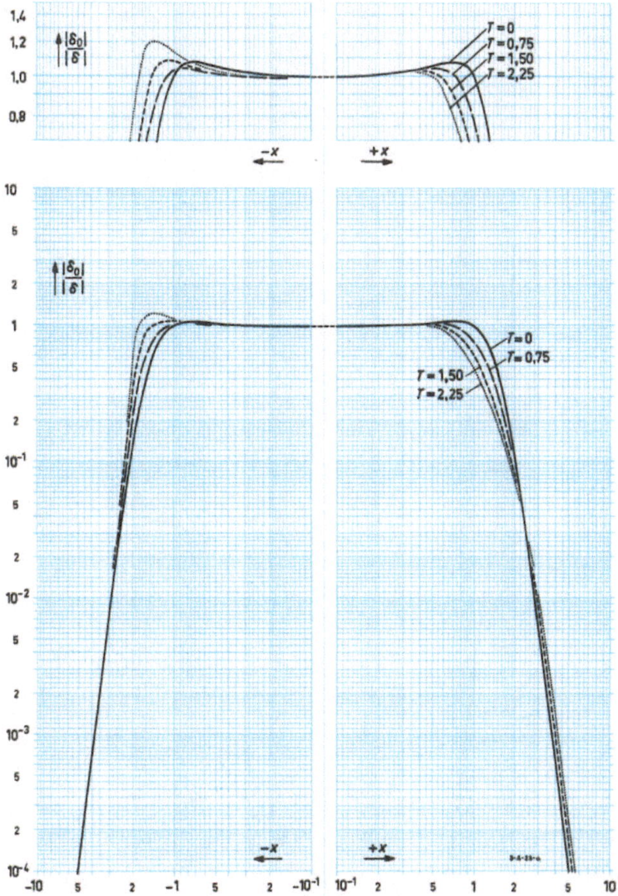

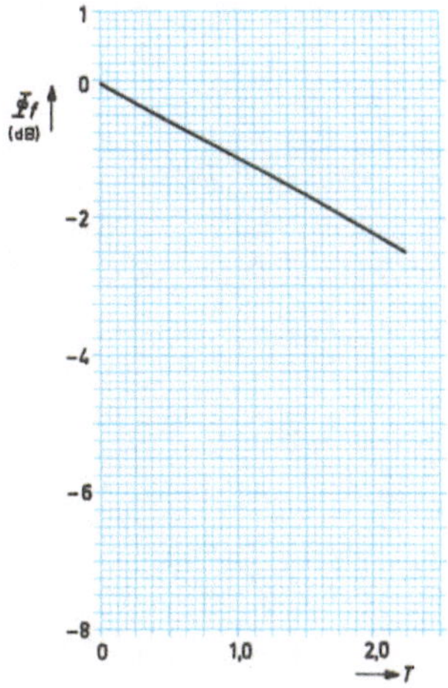

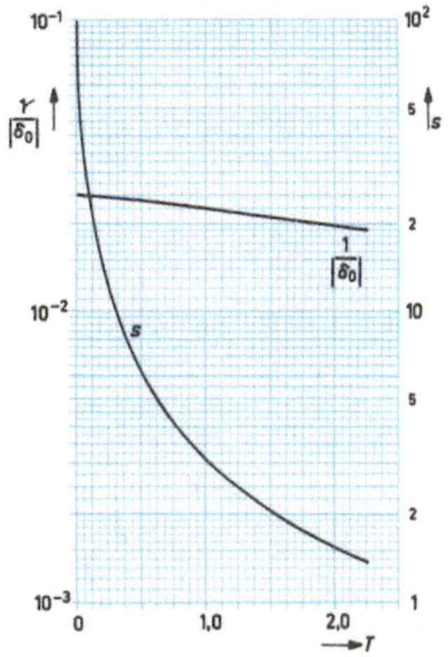

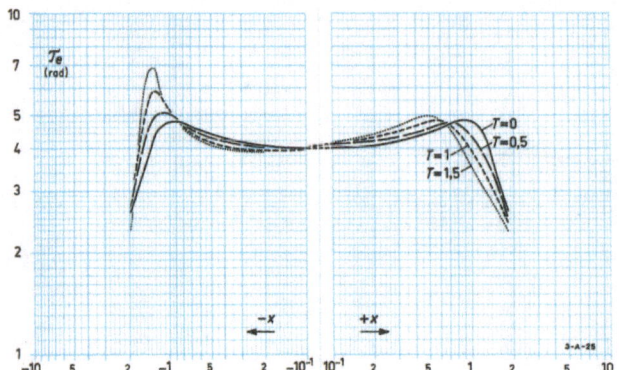

Fig. 15.45
Three-stage amplifier
Configuration II
Tuning method B
$\Theta = 270°$
$q^2 = 1, r = 1$

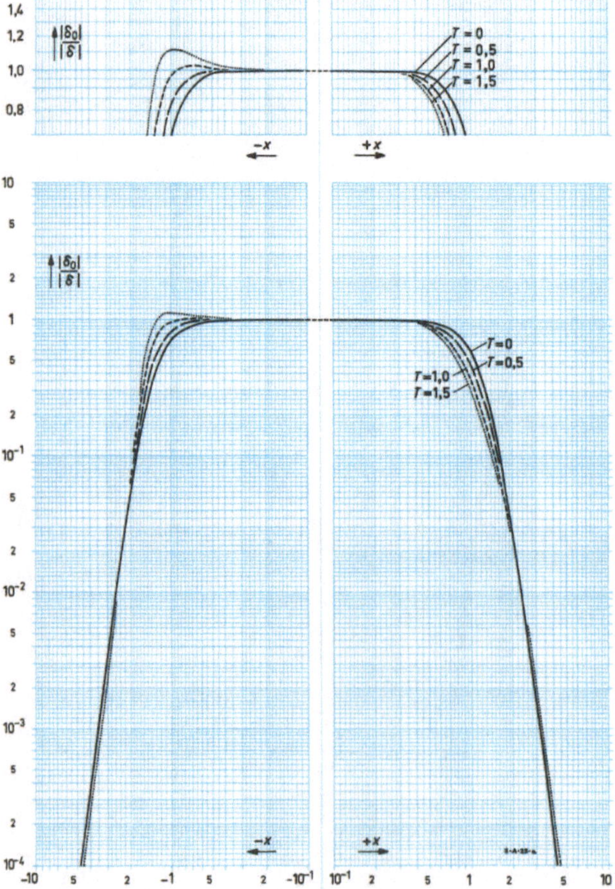

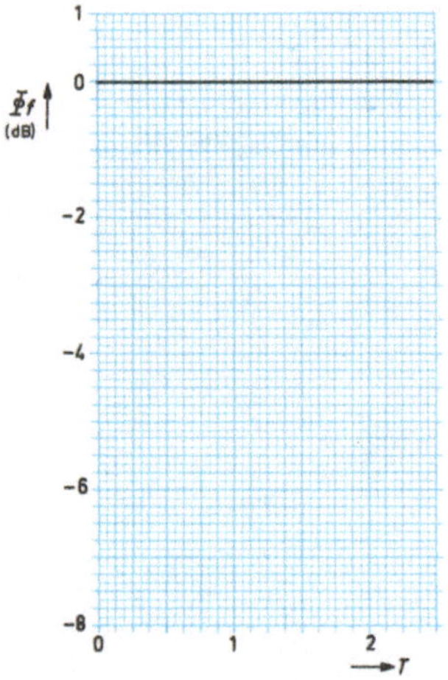

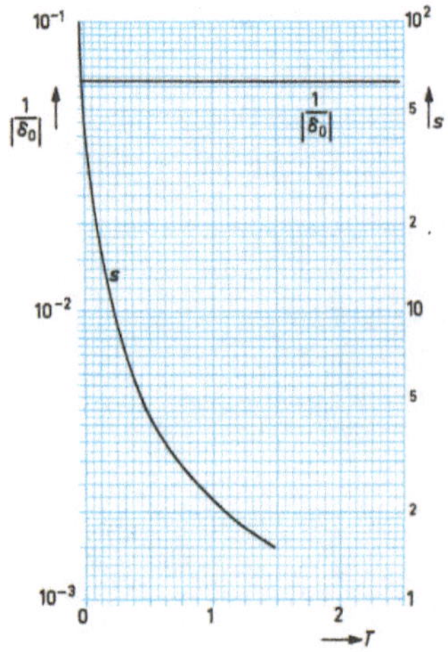

Four-Stage Amplifier with Five Double-Tuned Bandpass Filters

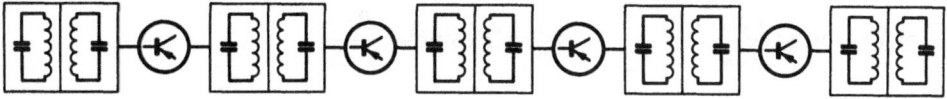

15.8   Four-stage amplifier with five double-tuned bandpass filters

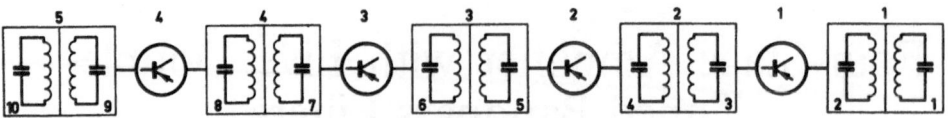

Fig. 15.46 Schematic representation of a four-stage amplifier with five double-tuned bandpass filters (Configuration II).

Reduced amplifier determinant:
See Determinant II on page 233 for $n = 4$.

Method of tuning: B
Reduced amplifier determinant at tuning frequency $(x = 0)$:
$$|_4\delta_0| = P_{10M}$$
Tuning correction terms:

$$x_1' = T_1 \sin\Theta_1 \cdot \frac{P_{1M}}{P_{2M}} \qquad\qquad x_3' = T_3 \sin\Theta_3 \cdot \frac{P_{5M}}{P_{6M}}$$

$$x_2' = T_2 \sin\Theta_2 \cdot \frac{P_{3M}}{P_{4M}} \qquad\qquad x_4' = T_4 \sin\Theta_4 \cdot \frac{P_{7M}}{P_{8M}}$$

Minor determinant values at tuning frequency $(x = 0)$:

$$P_{1M} = 1$$

$$P_{2M} = 1 + q_1^2$$

$$P_{3M} = 1 + q_1^2 - T_1 \cos\Theta_1$$

$$P_{4M} = (1 + q_1^2)(1 + q_2^2) - T_1 \cos\Theta_1$$

$$P_{5M} = (1 + q_1^2)(1 + q_2^2 - T_2 \cos\Theta_2) - T_1 \cos\Theta_1 (1 - T_2 \cos\Theta_2)$$

$$P_{6M} = (1 + q_1^2)\left\{(1 + q_2^2)(1 + q_3^2) - T_2 \cos\Theta_2\right\} - T_1 \cos\Theta_1 (1 + q_3^2 - T_2 \cos\Theta_2)$$

$$P_{7M} = (1 + q_1^2)\left\{(1 + q_2^2)(1 + q_3^2 - T_3 \cos\Theta_3) - T_2 \cos\Theta_2 (1 - T_3 \cos\Theta_3)\right\}$$
$$\qquad - T_1 \cos\Theta_1 \left\{- T_2 \cos\Theta_2 (1 - T_3 \cos\Theta_3) + 1 + q_3^2 - T_3 \cos\Theta_3\right\}$$

$$P_{8M} = (1 + q_1^2)\left[(1 + q_2^2)\left\{(1 + q_3^2)(1 + q_4^2) - T_3 \cos\Theta_3\right\}\right.$$
$$\qquad \left. - T_2 \cos\Theta_2 (1 + q_4^2 - T_3 \cos\Theta_3)\right]$$
$$\qquad - T_1 \cos\Theta_1 \left\{- T_2 \cos\Theta_2 (1 + q_4^2 - T_3 \cos\Theta_3)\right.$$
$$\qquad \left. + (1 + q_3^2)(1 + q_4^2) - T_3 \cos\Theta_3\right\}$$

$$P_{9M} = (1 + q_1^2) \left[(1 + q_2^2) \left\{ (1 + q_3^2)(1 + q_4^2 - T_4 \cos\Theta_4) \right. \right.$$
$$- T_3 \cos\Theta_3 (1 - T_4 \cos\Theta_4) \left. \right\}$$
$$- T_2 \cos\Theta_2 (- T_3 \cos\Theta_3 (1 - T_4 \cos\Theta_4) + 1 + q_4^2 - T_4 \cos\Theta_4) \left. \right]$$
$$- T_1 \cos\Theta_1 [- T_2 \cos\Theta_2 \left\{ - T_3 \cos\Theta_3 (1 - T_4 \cos\Theta_4) \right.$$
$$+ 1 + q_4^2 - T_4 \cos\Theta_4 \left. \right\} + (1 + q_3^2)(1 + q_4^2 - T_4 \cos\Theta_4)$$
$$- T_3 \cos\Theta_3 (1 - T_4 \cos\Theta_4)]$$

$$P_{10M} = (1 + q_1^2) \left[[(1 + q_2^2) [(1 + q_3^2) \left\{ (1 + q_4^2)(1 + q_5^2) - T_4 \cos\Theta_4 \right\} \right.$$
$$- T_3 \cos\Theta_3 (1 + q_5^2 - T_4 \cos\Theta_4)]$$
$$- T_2 \cos\Theta_2 \left\{ (1 + q_4^2)(1 + q_5^2) \right.$$
$$- T_2 \cos\Theta_2 (1 + q_5^2 - T_4 \cos\Theta_4) - T_4 \cos\Theta_4 \left. \right\}]]$$
$$- T_1 \cos\Theta_1 [- T_2 \cos\Theta_2 \left\{ - T_3 \cos\Theta_3 (1 + q_5^2) - T_4 \cos\Theta_4 \right.$$
$$+ (1 + q_4^2)(1 + q_5^2) - T_4 \cos\Theta_4 \left. \right\}$$
$$+ (1 + q_3^2)(1 + q_4^2)(1 + q_5^2)$$
$$- T_3 \cos\Theta_3 (1 + q_5^2 - T_4 \cos\Theta_4) - T_4 \cos\Theta_4 (1 + q_3^2)]$$

Minor determinant values at tuning frequency ($x = 0$) for identical stages:

$$P_{1M} = 1$$
$$P_{2M} = 1 + q^2$$
$$P_{3M} = 1 + q^2 - T\cos\Theta$$
$$P_{4M} = (1 + q^2)^2 - T\cos\Theta$$
$$P_{5M} = (1 + q^2)^2 - T\cos\Theta (2 + q^2 - T\cos\Theta)$$
$$P_{6M} = (1 + q^2)^3 - T\cos\Theta (2 + 2q^2 - T\cos\Theta)$$
$$P_{7M} = (1 + q^2)^3 - T\cos\Theta \left\{ 3 + 4q^2 + q^4 - (3 + q^2) T\cos\Theta + T^2 \cos^2\Theta \right\}$$
$$P_{8M} = (1 + q^2)^4 - T\cos\Theta \left\{ 3 + 6q^2 + 3q^4 - (3 + 2q^2) T\cos\Theta + T^2 \cos^2\Theta \right\}$$
$$P_{9M} = (1 + q^2)^4 - T\cos\Theta \left\{ 4 + 9q^2 + 6q^4 + q^6 - (6 + 6q^2 + q^4) T\cos\Theta \right.$$
$$+ (4 + q^2) T^2 \cos\Theta^2 - T^3 \cos^3\Theta \left. \right\}$$

$$P_{10M} = (1 + q^2)^5 - T\cos\Theta \left\{ 4 + 12q^2 + 12q^4 + 4q^6 \right.$$
$$- (6 + 9q^2 + 3q^4) T\cos\Theta + (4 + 2q^2) T^2 \cos^2\Theta - T^3 \cos^3\Theta$$

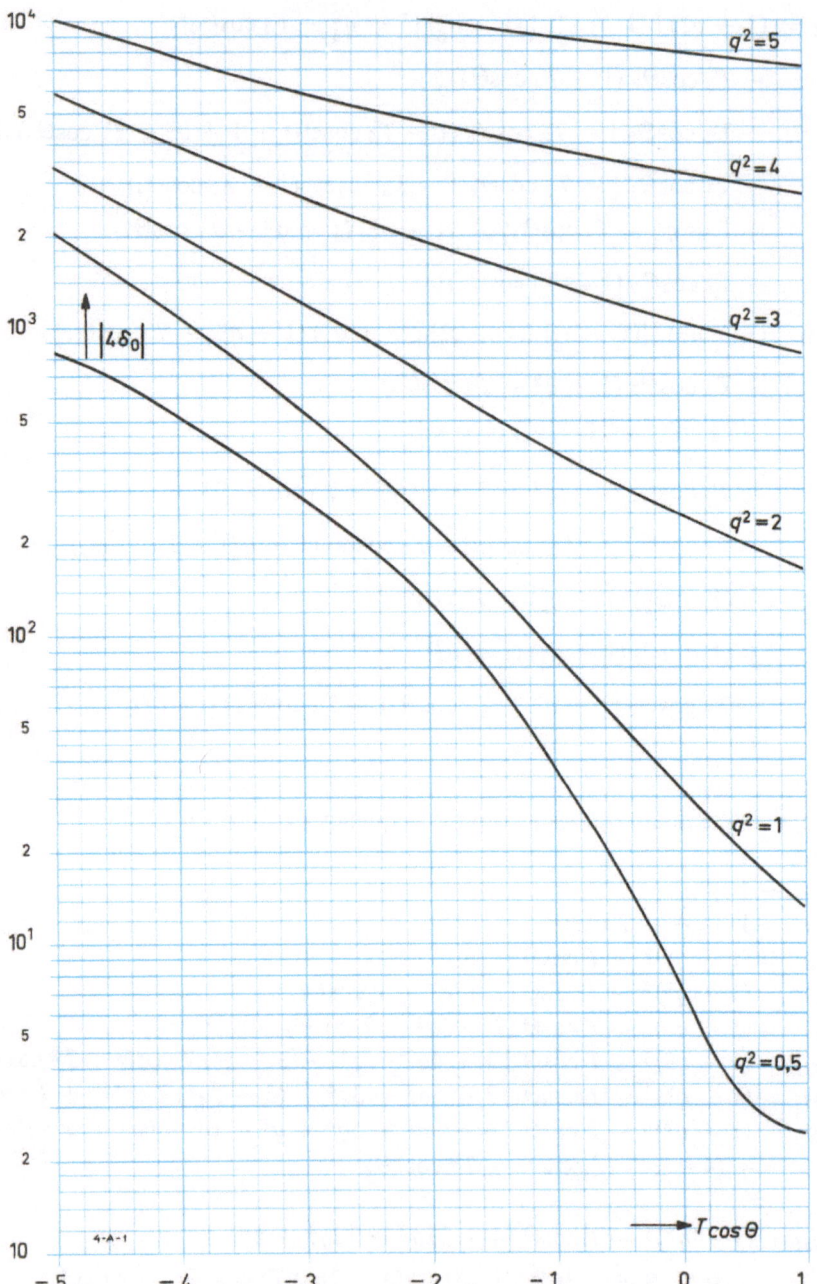

Fig. 15.47 Graph representing the value of the reduced determinant $|1/\delta_0|$ as a function of $T\cos\Theta$ for a four-stage amplifier with five double-tuned bandpass filters (Configuration II, tuning method B).

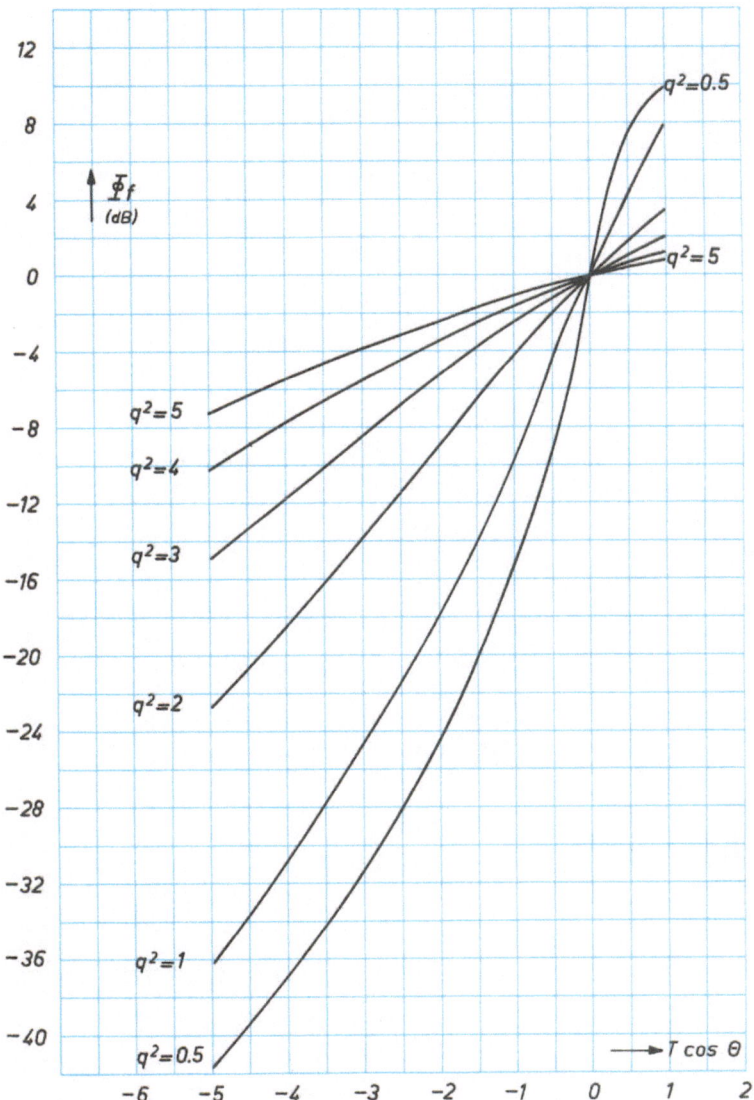

Fig. 15.48 Graph representing the feedback losses Φ_f as a function of $T\cos\Theta$ for a four-stage amplifier with five double-tuned bandpass filters (Configuration II, tuning method B).

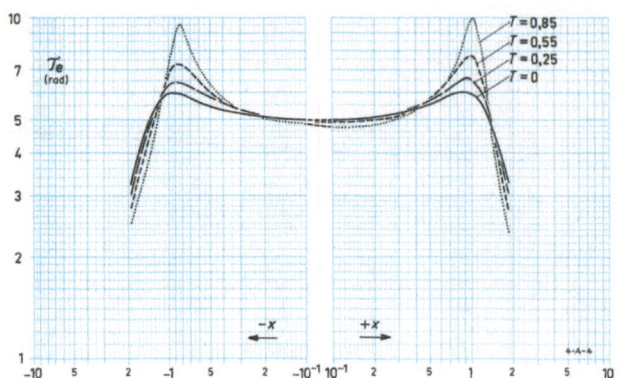

Fig. 15.49
Four-stage amplifier
Configuration II
Tuning method B
$\Theta = 30°$
$q^2 = 1, r = 1$

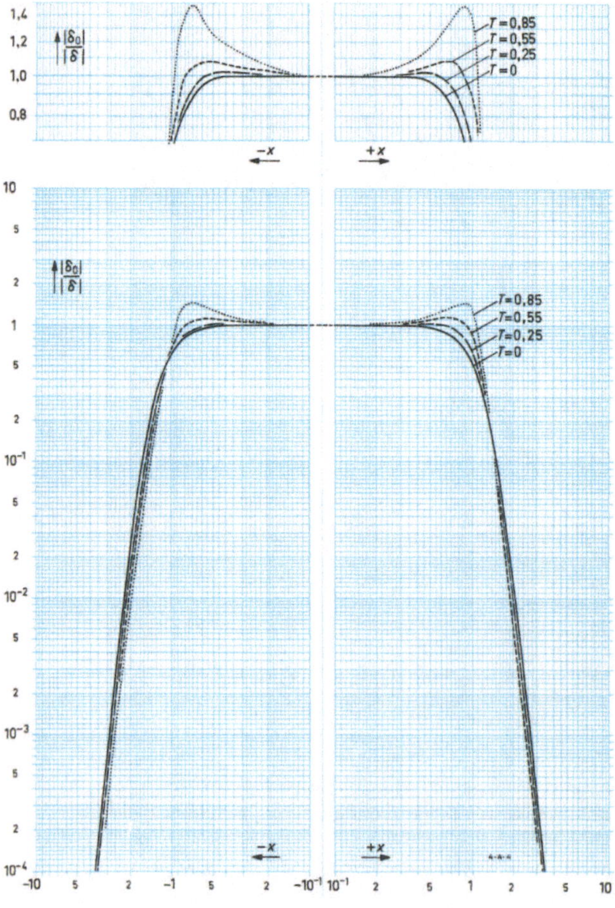

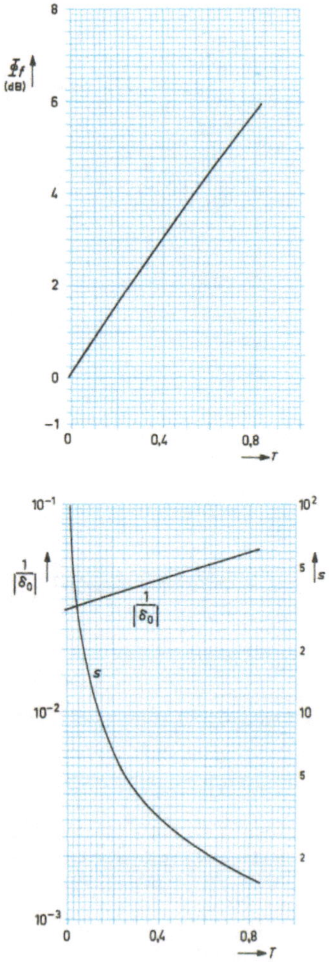

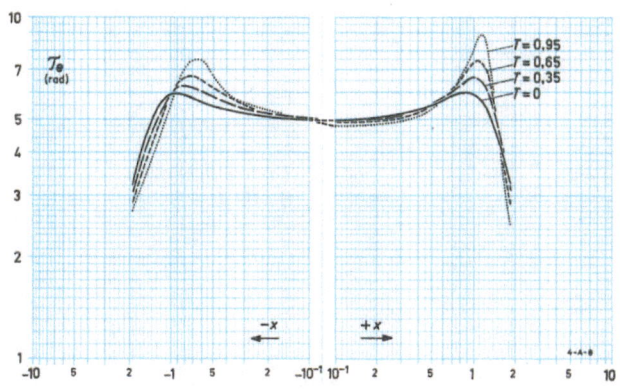

Fig. 15.50
Four-stage amplifier
Configuration II
Tuning method B
$\Theta = 60°$
$q^2 = 1, r = 1$

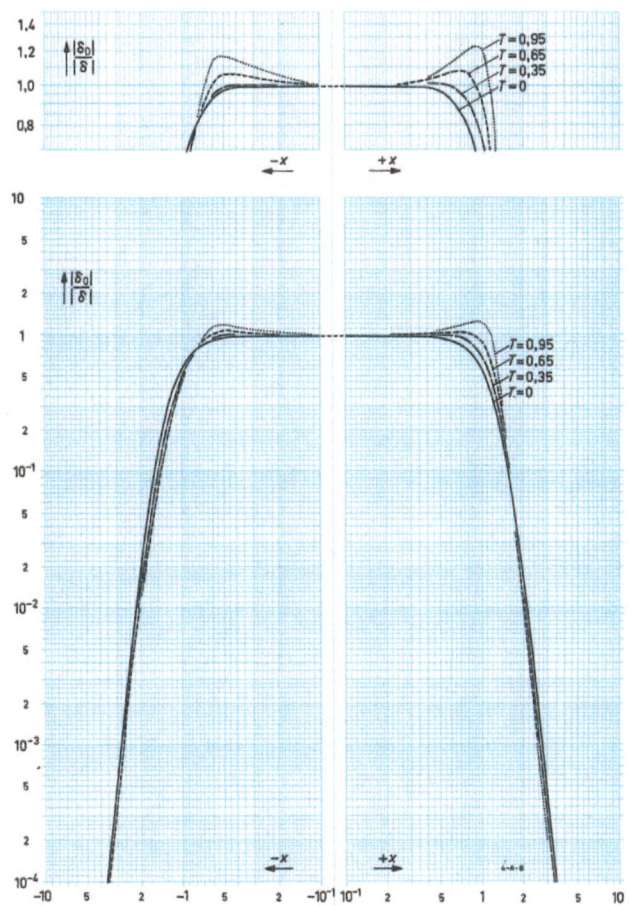

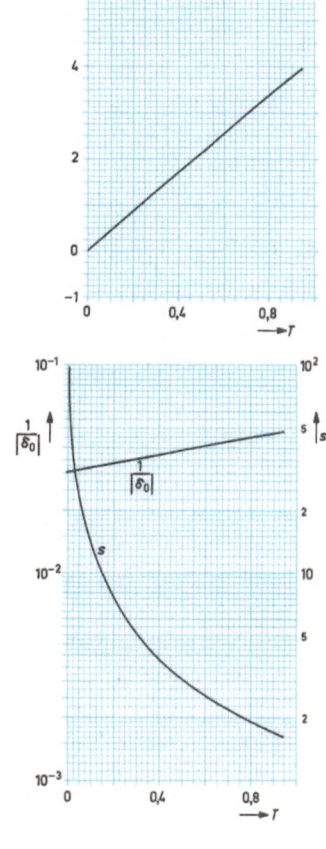

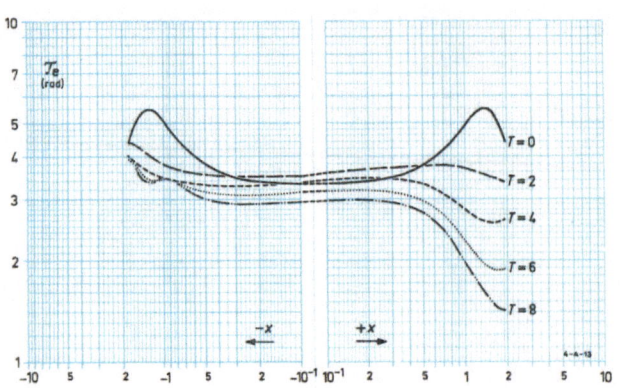

Fig. 15.51
Four-stage amplifier
Configuration II
Tuning method B
$\Theta = 210°$
$q^2 = 2, r = 1$

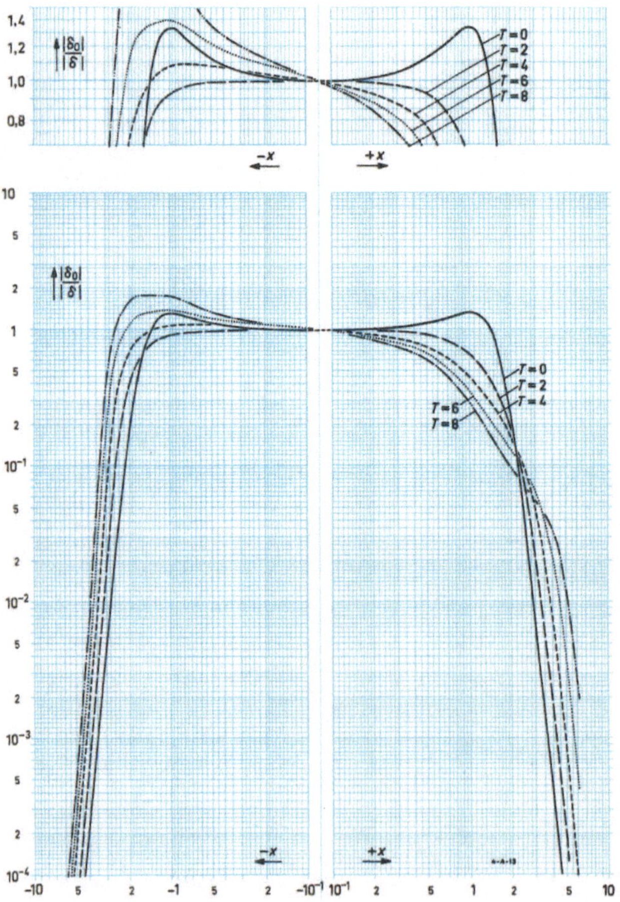

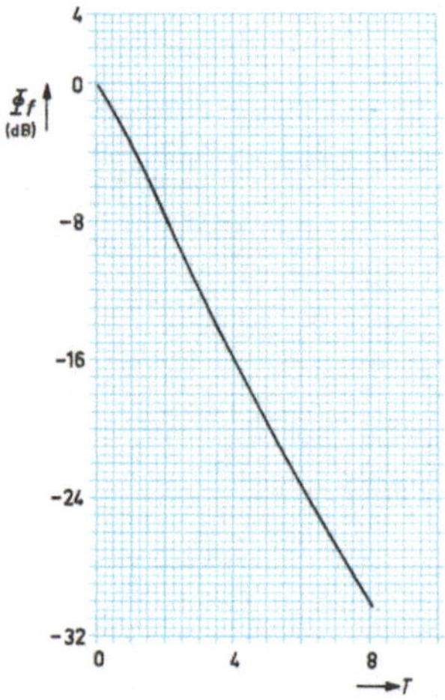

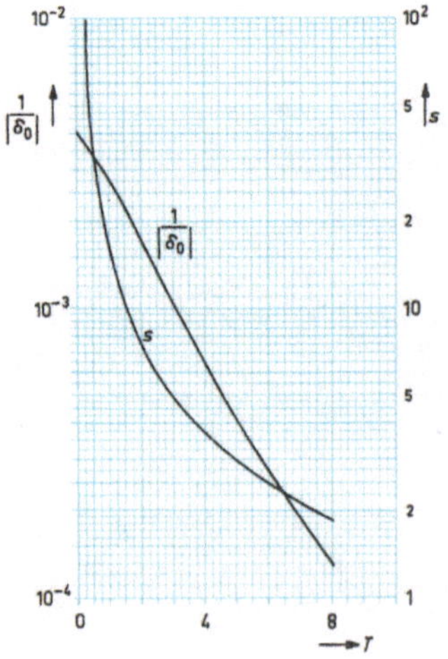

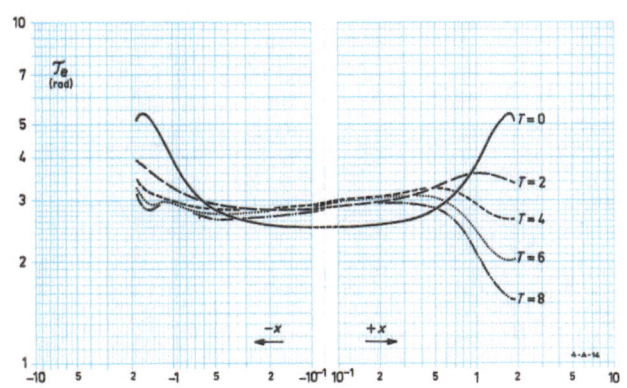

Fig. 15.52
Four-stage amplifier
Configuration II
Tuning method B
$\Theta = 210°$
$q^2 = 3$, $r = 1$

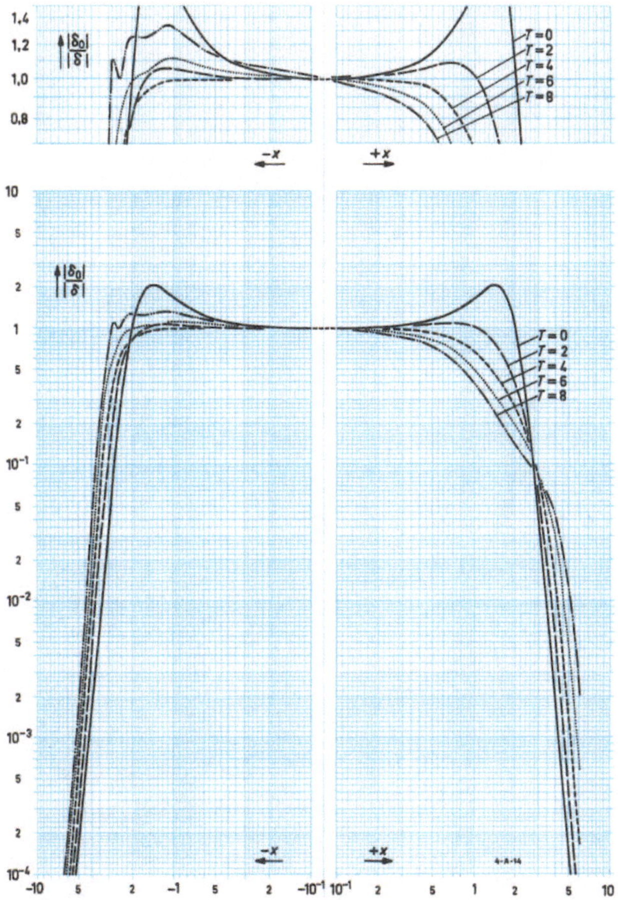

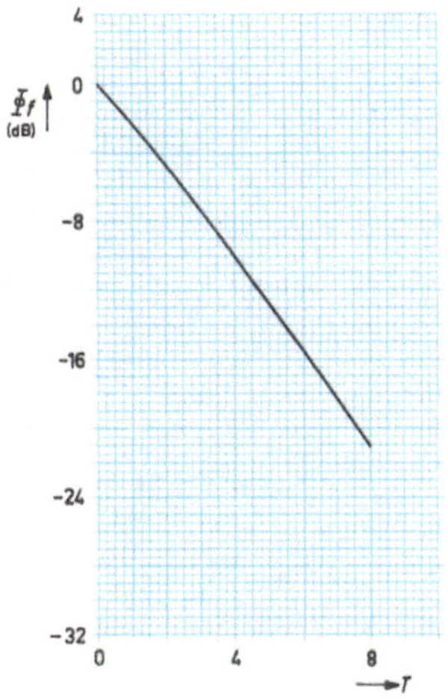

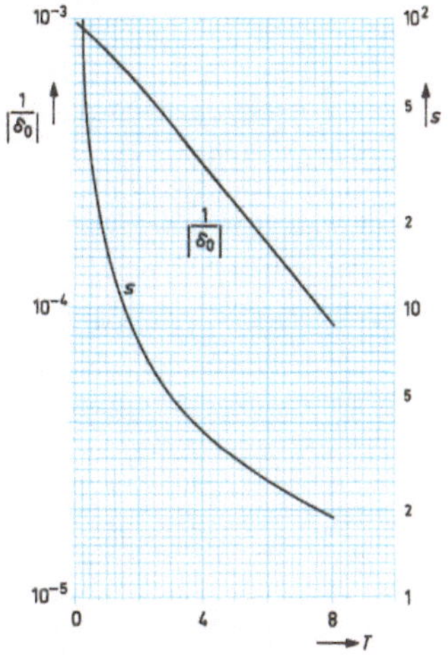

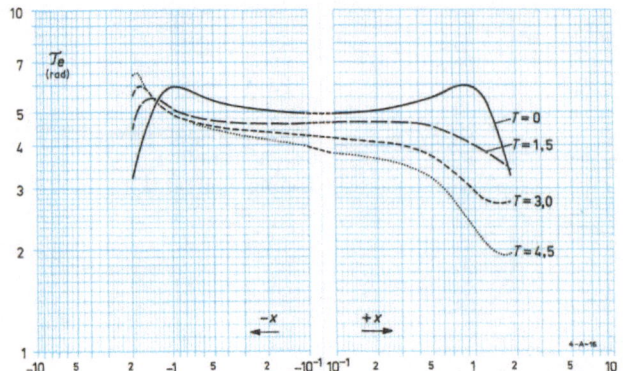

Fig. 15.53
Four-stage amplifier
Configuration II
Tuning method B
$\Theta = 225°$
$q^2 = 1, r = 1$

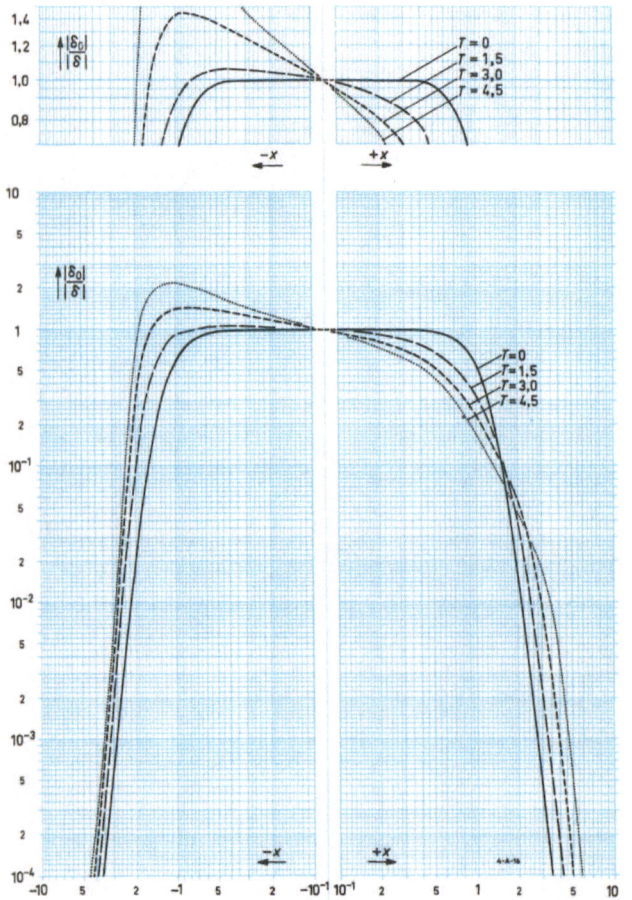

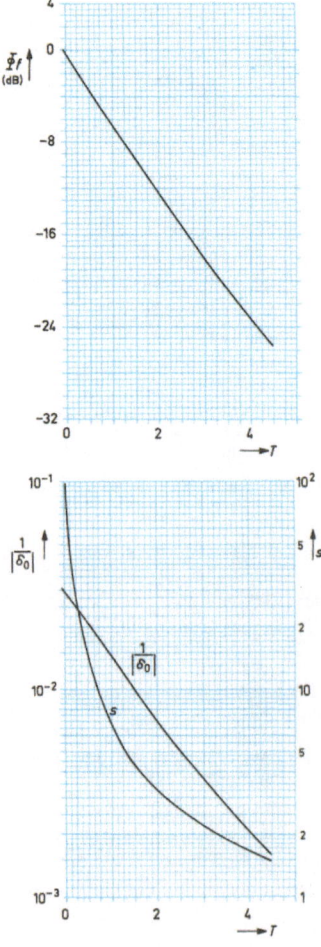

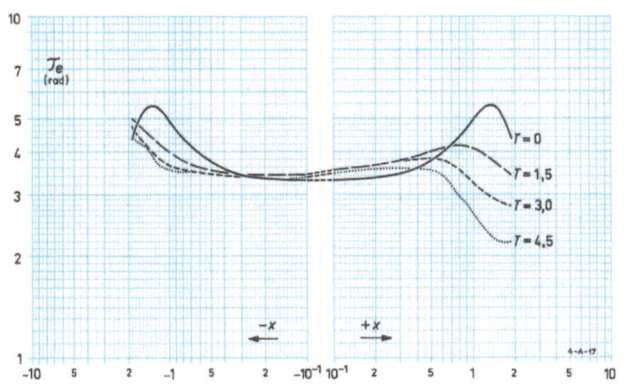

Fig. 15.54
Four-stage amplifier
Configuration II
Tuning method B
$\Theta = 225°$
$q^2 = 2, r = 1$

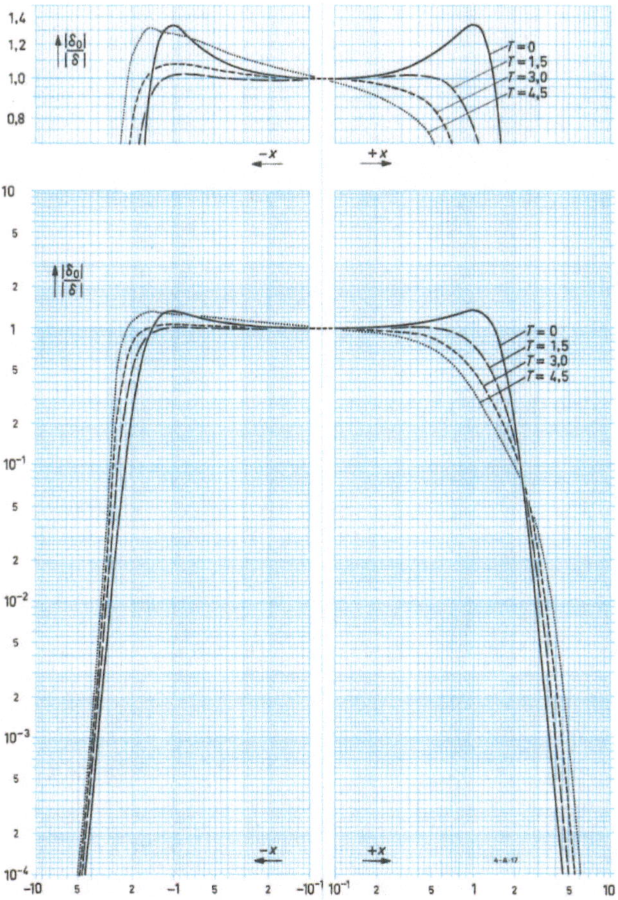

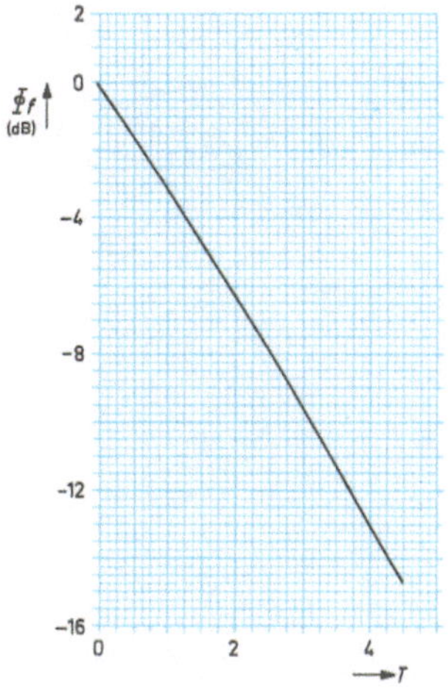

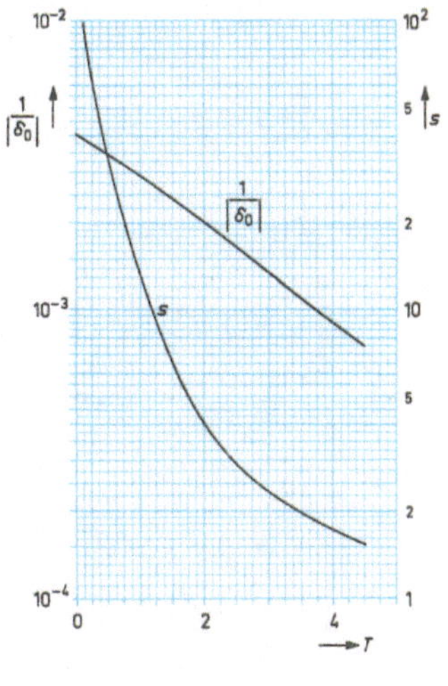

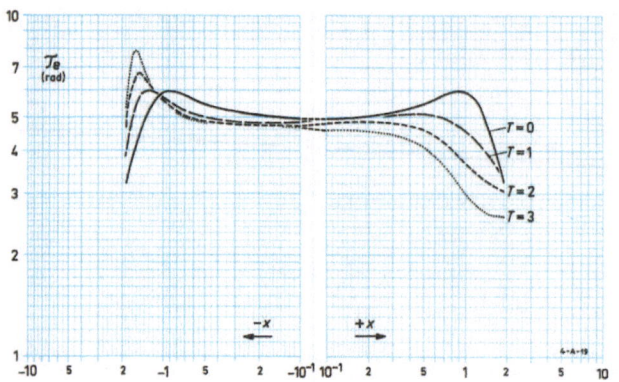

Fig. 15.55
Four-stage amplifier
Configuration II
Tuning method B
$\Theta = 240°$
$q^2 = 1, r = 1$

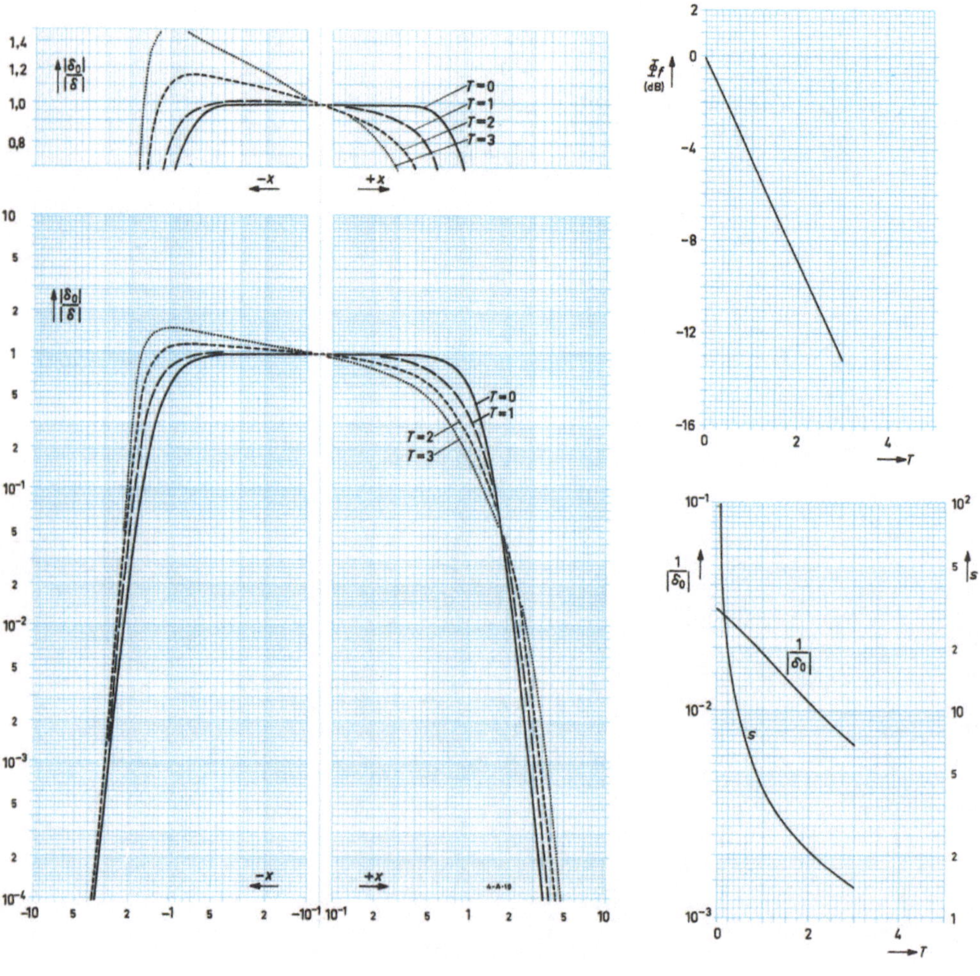

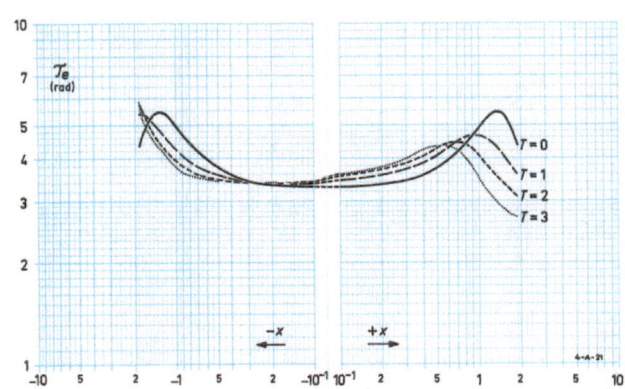

Fig. 15.56
Four-stage amplifier
Configuration II
Tuning method B
$\Theta = 240°$
$q^2 = 2, r = 1$

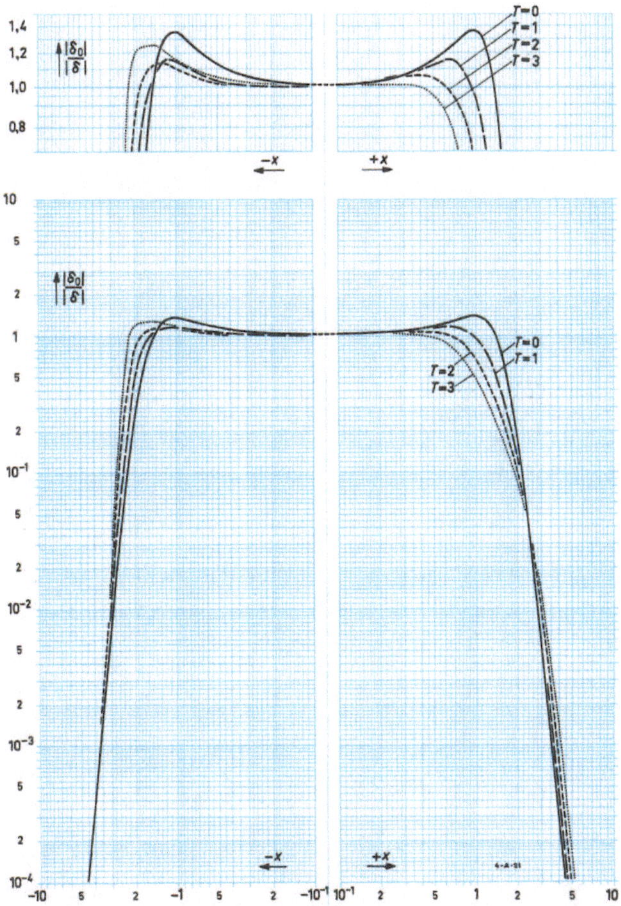

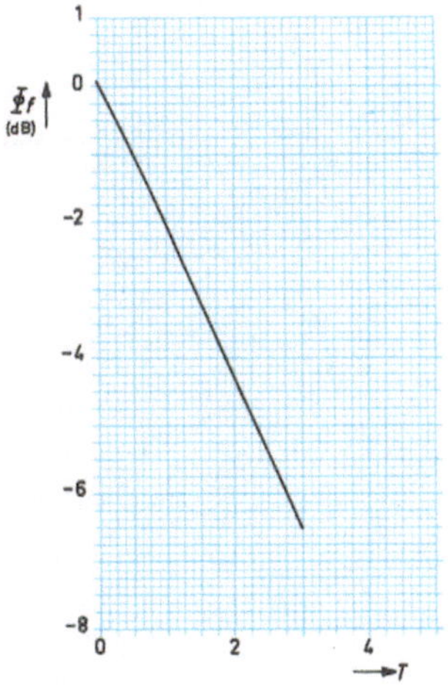

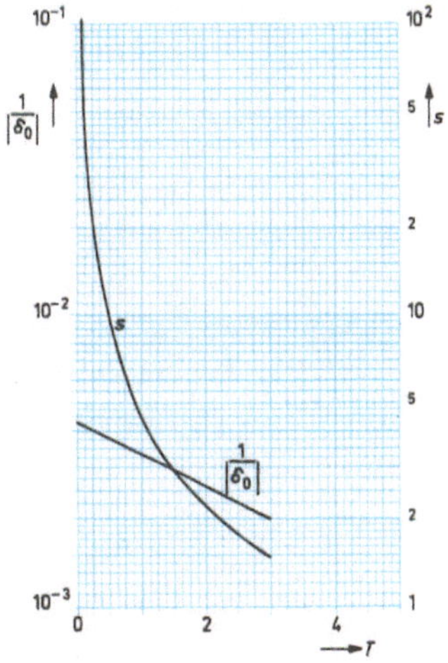

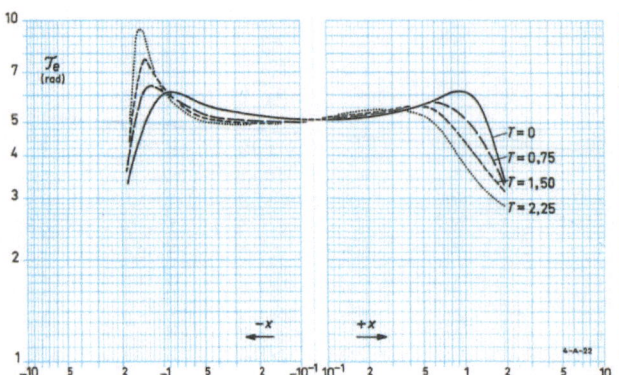

Fig. 15.57
Four-stage amplifier
Configuration II
Tuning method B
$\Theta = 255°$
$q^2 = 1, r = 1$

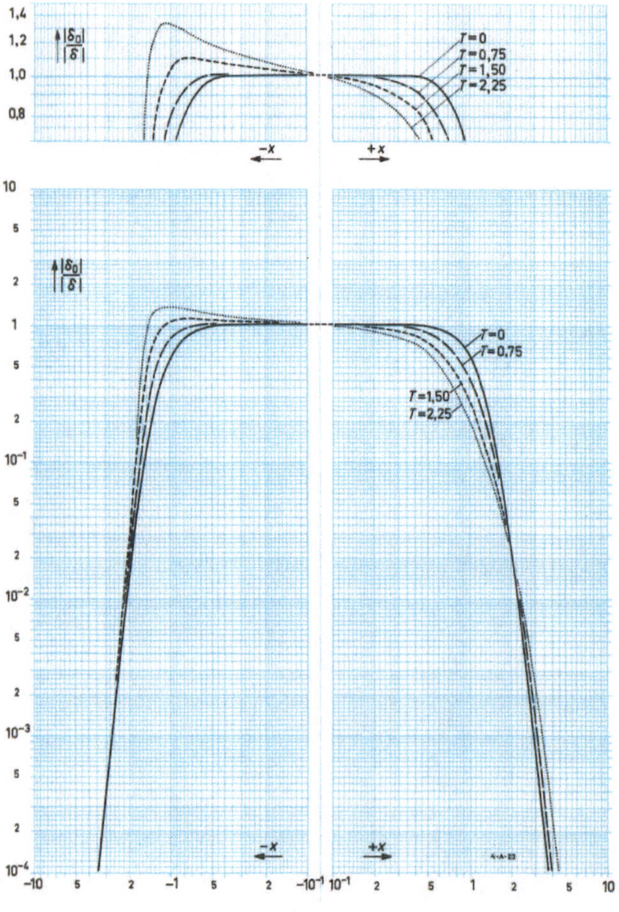

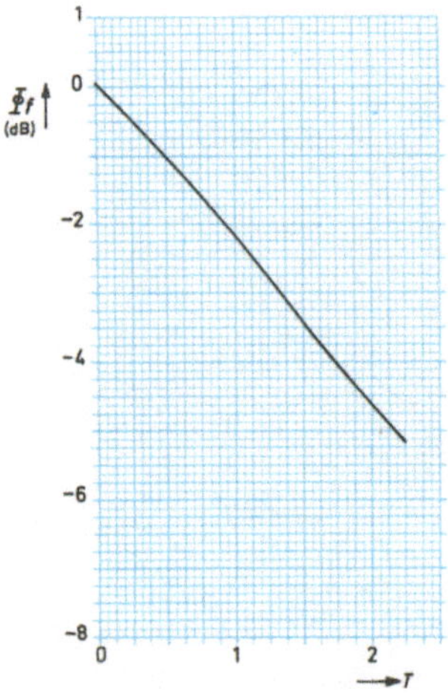

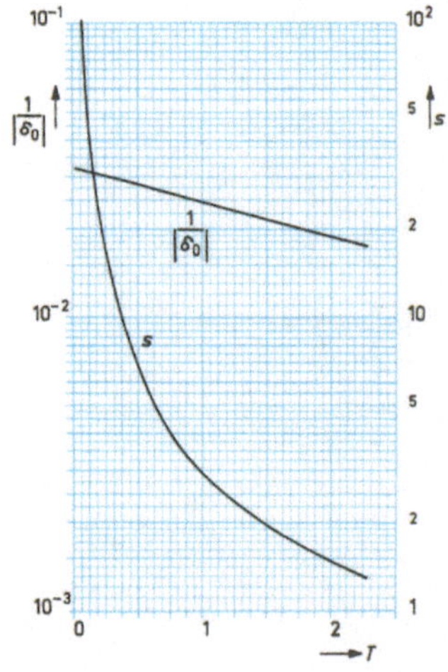

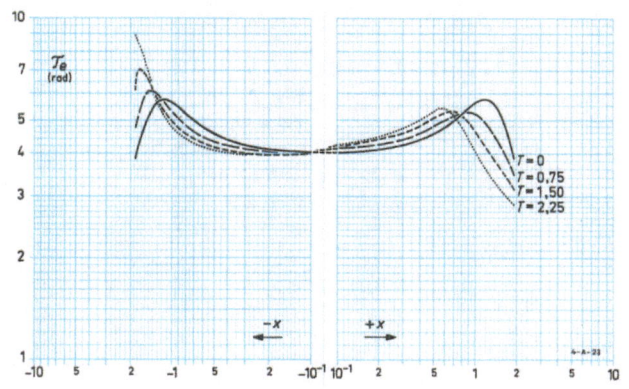

Fig. 15.58
Four-stage amplifier
Configuration II
Tuning method B
$\Theta = 255°$
$q^2 = 1.5, r = 1$

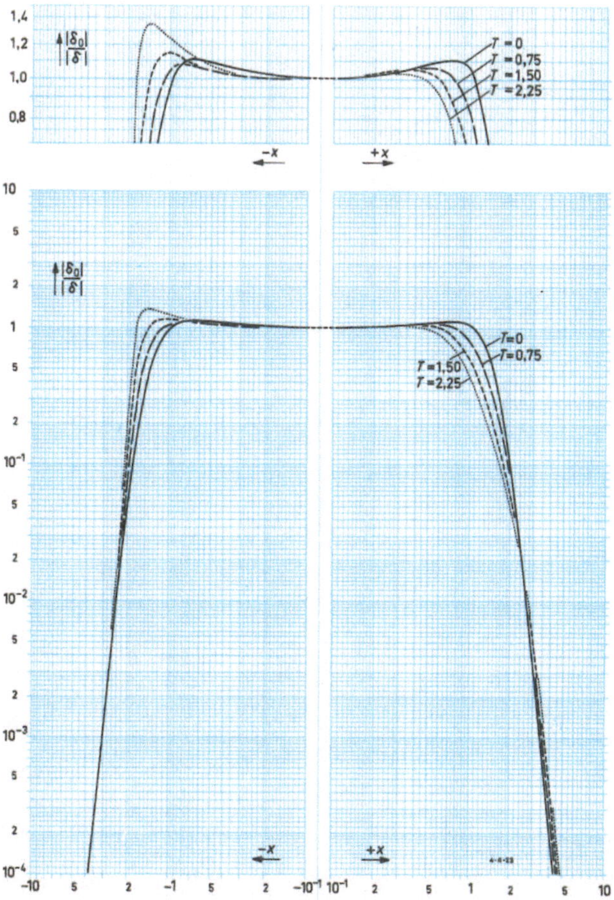

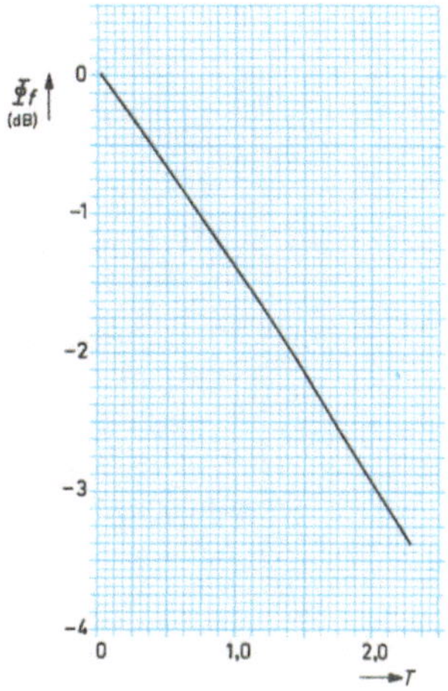

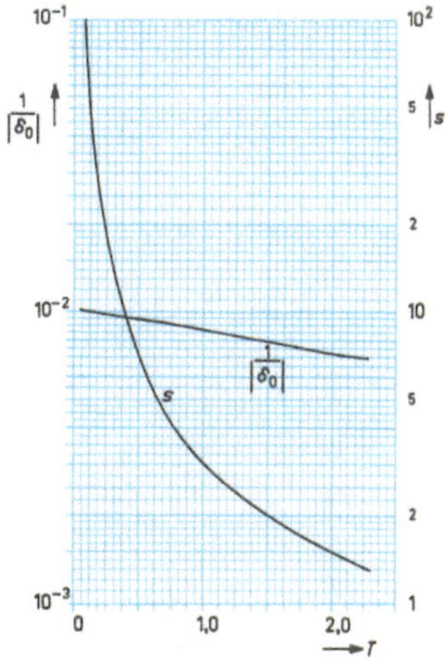

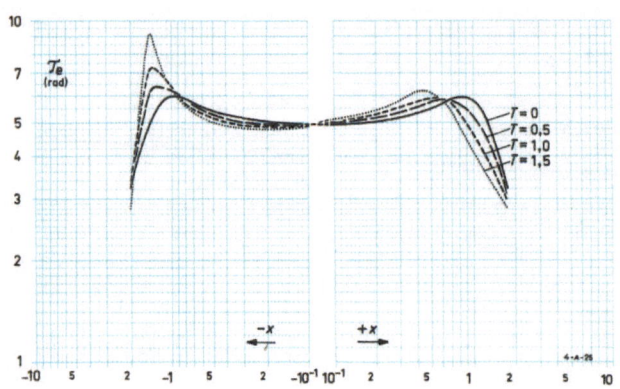

Fig. 15.59
Four-stage amplifier
Configuration II
Tuning method B
$\Theta = 270°$
$q^2 = 1, r = 1$

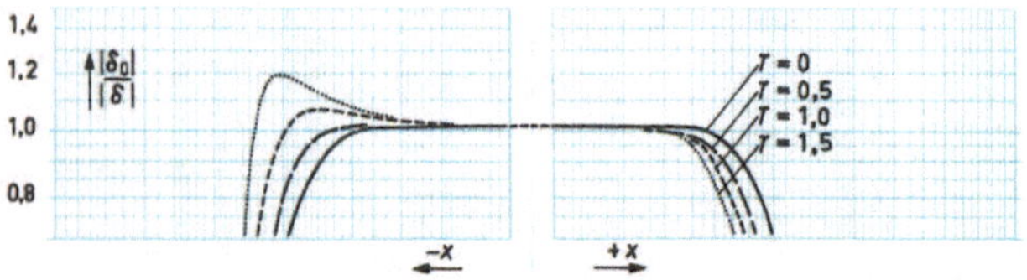

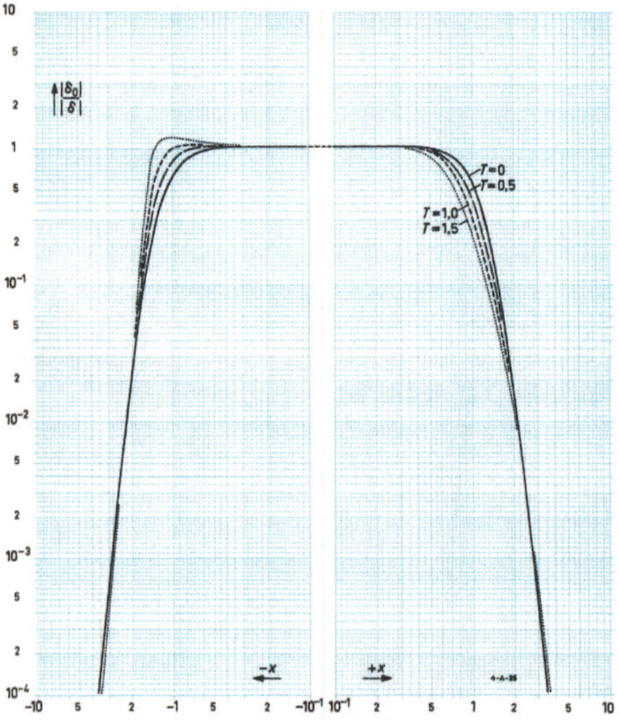

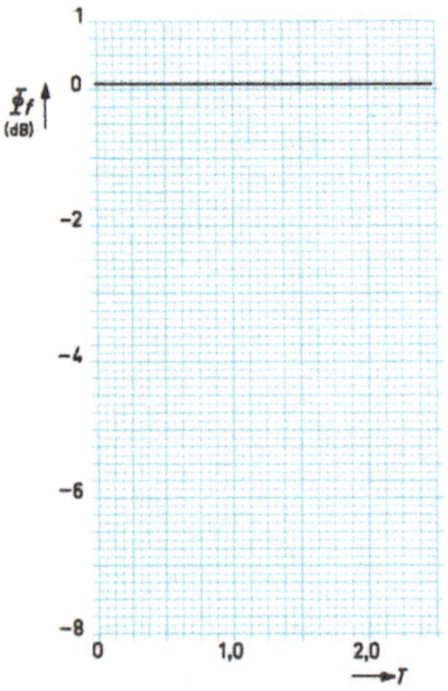

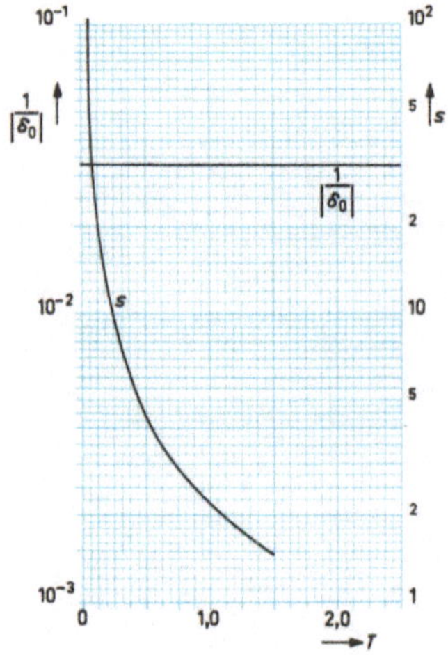

Three-Stage Amplifier with Three Double-Tuned and One Single-Tuned Bandpass Filters

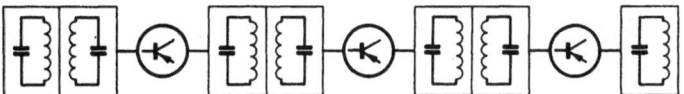

15.9 Three-stage amplifier with three double-tuned and one single-tuned bandpass filters

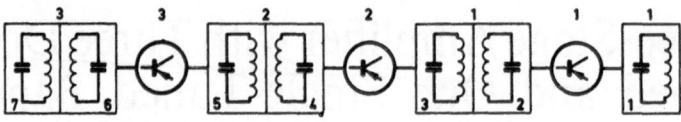

Fig. 15.60 Schematic representation of a three-stage amplifier with three double-tuned and one single-tuned bandpass filters (Configuration II).

Reduced amplifier determinant:

See Determinant III on page 234 for $n = 3$.

Method of tuning: B

Reduced amplifier determinant at tuning frequency ($x = 0$):

$$|{}_3\delta_0| = P_{7M}$$

Tuning correction terms:

$$x'_1 = T_1 \sin\Theta_1 \cdot \frac{1}{P_{1M}}$$

$$x'_2 = T_2 \sin\Theta_2 \cdot \frac{P_{2M}}{P_{3M}}$$

$$x'_3 = T_3 \sin\Theta_3 \cdot \frac{P_{4M}}{P_{5M}}$$

Minor determinant values at tuning frequency ($x = 0$):

$$P_{1M} = 1$$

$$P_{2M} = 1 - T_1 \cos\Theta_1$$

$$P_{3M} = 1 + q_1^2 - T_1 \cos\Theta_1$$

$$P_{4M} = 1 + q_1^2 - T_1 \cos\Theta_1 (1 - T_2 \cos\Theta_2) - T_2 \cos\Theta_2$$

$$P_{5M} = (1 + q_1^2)(1 + q_2^2) - T_1 \cos\Theta_1 (1 + q_2^2 - T_2 \cos\Theta_2) - T_2 \cos\Theta_2$$

$$P_{6M} = (1 + q_1^2)(1 + q_2^2 - T_3 \cos\Theta_3) - T_1 \cos\Theta_1 \{1 + q_2^2 - T_3 \cos\Theta_3$$
$$- T_2 \cos\Theta_2 (1 - T_3 \cos\Theta_3)\}$$
$$- T_2 \cos\Theta_2 (1 - T_3 \cos\Theta_3)$$

$$P_{7M} = (1 + q_1^2) \{(1 + q_2^2)(1 + q_3^2) - T_3 \cos\Theta_3\}$$
$$- T_1 \cos\Theta_1 \{(1 + q_2^2)(1 + q_3^2)$$
$$- T_2 \cos\Theta_2 (1 + q_3^2 - T_3 \cos\Theta_3) - T_3 \cos\Theta_3\}$$
$$- T_2 \cos\Theta_2 (1 + q_3^2 - T_3 \cos\Theta_3)$$

Minor determinant values at tuning frequency ($x = 0$) for identical stages:

$$P_{1M} = 1$$
$$P_{2M} = 1 - T \cos\Theta$$
$$P_{3M} = 1 + q^2 - T \cos\Theta$$
$$P_{4M} = 1 + q^2 - T \cos\Theta (2 - T \cos\Theta)$$
$$P_{5M} = (1 + q^2)^2 - T \cos\Theta (2 + q^2 - T \cos\Theta)$$
$$P_{6M} = (1 + q^2)^2 - T \cos\Theta (3 + 2q^2 - 3T \cos\Theta + T^2 \cos^2\Theta)$$
$$P_{7M} = (1 + q^2)^3 - T \cos\Theta \{3 + 4q^2 + q^4 - (3 + q^2) T \cos\Theta + T^2 \cos^2\Theta\}$$

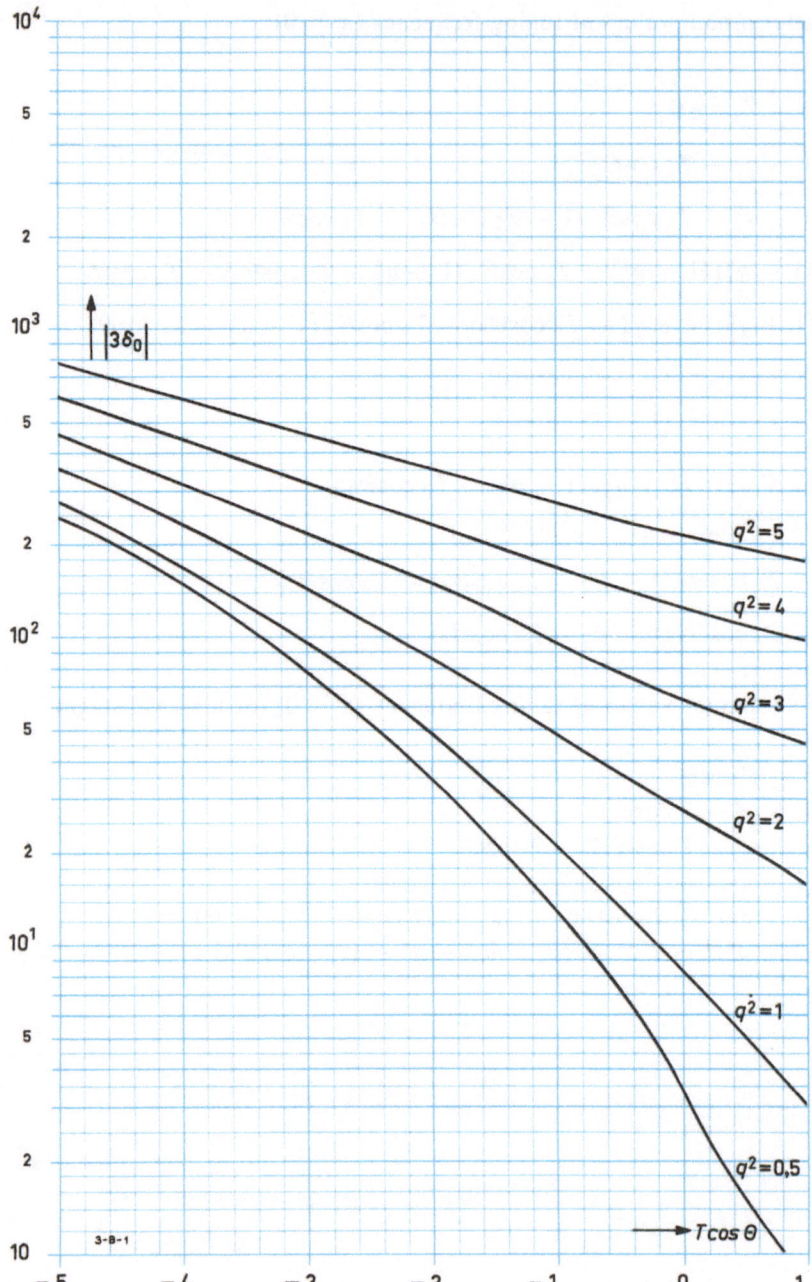

Fig. 15.61 Graph representing the value of the reduced determinant $|\,1/\delta_0\,|$ as a function of $T\cos\Theta$ for a three-stage amplifier with three double-tuned and one single-tuned band-pass filters (Configuration III, tuning method B).

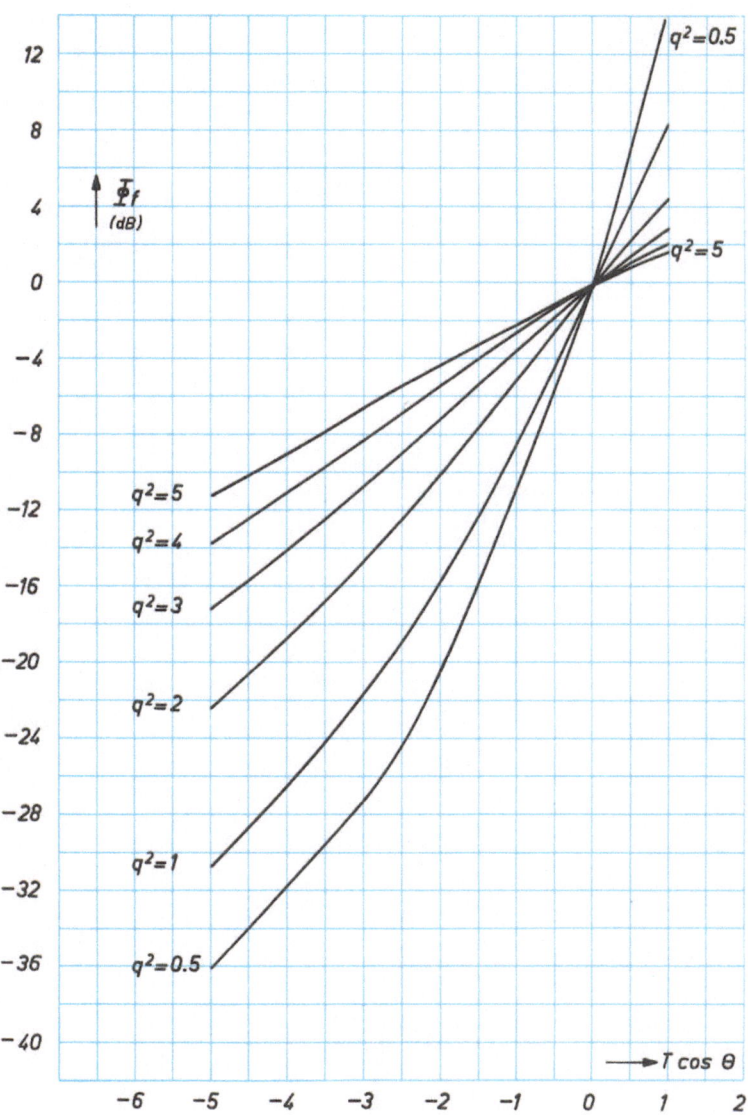

Fig. 15.62 Graph representing the feedback losses Φ_f as a function of $T\cos\Theta$ for a three-stage amplifier with three double-tuned and one single-tuned bandpass filters (Configuration III, tuning method B).

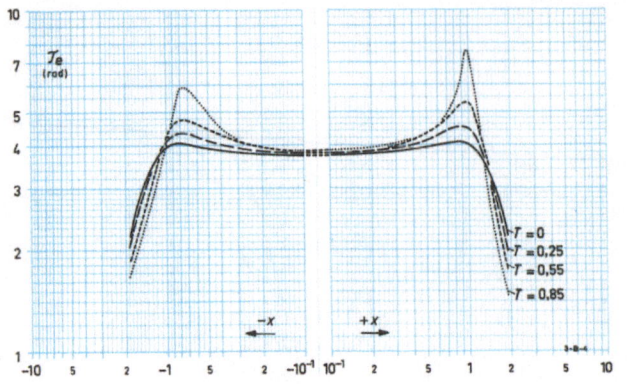

Fig. 15.63
Three-stage amplifier
Configuration III
Tuning method B
$\Theta = 30°$
$q^2 = 1, r = 1, m = 0.75$

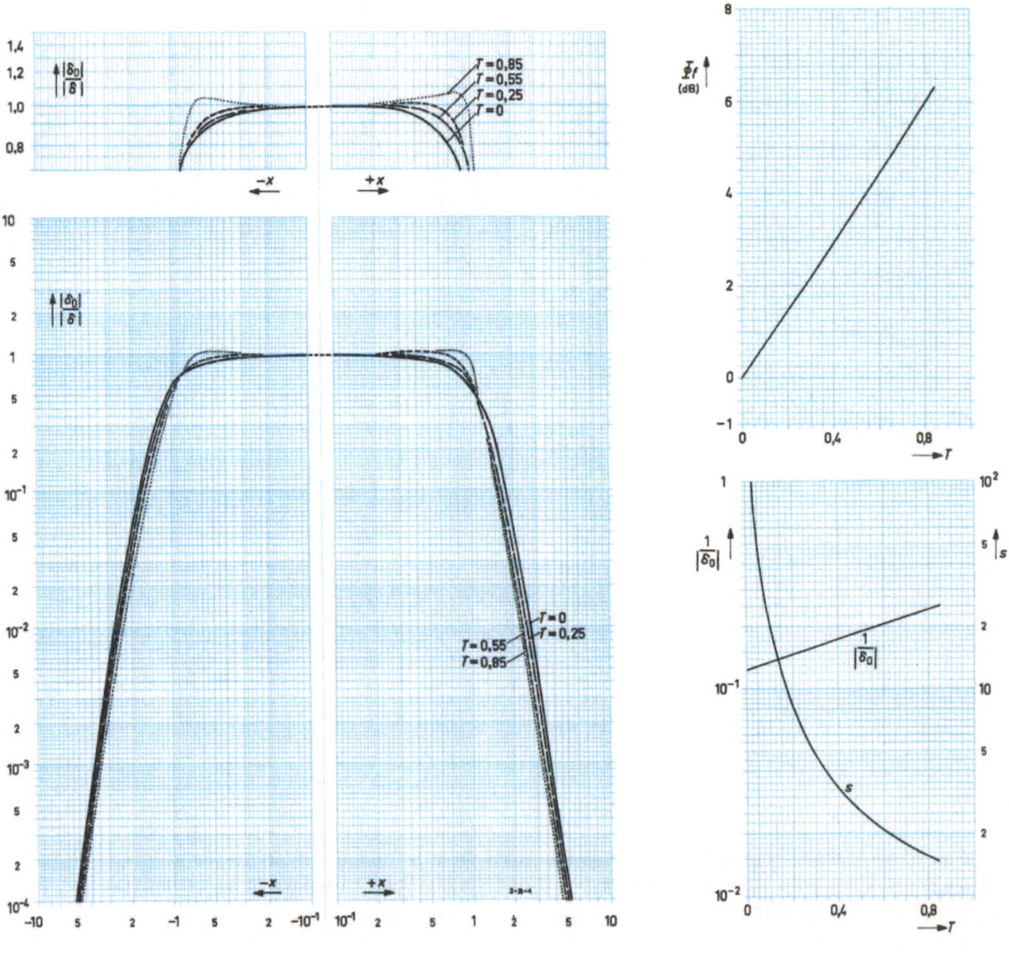

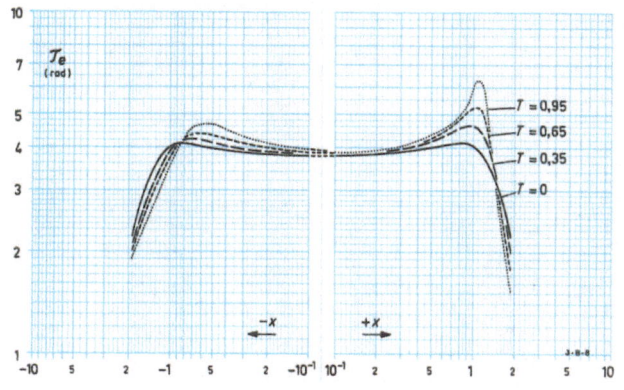

Fig. 15.64
Three-stage amplifier
Configuration III
Tuning method B
$\Theta = 60°$
$q^2 = 1, r = 1, m = 0.75$

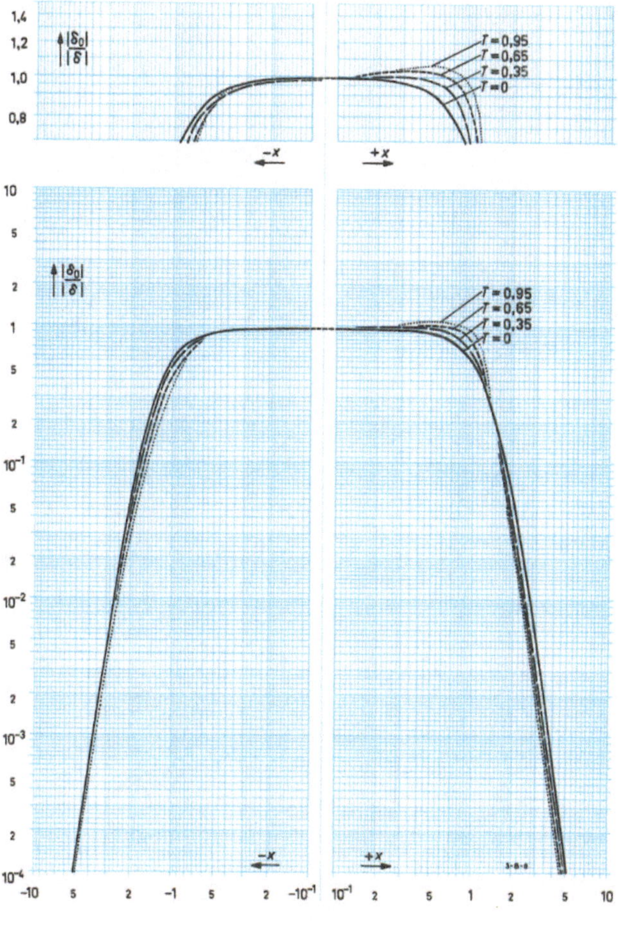

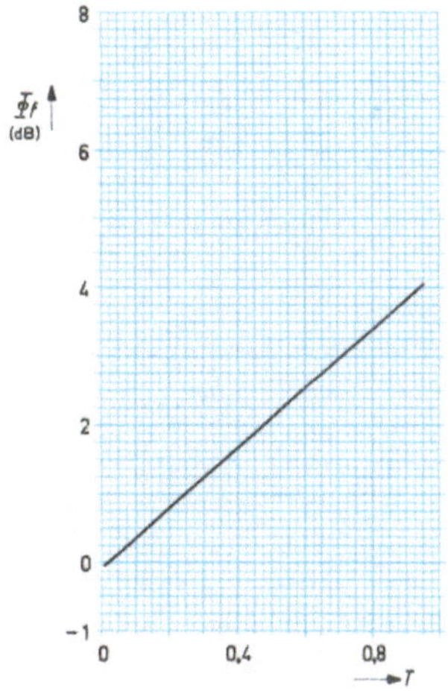

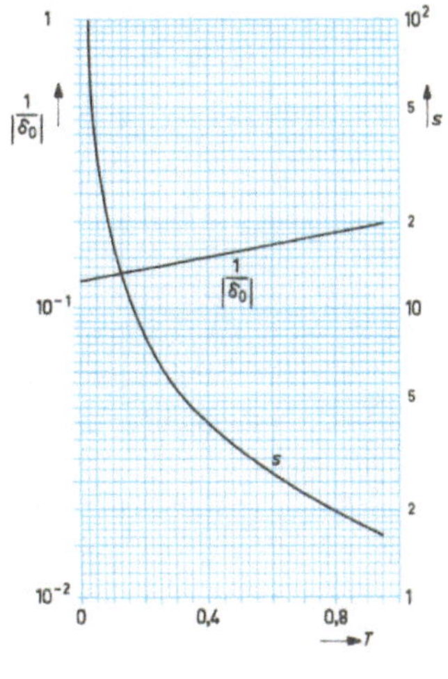

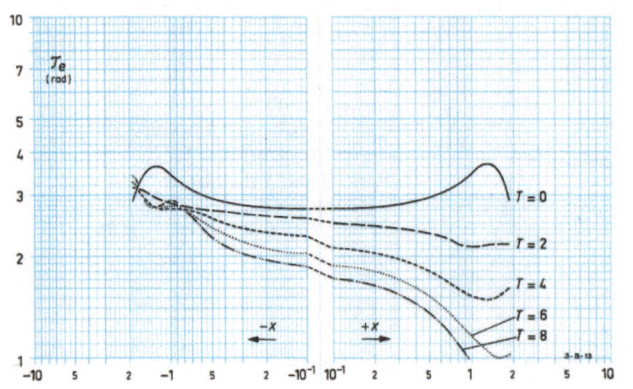

Fig. 15.65
Three-stage maplifier
Configuration III
Tuning method B
$\Theta = 210°$
$q^2 = 2, r = 1, m = 0.75$

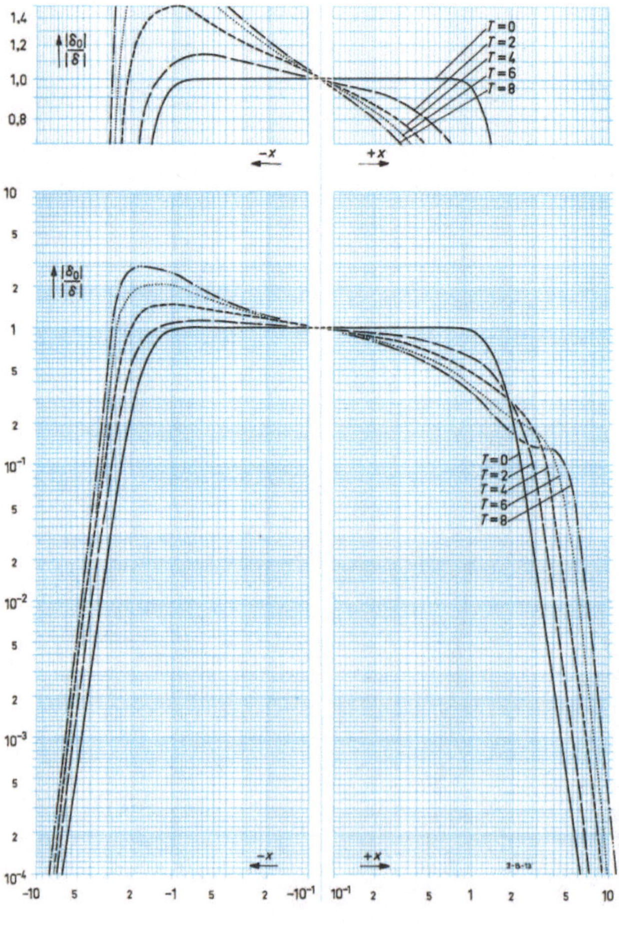

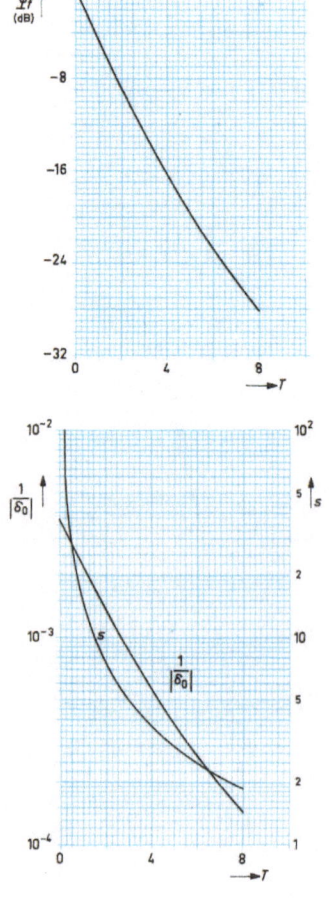

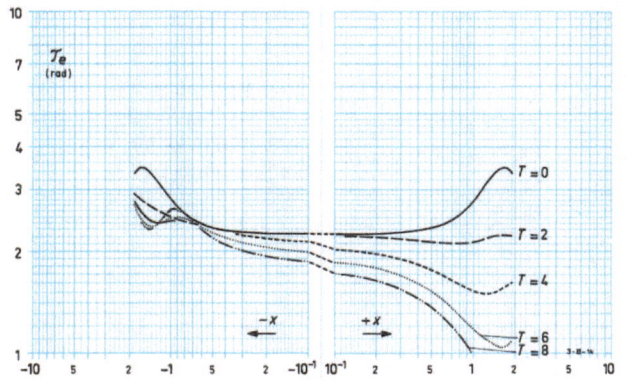

Fig. 15.66
Three-stage amplifier
Configuration III
Tuning method B
$\Theta = 210°$
$q^2 = 3, r = 1, m = 0.75$

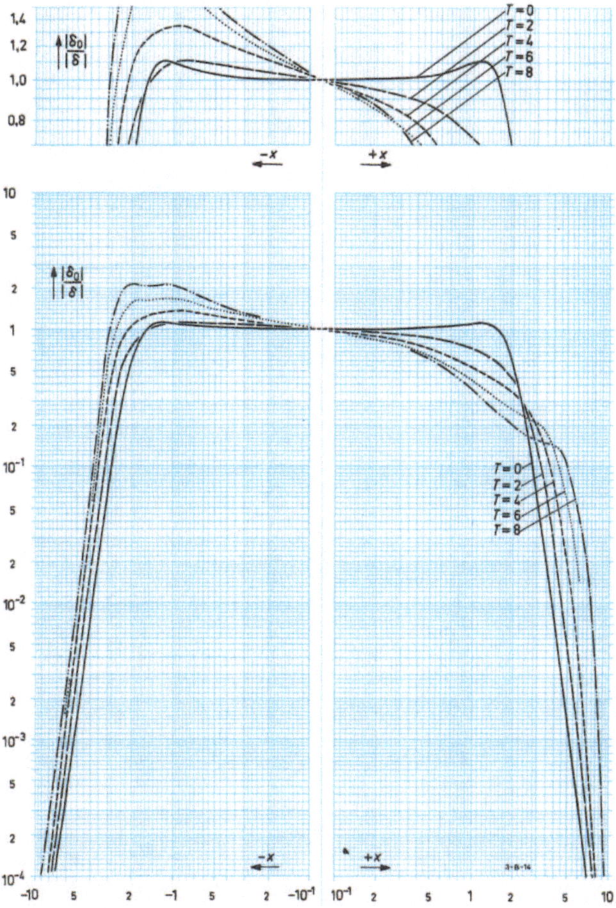

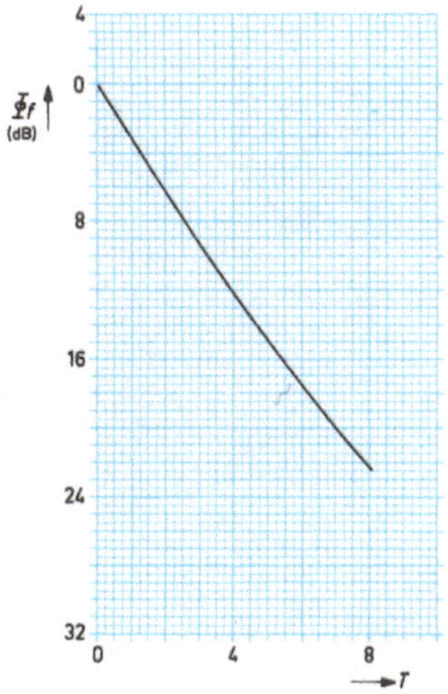

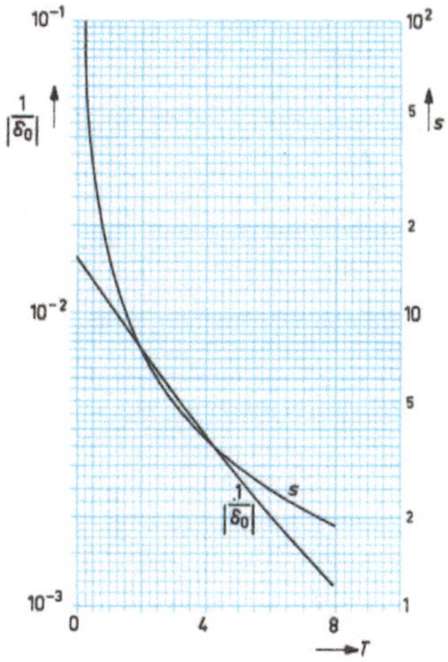

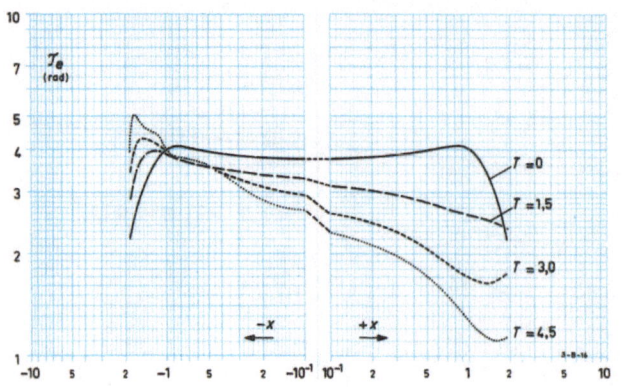

Fig. 15.67
Three-stage amplifier
Configuration III
Tuning method B
$\Theta = 225°$
$q^2 = 1, r = 1, m = 0.75$

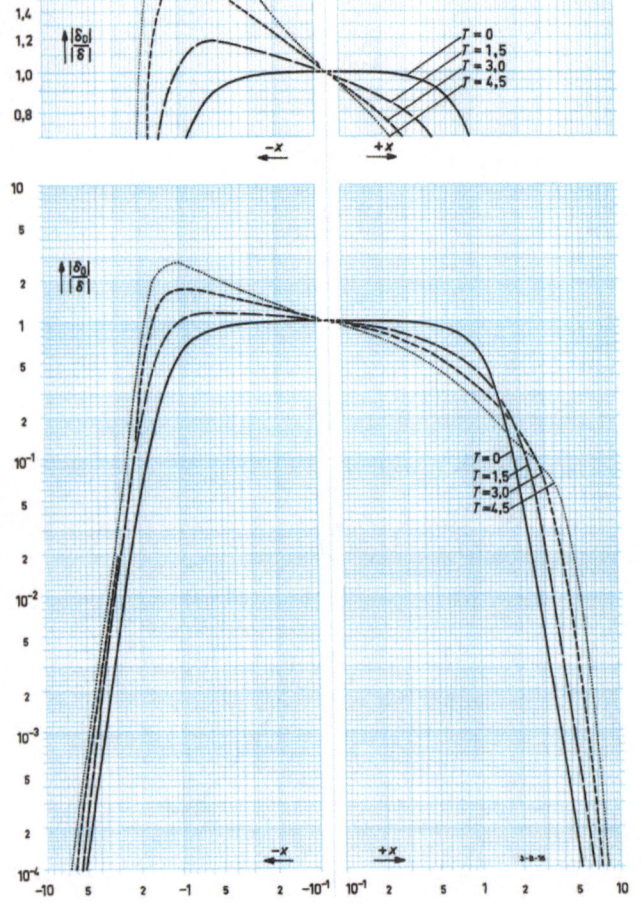

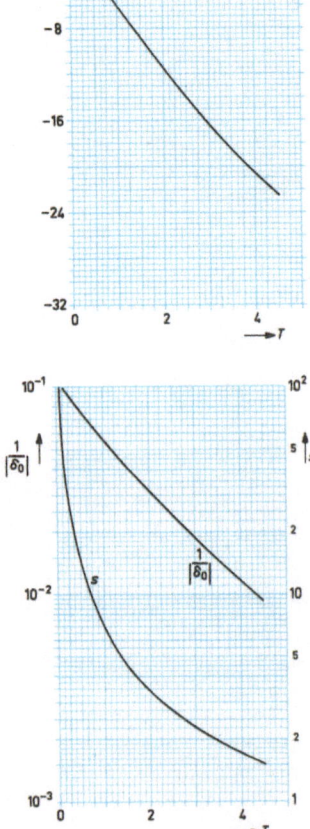

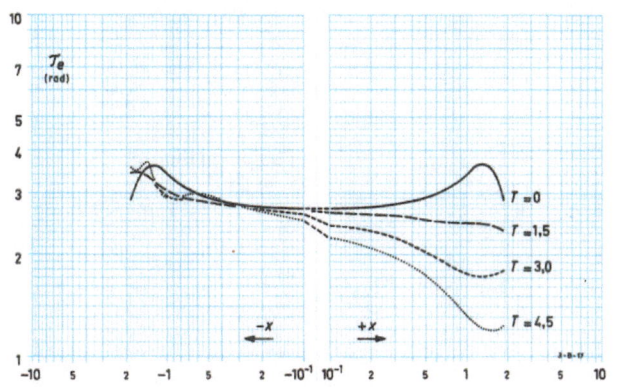

Fig. 15.68
Three-stage amplifier
Configuration III
Tuning method B
$\Theta = 225°$
$q^2 = 2, r = 1, m = 0.75$

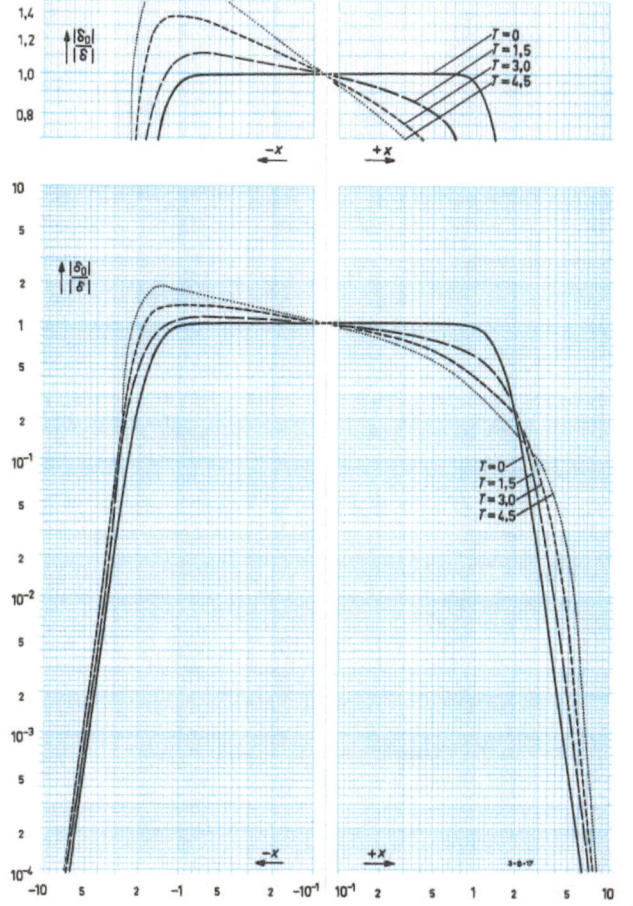

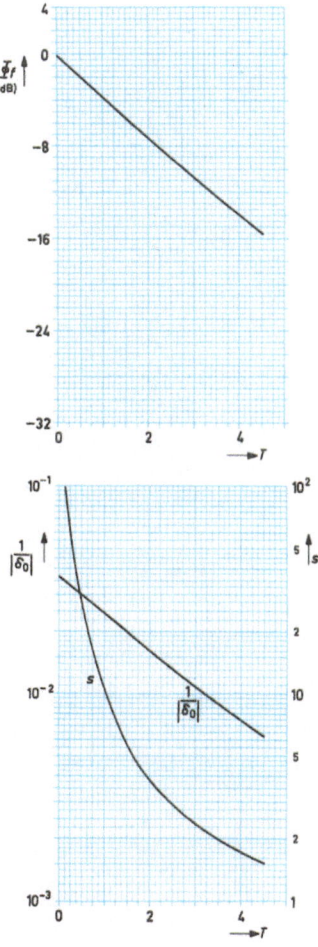

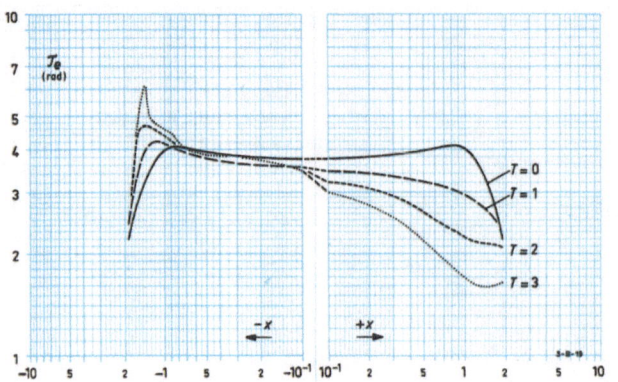

Fig. 15.69
Three-stage amplifier
Configuration III
Tuning method B
$\Theta = 240°$
$q^2 = 1, r = 1, m = 0.75$

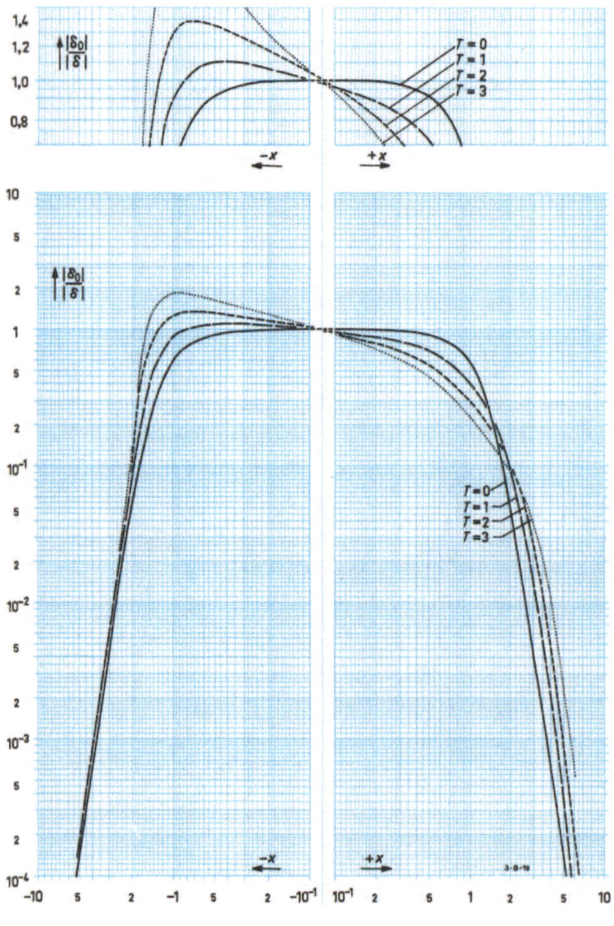

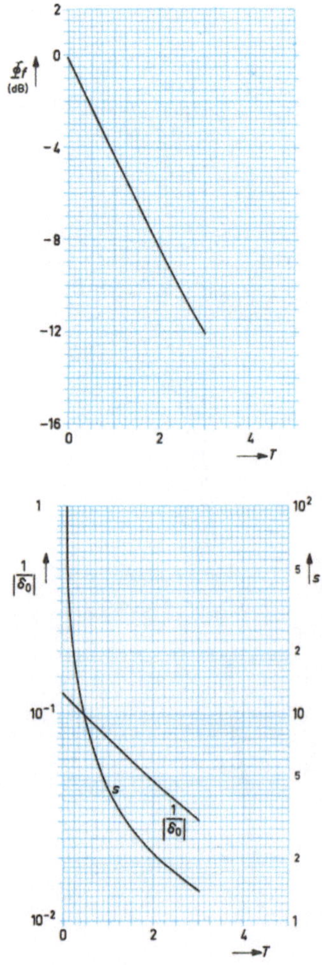

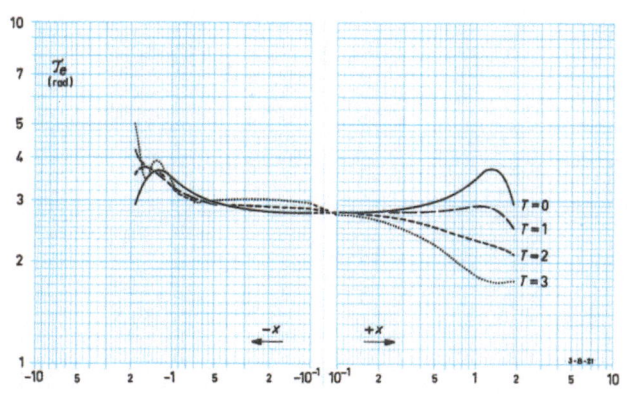

Fig. 15.70
Three-stage amplifier
Configuration III
Tuning method B
$\Theta = 240°$
$q^2 = 2$, $r = 1$, $m = 0.75$

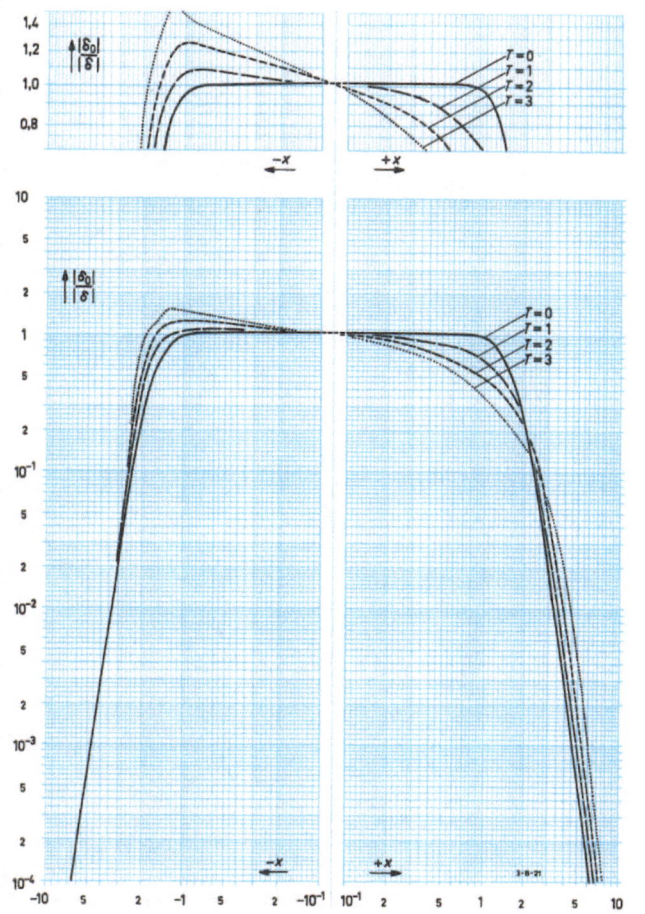

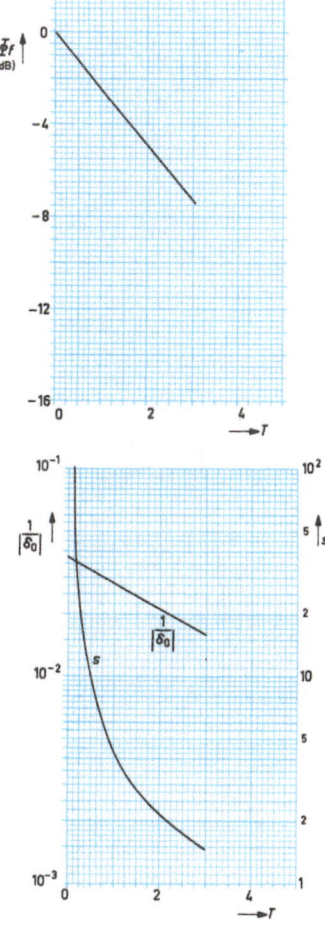

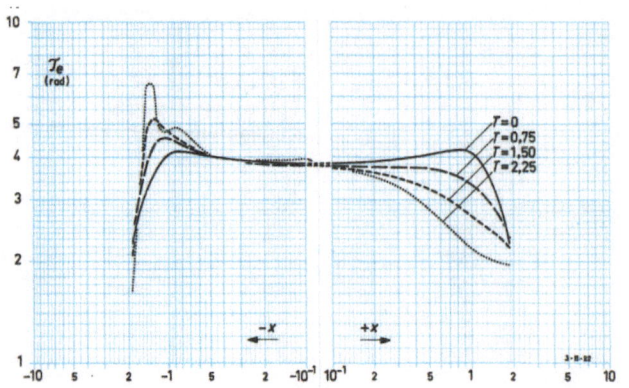

Fig. 15.71
Three-stage amplifier
Configuration III
Tuning method B
$\Theta = 255°$
$q^2 = 1, r = 1, m = 0.75$

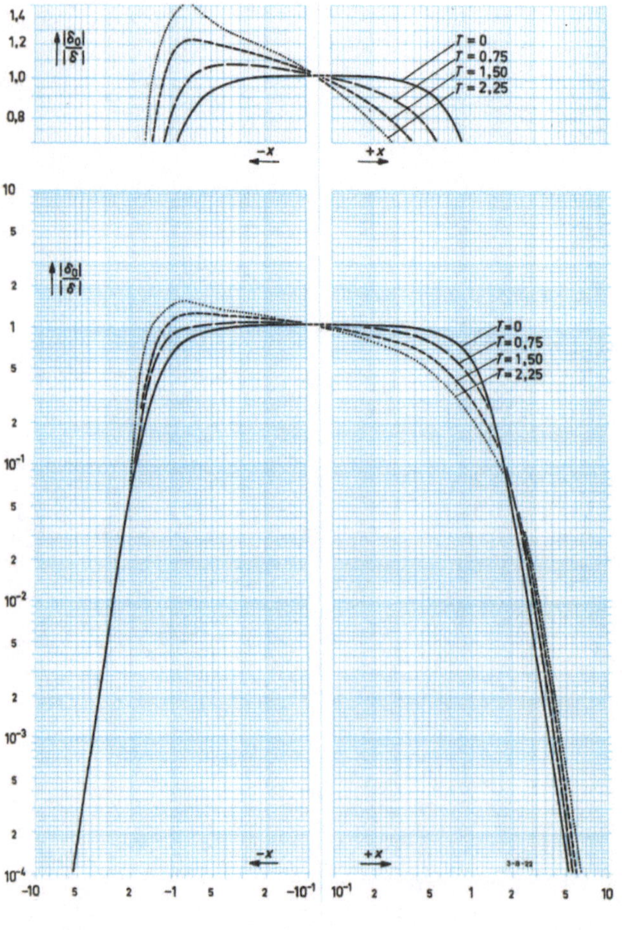

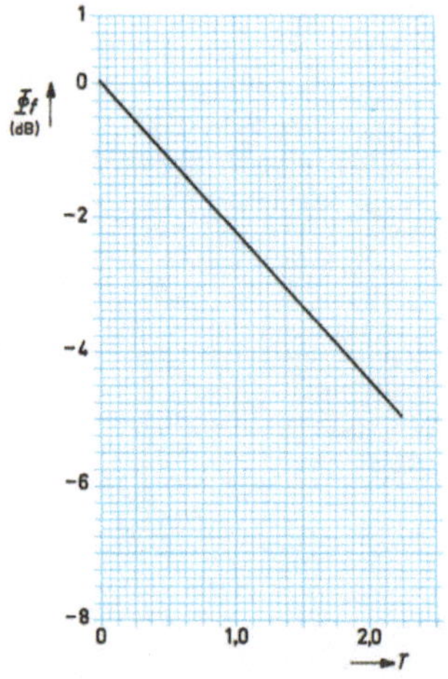

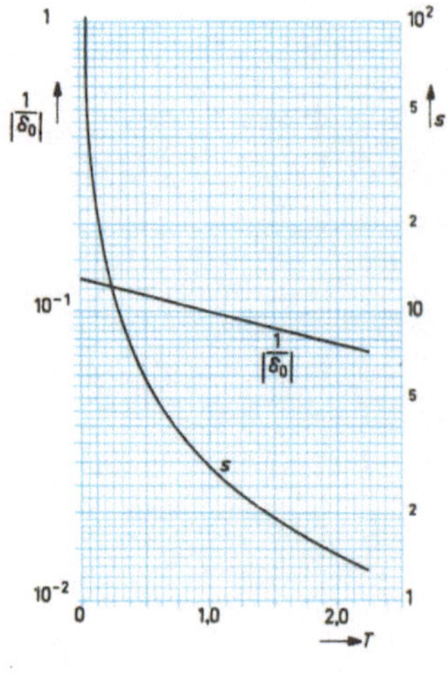

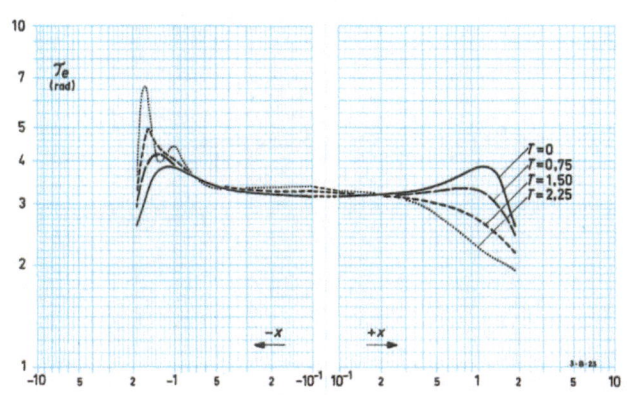

Fig. 15.72
Three-stage amplifier
Configuration III
Tuning method B
$\Theta = 255°$
$q^2 = 1.5$, $r = 1$, $m = 0.75$

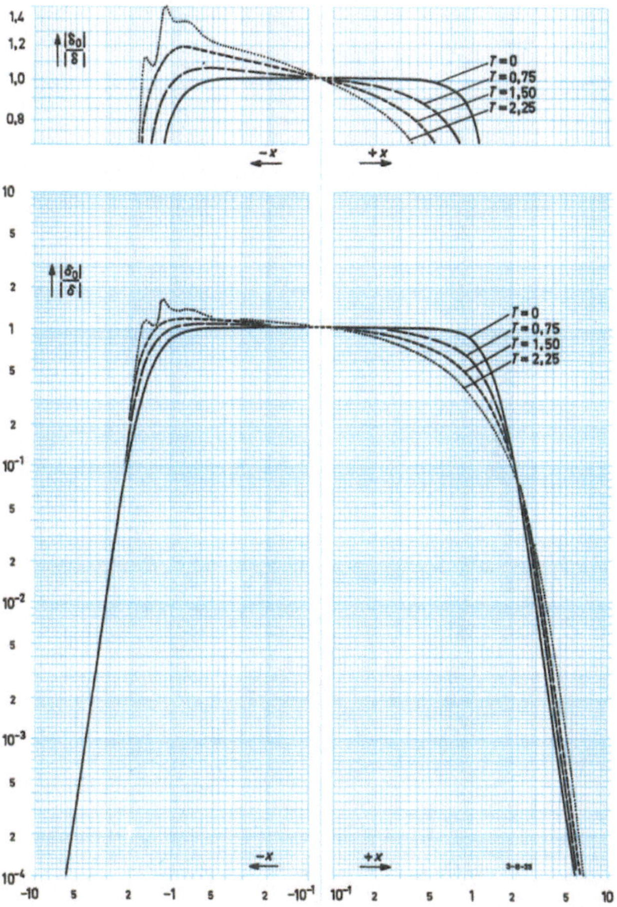

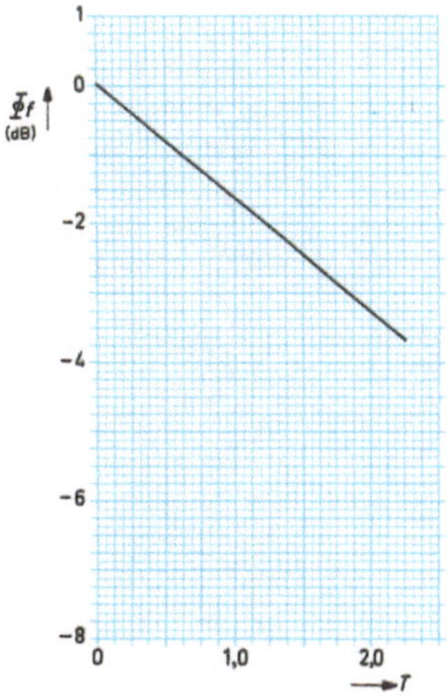

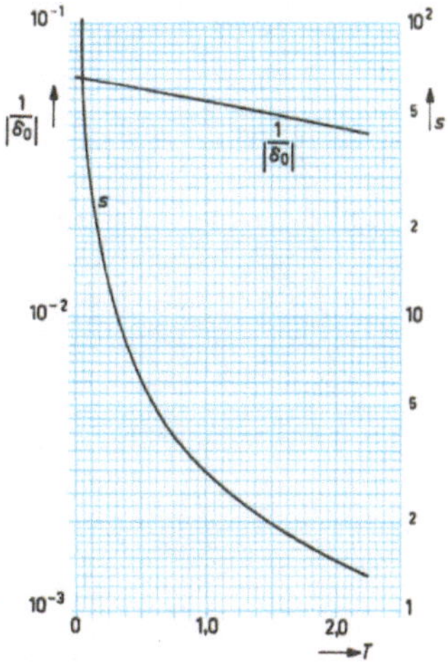

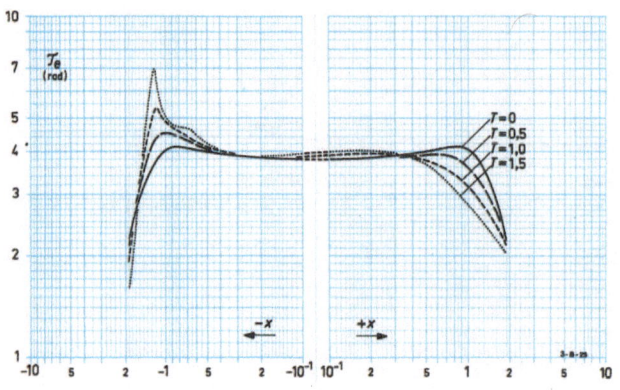

Fig. 15.73
Three-stage amplifier
Configuration III
Tuning method B
$\Theta = 270°$
$q^2 = 1, r = 1, m = 0.75$

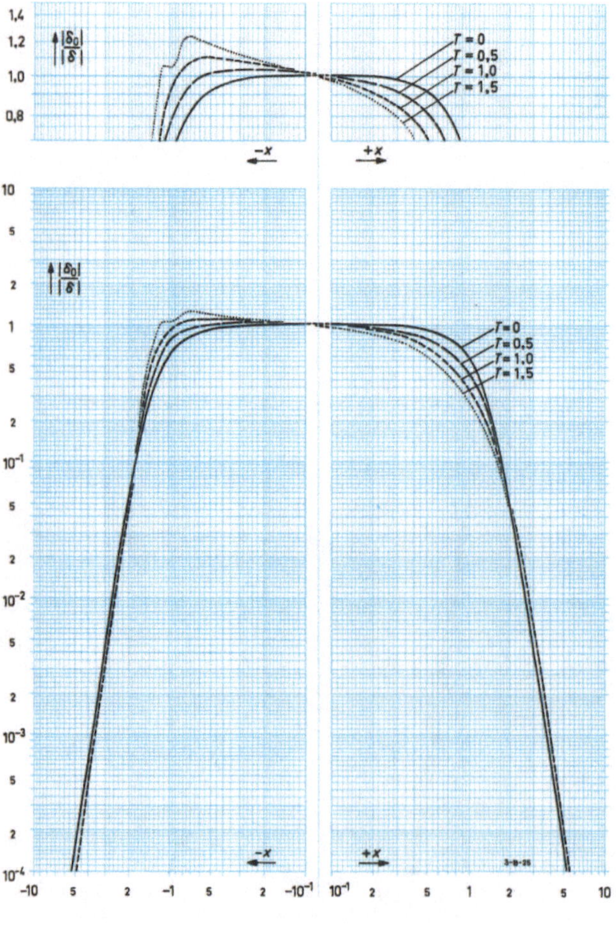

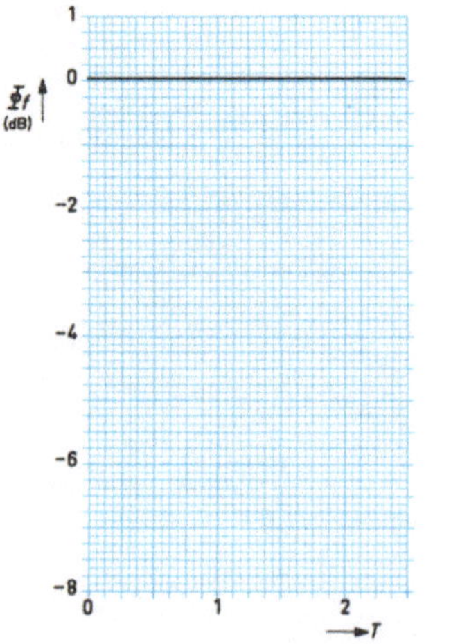

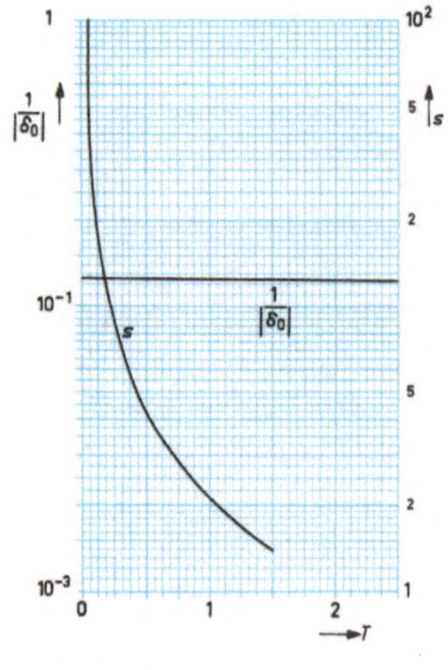

Four-Stage Amplifier with Four Double-Tuned and One Single-Tuned Bandpass Filters

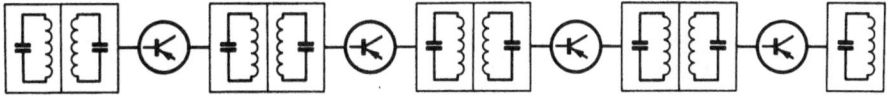

15.10 Four-stage amplifier with four double-tuned and one single-tuned bandpass filters

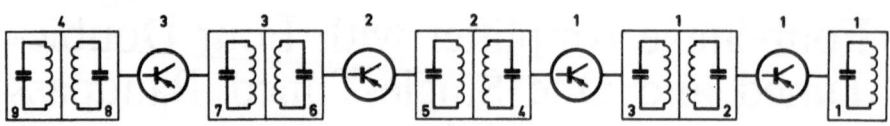

Fig. 15.74 Schematic representation of a four-stage amplifier with four double-tuned and one single-tuned bandpass filters (Configuration III).

Reduced amplifier determinant:

See Determinant III on page 234 for $n = 4$.

Method of tuning: **B**

Reduced amplifier determinant at tuning frequency ($x = 0$):

$$|_4\delta_0| = P_{9M}$$

Tuning correction terms:

$$x_1' = T_1 \sin\Theta_1 \cdot \frac{1}{P_{1M}}$$

$$x_2' = T_2 \sin\Theta_2 \cdot \frac{P_{2M}}{P_{3M}}$$

$$x_3' = T_3 \sin\Theta_3 \cdot \frac{P_{4M}}{P_{5M}}$$

$$x_4' = T_4 \sin\Theta_4 \cdot \frac{P_{6M}}{P_{7M}}$$

Minor determinant values at tuning frequency ($x = 0$):

$$P_{1M} = 1$$
$$P_{2M} = 1 - T_1 \cos\Theta_1$$
$$P_{3M} = 1 + q_1^2 - T_1 \cos\Theta_1$$
$$P_{4M} = 1 + q_1^2 - T_1 \cos\Theta_1 (1 - T_2 \cos\Theta_2) - T_2 \cos\Theta_2$$
$$P_{5M} = (1 + q_1^2)(1 + q_2^2) - T_1 \cos\Theta_1 (1 + q_2^2 - T_2 \cos\Theta_2) - T_2 \cos\Theta_2$$
$$\begin{aligned} P_{6M} = {}&(1 + q_1^2)(1 + q_2^2 - T_3 \cos\Theta_3) - T_1 \cos\Theta_1 \{1 + q_2^2 - T_3 \cos\Theta_3 \\ &- T_2 \cos\Theta_2 (1 - T_3 \cos\Theta_3)\} \\ &- T_2 \cos\Theta_2 (1 - T_3 \cos\Theta_3) \end{aligned}$$
$$\begin{aligned} P_{7M} = {}&(1 + q_1^2)\{(1 + q_2^2)(1 + q_3^2) - T_3 \cos\Theta_3\} \\ &- T_1 \cos\Theta_1 \{(1 + q_2^2)(1 + q_3^2) - T_2 \cos\Theta_2 (1 + q_3^2 - T_3 \cos\Theta_3) \\ &- T_3 \cos\Theta_3\} - T_2 \cos\Theta_2 (1 + q_3^2 - T_3 \cos\Theta_3) \end{aligned}$$

Minor determinant values at tuning frequency ($x = 0$) for identical stages:

$P_{1M} = 1$

$P_{2M} = 1 - T\cos\Theta$

$P_{3M} = 1 + q^2 - T\cos\Theta$

$P_{4M} = 1 + q^2 - T\cos\Theta\,(2 - T\cos\Theta)$

$P_{5M} = (1 + q^2)^2 - T\cos\Theta\,(2 + q^2 - T\cos\Theta)$

$P_{6M} = (1 + q^2)^2 - T\cos\Theta\,(3 + 2q^2 - 3T\cos\Theta + T^2\cos^2\Theta)$

$P_{7M} = (1 + q^2)^3 - T\cos\Theta\,\{3 + 4q^2 + q^4 - (3 + q^2)\,T\cos\Theta + T^2\cos^2\Theta\}$

$P_{8M} = (1 + q^2)^3 - T\cos\Theta\,\{4 + 6q^2 + 2q^4 - (6 + 3q^2)\,T\cos\Theta + 4T^2\cos^2\Theta$
$\qquad - T^3\cos^3\Theta\}$

$P_{9M} = (1 + q^2)^4 - T\cos\Theta\,\{4 + 9q^2 + 6q^4 + q^6 - (6 + 6q^2 + q^4)\,T\cos\Theta$
$\qquad + (4 + q^2)\,T^3\cos^3\Theta - T^4\cos^4\Theta\}$

$P_{8M} = (1 + q_1^2)\,\{(1 + q_2^2)\,(1 + q_3^2 - T_4\cos\Theta_4) - T_3\cos\Theta_3\,(1 - T_4\cos\Theta_4)\}$

$\qquad - T_1\cos\Theta_1\,[(1 + q_2^2)\,(1 + q_3^2 - T_4\cos\Theta_4)$

$\qquad\quad - T_2\cos\Theta_2\,\{1 + q_3^2 - T_3\cos\Theta_3 - T_4\cos\Theta_4\,(1 - T_3\cos\Theta_3)\}$

$\qquad\quad - T_3\cos\Theta_3\,(1 - T_4\cos\Theta_4)]$

$\qquad - T_2\cos\Theta_2\,\{1 + q_3^2 - T_3\cos\Theta_3 - T_4\cos\Theta_4\,(1 - T_3\cos\Theta_3)\}$

$P_{9M} = (1 + q_1^2)\,[(1 + q_2^2)\,\{(1 + q_3^2)\,(1 + q_4^2) - T_4\cos\Theta_4\}$

$\qquad - T_3\cos\Theta_3\,(1 + q_4^2 - T_4\cos\Theta_4)]$

$\qquad - T_1\cos\Theta_1\,[\,(1 + q_2^2)\,(1 + q_3^2)\,(1 + q_4^2)$

$\qquad\quad - T_2\cos\Theta_2\,\{(1 + q_3^2 - T_3\cos\Theta_3)\,(1 + q_4^2) - T_4\cos\Theta_4\,(1 - T_3\cos\Theta_3)\}$

$\qquad\quad - T_3\cos\Theta_3\,(1 + q_4^2 - T_4\cos\Theta_4)\,]$

$\qquad - T_2\cos\Theta_2\,\{(1 + q_4^2)\,(1 + q_3^2 - T_3\cos\Theta_3)$

$\qquad\quad - T_4\cos\Theta_4\,(1 - T_3\cos\Theta_3)\}$

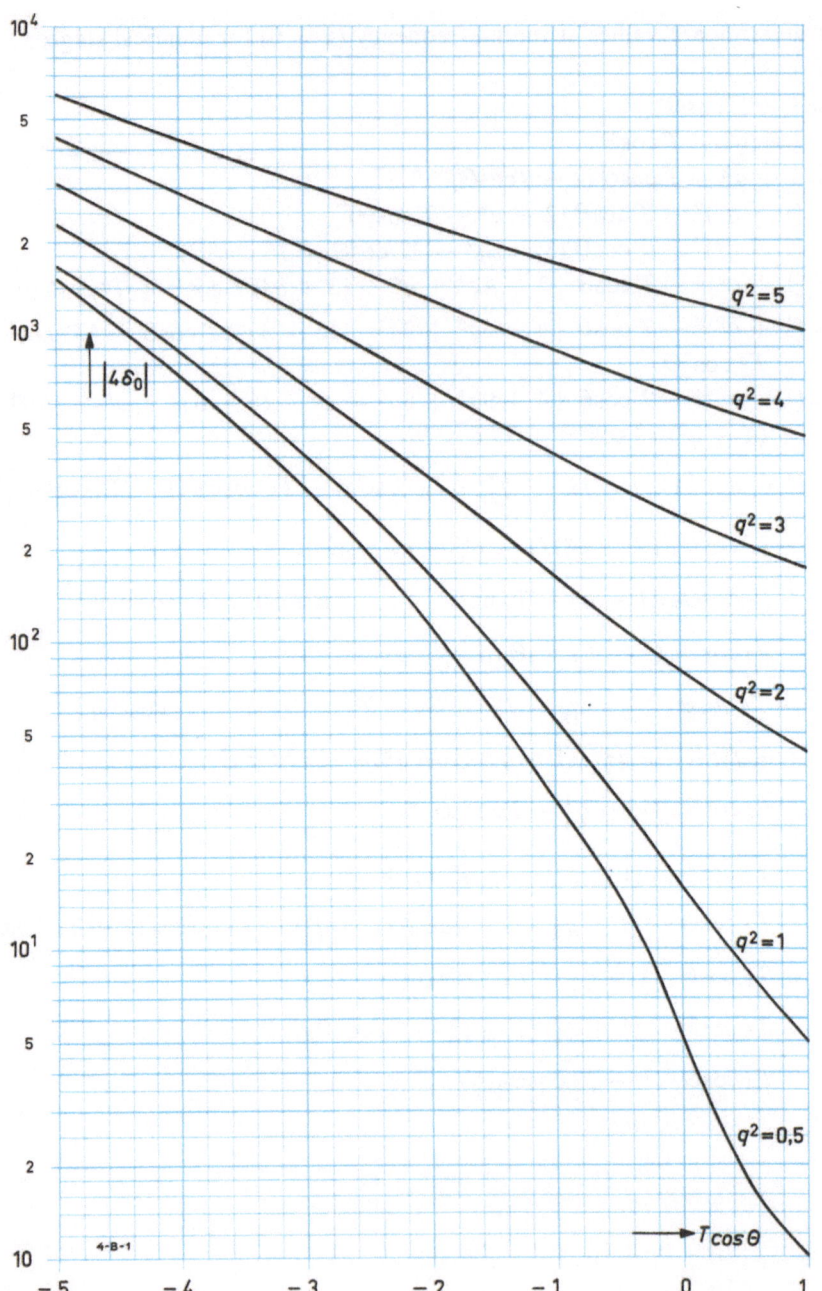

Fig. 15.75 Graph representing the value of the reduced determinant $|1/\delta_0|$ as a function of $T\cos\Theta$ for a four-stage amplifier with four double-tuned and one single-tuned bandpass filters (Configuration III, tuning method B).

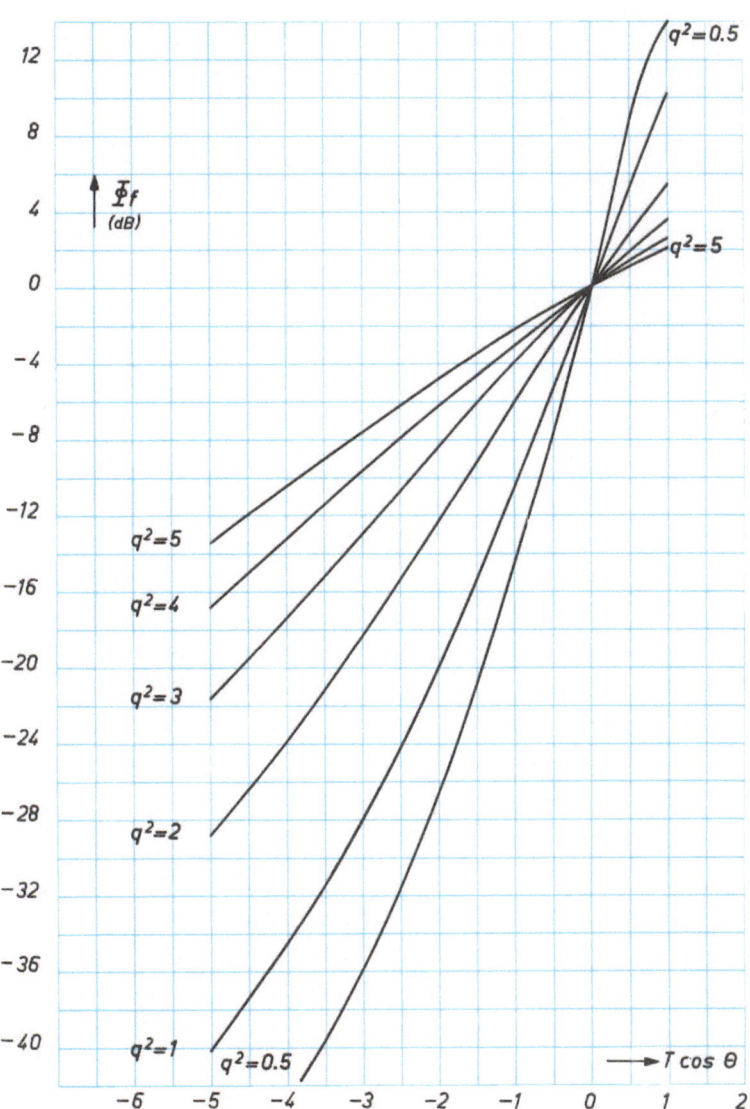

Fig. 15.76 Graph representing the feedback losses Φ_f as a function of $T\cos\Theta$ of a four-stage amplifier with four double-tuned and one single-tuned bandpass filters (Configuration III, tuning method B).

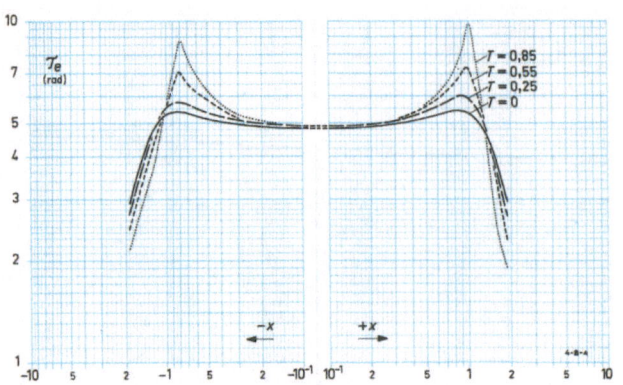

Fig. 15.77
Four-stage amplifier
Configuration III
Tuning method B
$\Theta = 30°$
$q^2 = 1$, $r = 1$, $m = 0.75$

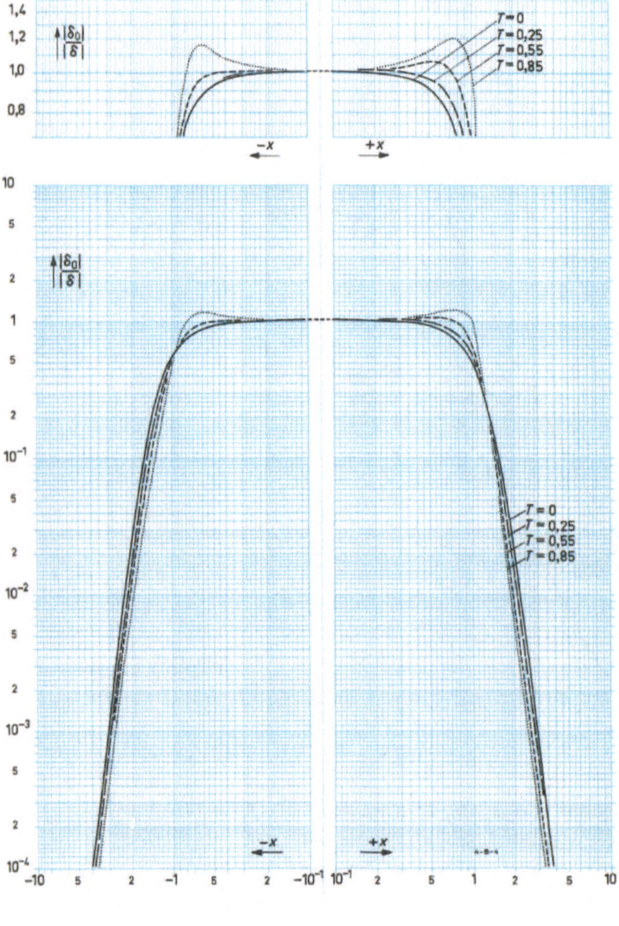

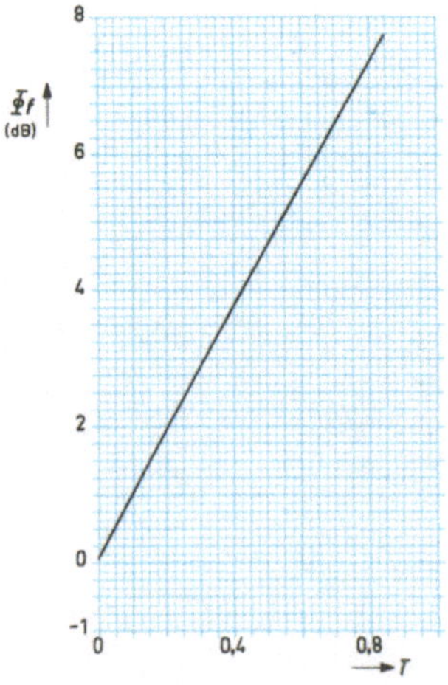

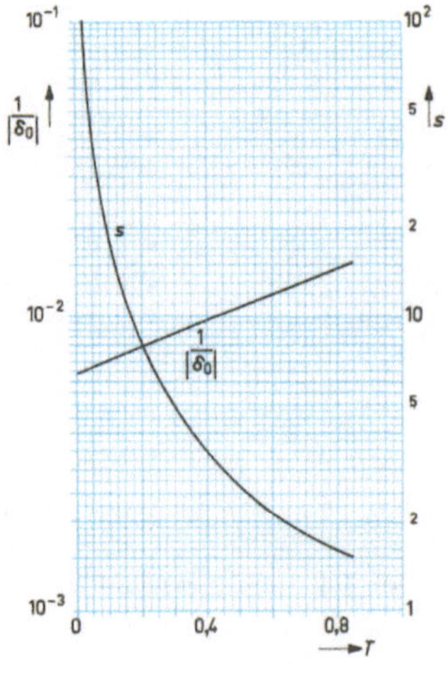

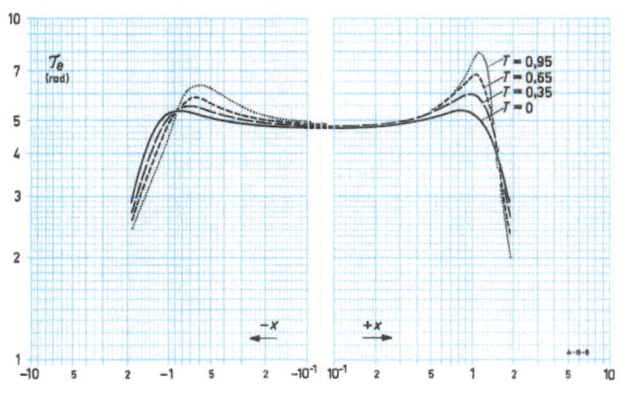

Fig. 15.78
Four-stage amplifier
Configuration III
Tuning method B
$\Theta = 60°$
$q^2 = 1, r = 1, m = 0.75$

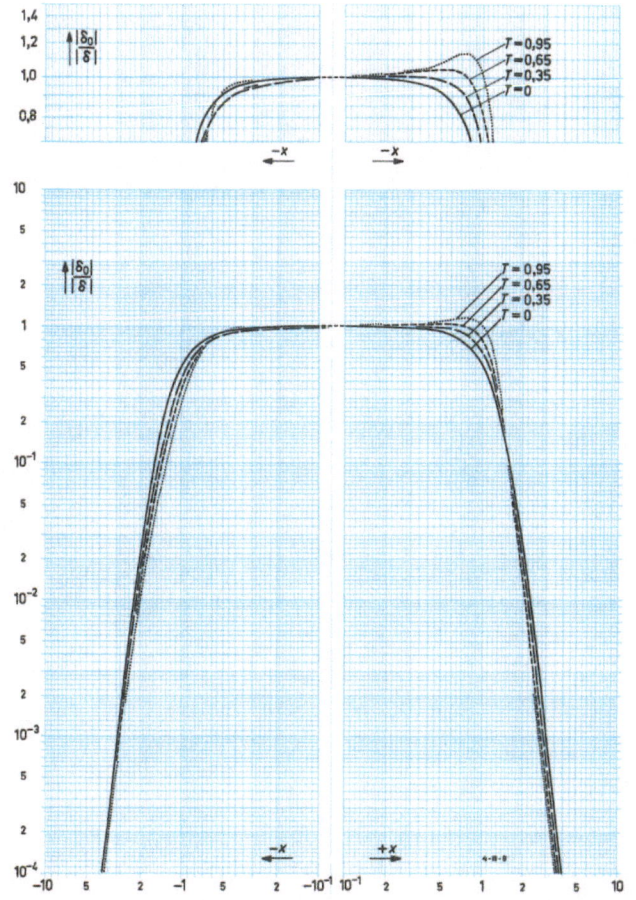

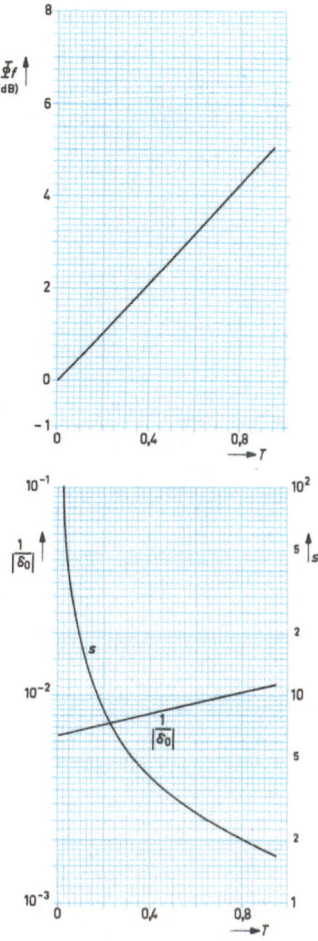

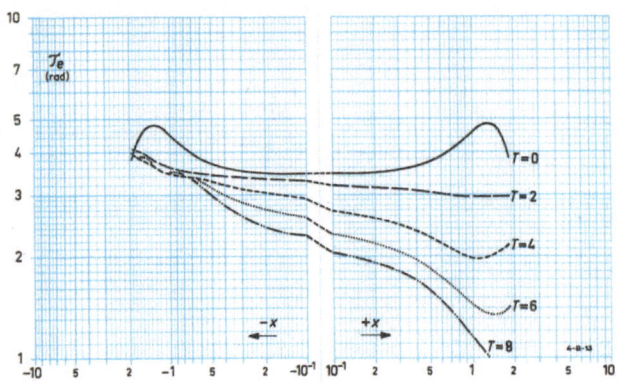

Fig. 15.79
Four-stage amplifier
Configuration III
Tuning method B
$\Theta = 210°$
$q^2 = 2, r = 1, m = 0.75$

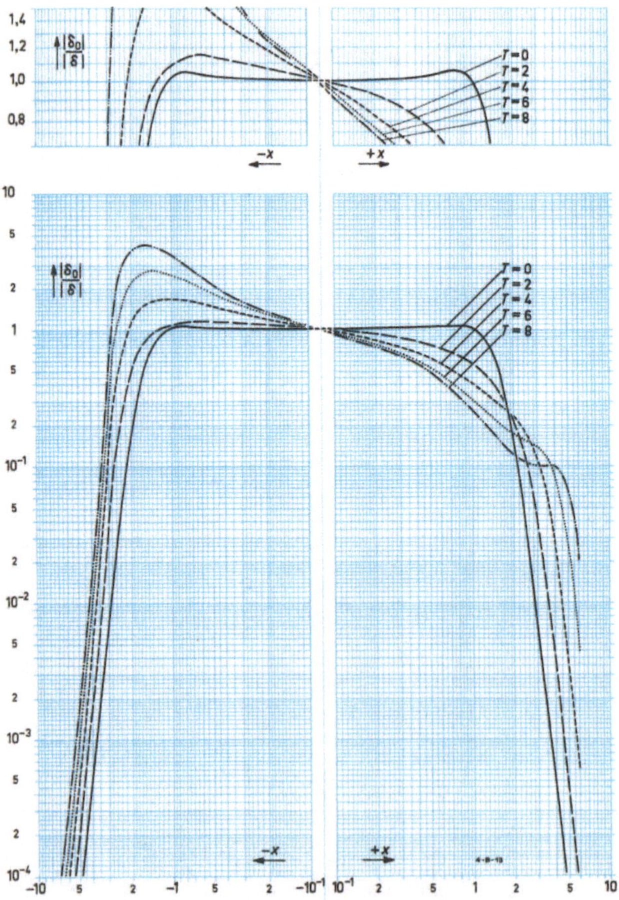

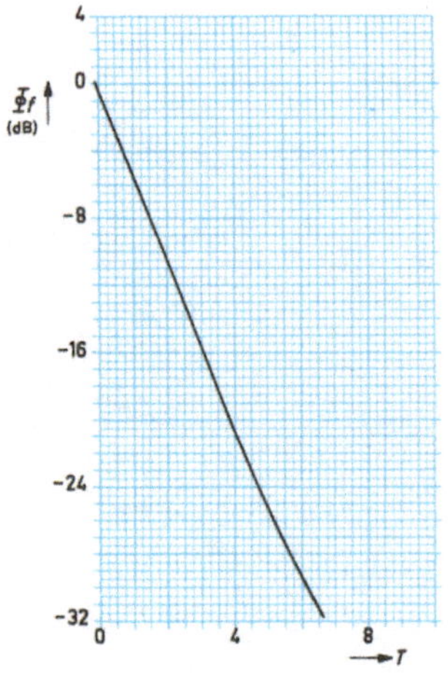

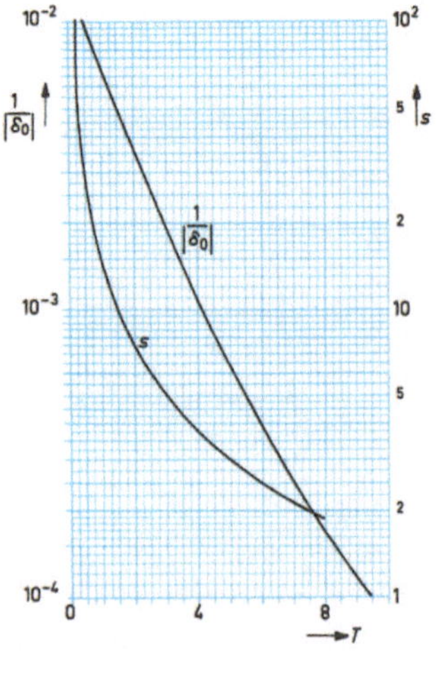

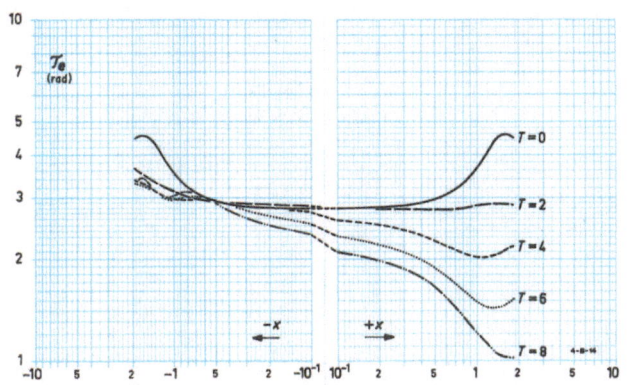

Fig. 15.80
Four-stage amplifier
Configuration III
Tuning method B
$\Theta = 210°$
$q^2 = 3, r = 1, m = 0.75$

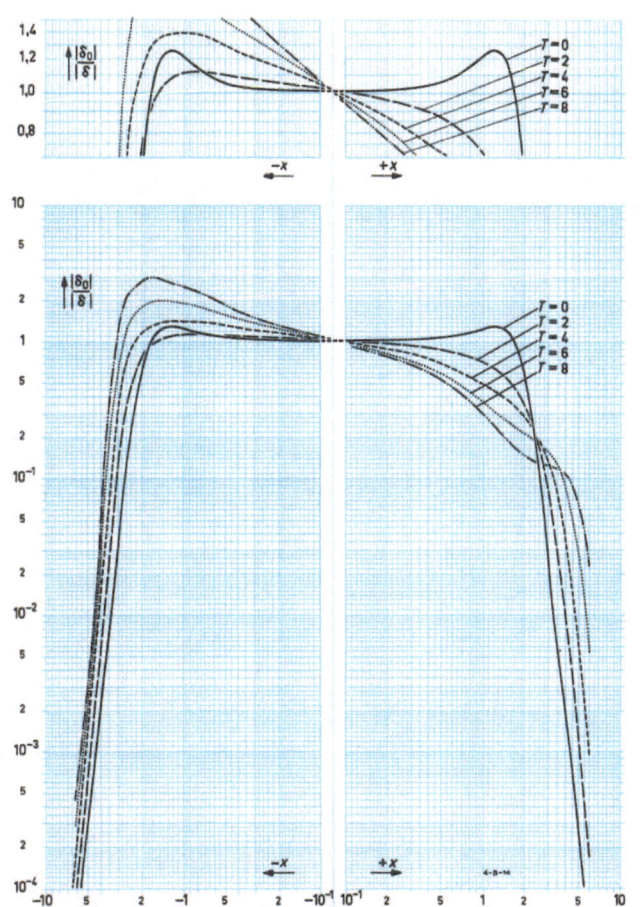

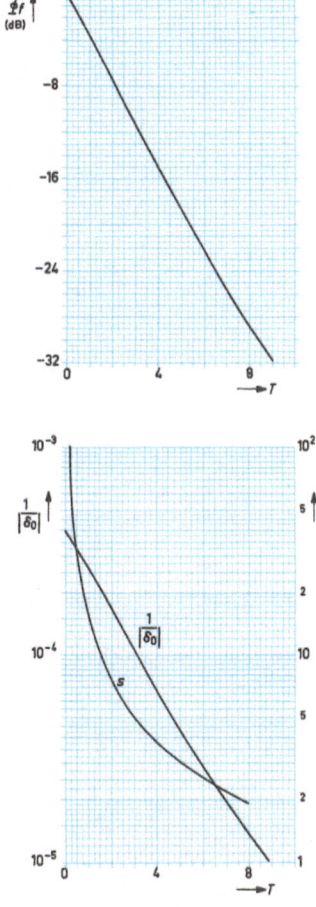

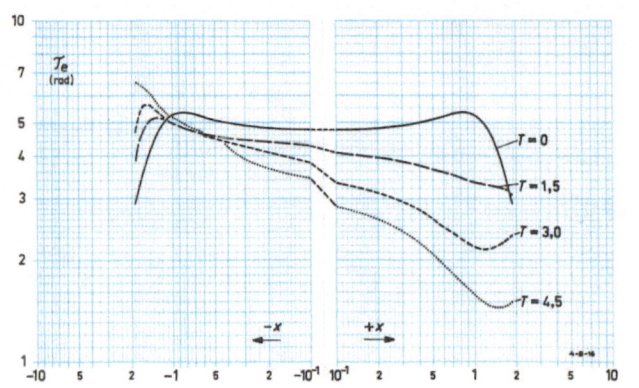

Fig. 15.81
Four-stage amplifier
Configuration III
Tuning method B
$\Theta = 225°$
$q^2 = 1, r = 1, m = 0.75$

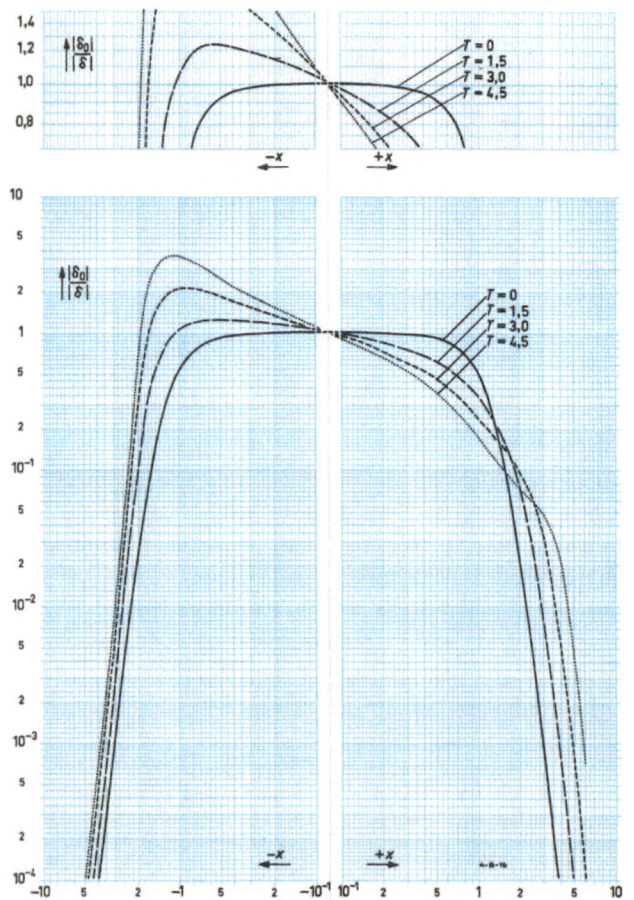

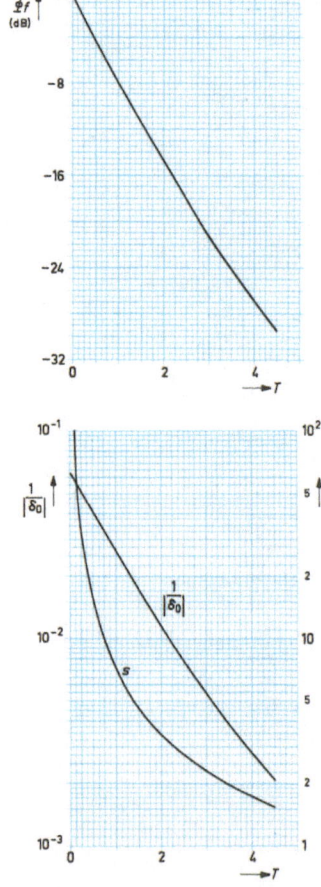

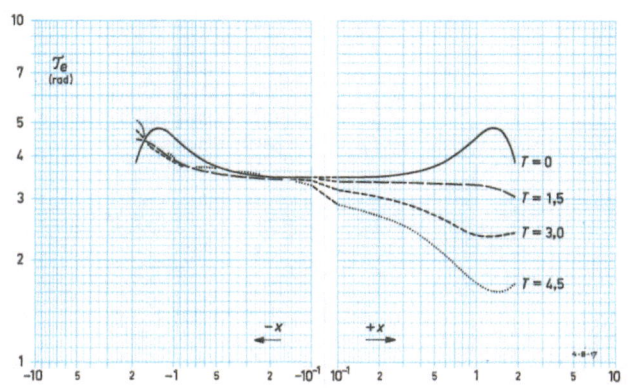

Fig. 15.82
Four-stage amplifier
Configuration III
Tuning method B
$\Theta = 225°$
$q^2 = 2, r = 1, m = 0.75$

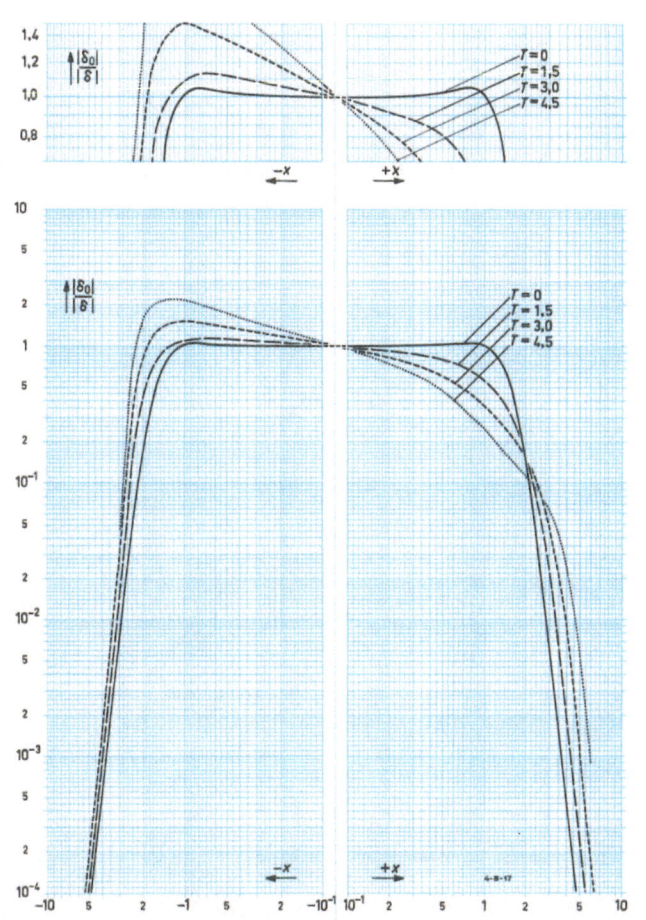

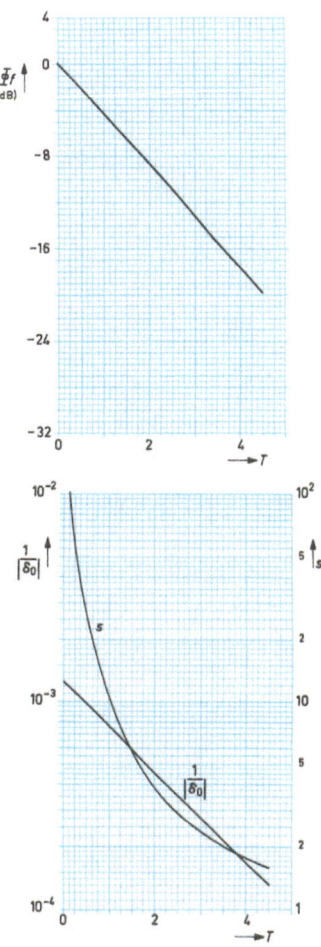

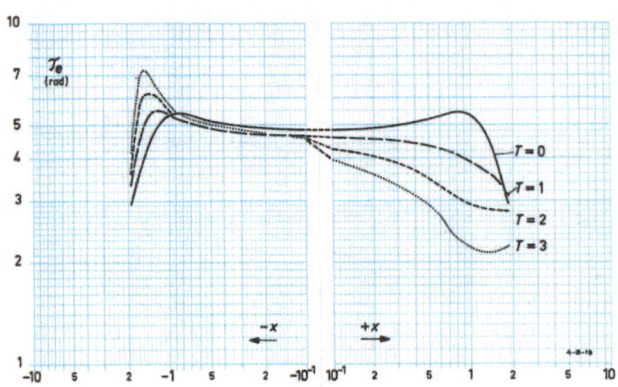

Fig. 15.83
Four-stage amplifier
Configuration III
Tuning method B
$\Theta = 240°$
$q^2 = 1, r = 1, m = 0.75$

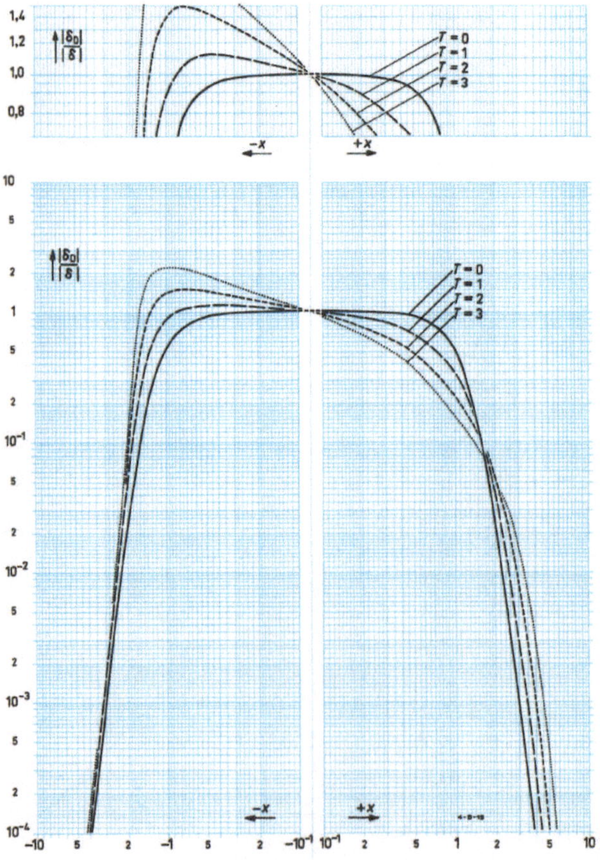

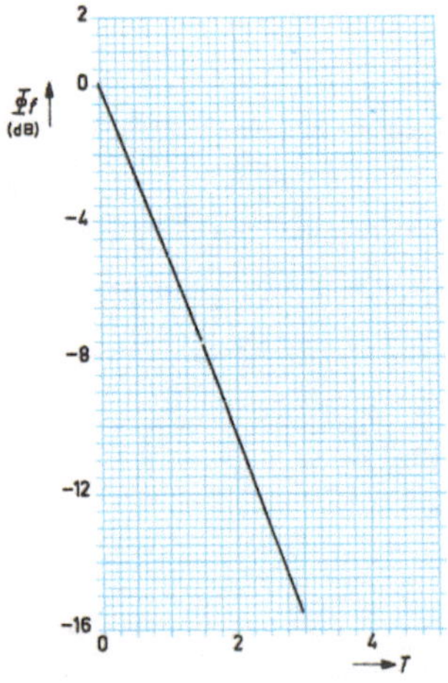

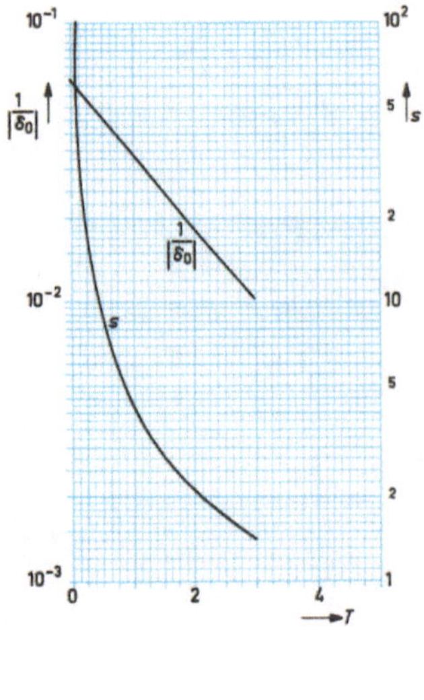

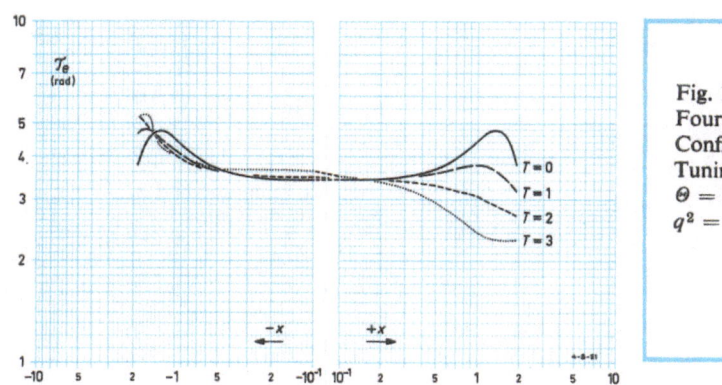

Fig. 15.84
Four-stage amplifier
Configuration III
Tuning method B
$\Theta = 240°$
$q^2 = 2, r = 1, m = 0.75$

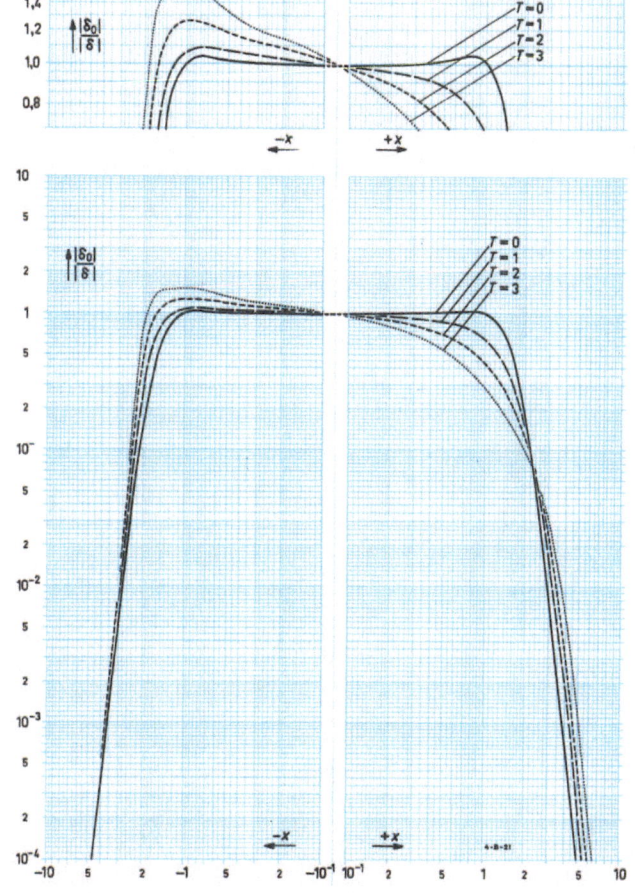

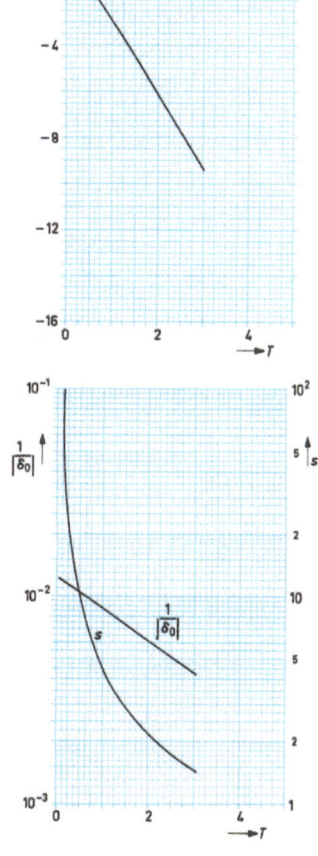

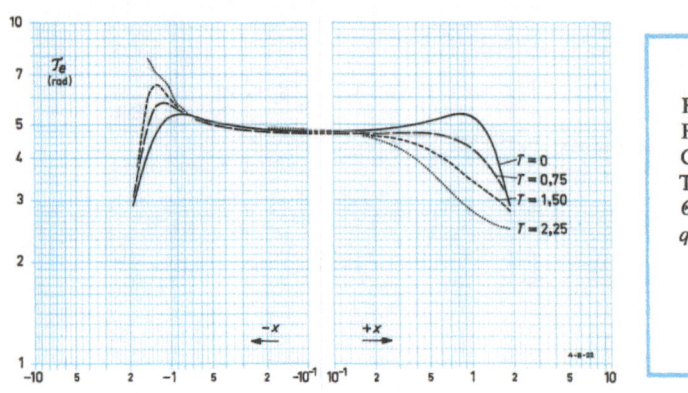

Fig. 15.85
Four-stage amplifier
Configuration III
Tuning method B
$\Theta = 255°$
$q^2 = 1, r = 1, m = 0.75$

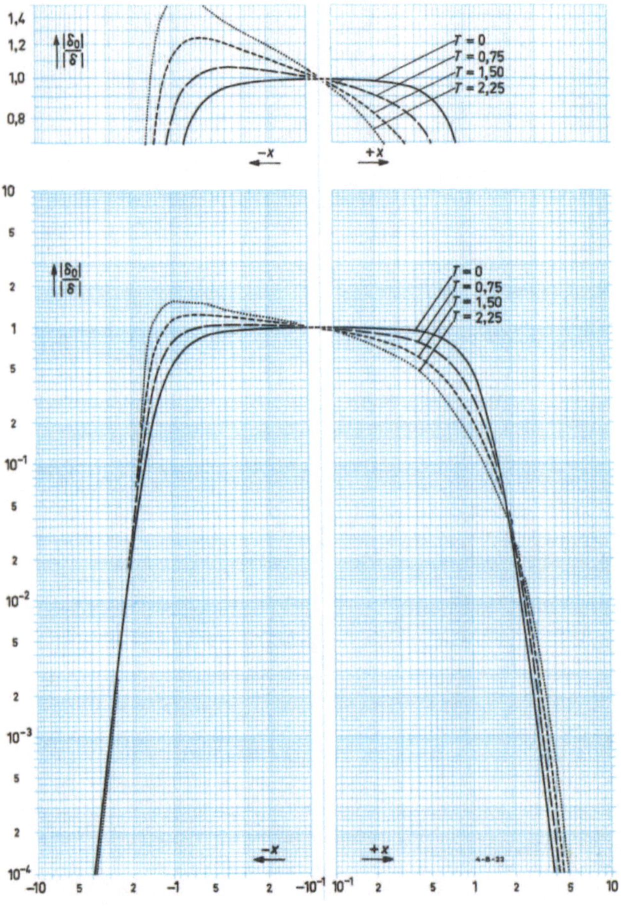

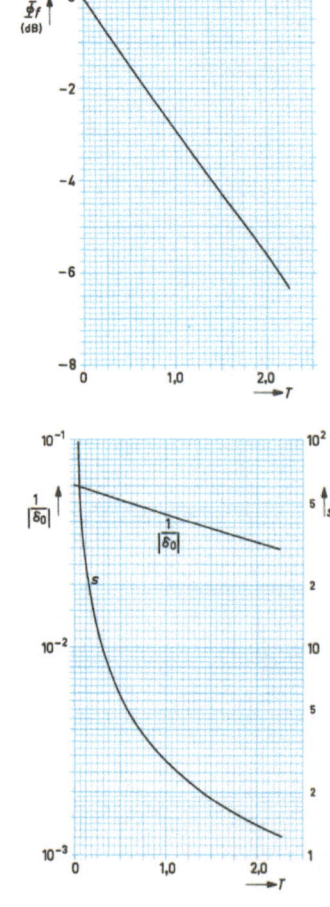

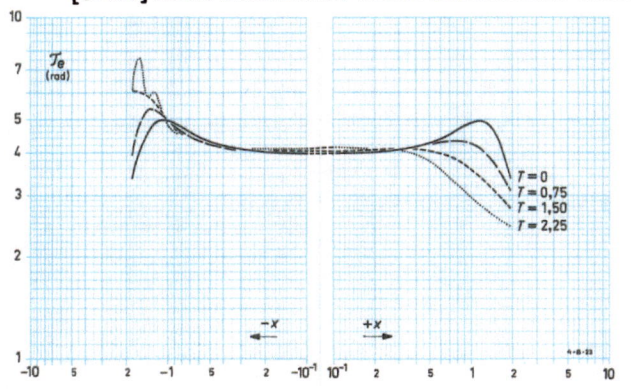

Fig. 15.86
Four-stage amplifier
Configuration III
Tuning method B
$\Theta = 255°$
$q^2 = 1.5, r = 1, m = 0.75$

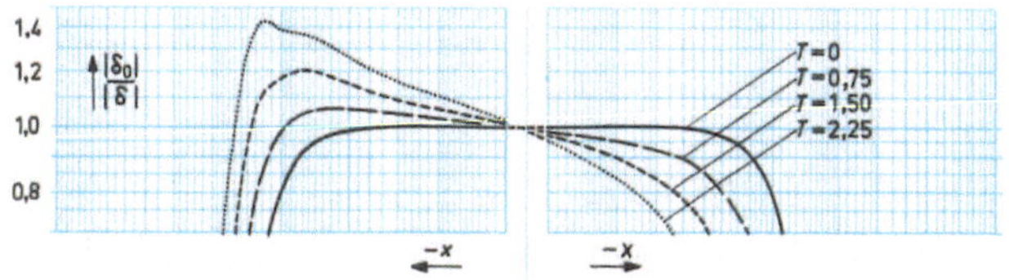

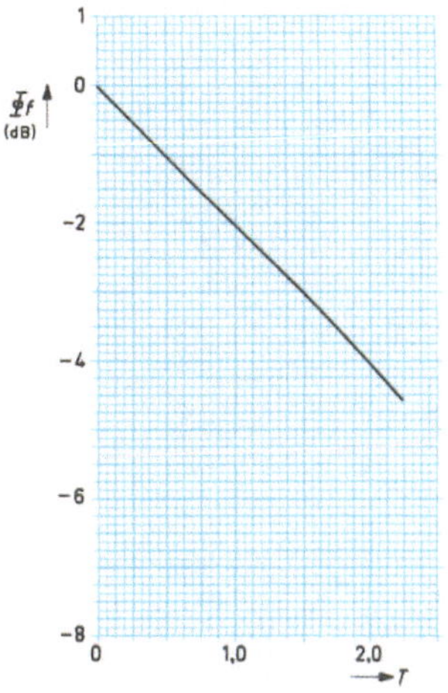

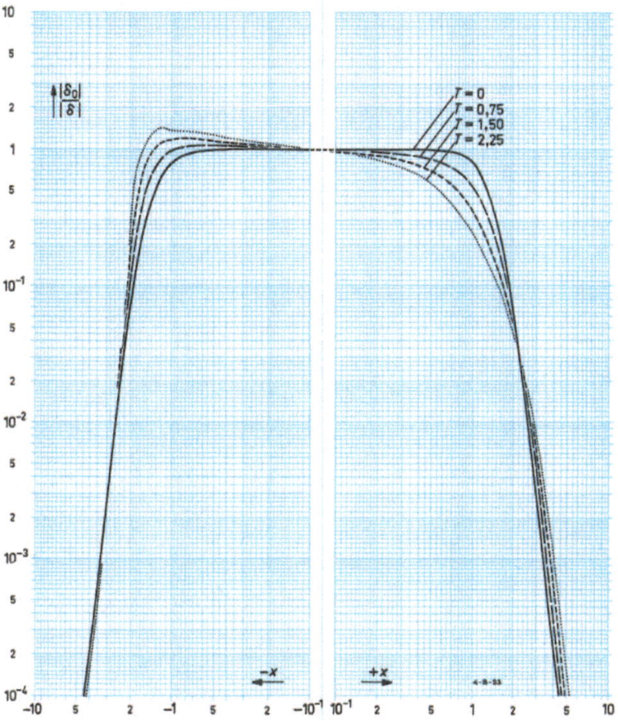

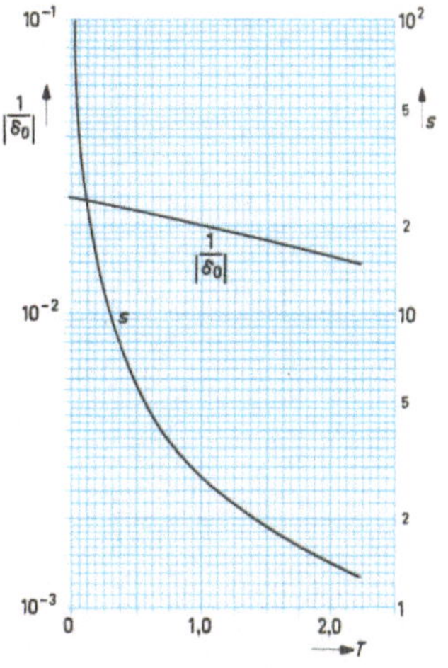

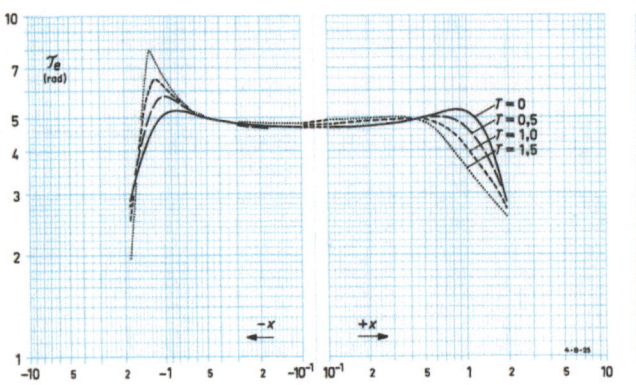

Fig. 15.87
Four-stage amplifier
Configuration III
Tuning method B
$\Theta = 270°$
$q^2 = 1, r = 1, m = 0.75$

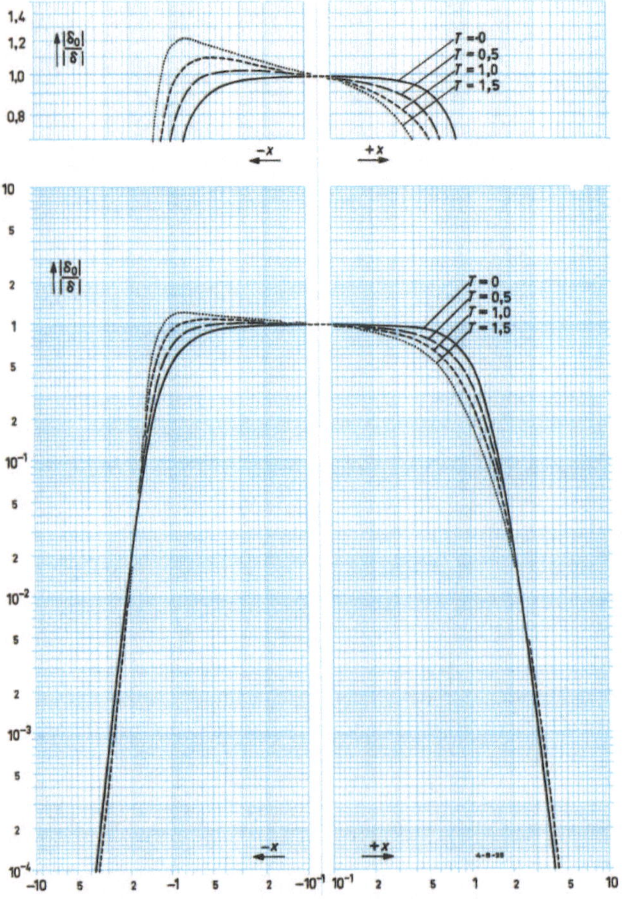

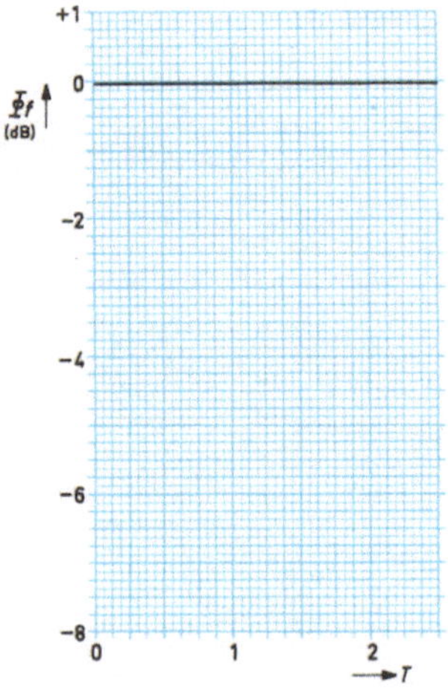

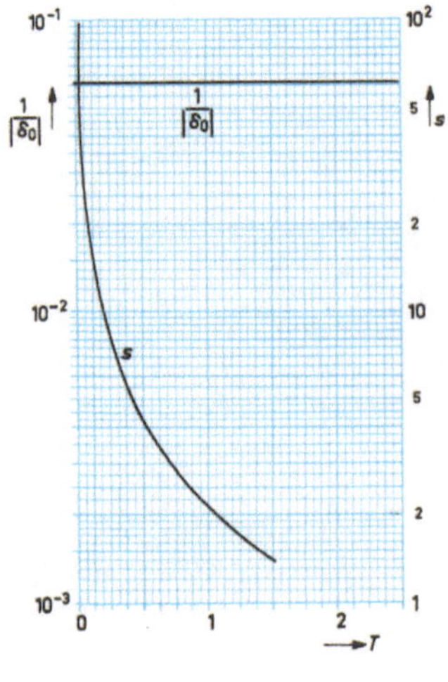

Additional material from *Designing Transistor I. F. Amplifiers,*

ISBN 978-3-662-38672-9 (978-3-662-38672-9_OSFO3)

is available at http://extras.springer.com